Adrenal Steroid Antagonism

Adrenal Steroid Antagonism

Proceedings
Satellite Workshop of the
VII. International Congress of Endocrinology
Quebec, Canada, July 7, 1984

Editor
M. K. Agarwal

Walter de Gruyter · Berlin · New York 1984

Editor

M. K. Agarwal, M. Sc.; Ph. D., M. D.
Maître de Recherche au CNRS
Scientific Director: Laboratoire de Physio-Hormono-Réceptérologie
Faculté de Médecine Broussais Hotel-Dieu
Université Pierre et Marie Curie
15, rue de l'Ecole de Médecine

F-75270 Paris Cédex 06
France

CIP-Kurztitelaufnahme der Deutschen Bibliothek

Adrenal steroid antagonism : proceedings, satellite workshop of the VII. Internat. Congress of Endocrinology, Quebec, Canada, July 7, 1984 / ed. M. K. Agarwal. – Berlin ; New York : de Gruyter, 1984.

ISBN 3-11-010090-8

NE: Agarwal, Manjul K. [Hrsg.]; International Congress of Endocrinology <07, 1984, Quebec, Quebec>

Library of Congress Cataloging in Publication Data

Main entry under title:

Adrenal steroid antagonism.

Bibliography: p.
Includes indexes.
1. Adrenocortical hormones--Antagonists--Congresses.
I. Agarwal, M. K.
QP572.A3A38 1984 599'.0142 84-19947
ISBN 3-11-010090-8

 Printing: Gerike GmbH, Berlin. Binding: Lüderitz & Bauer GmbH, Berlin. Printed in Germany.

Foreword

Molecular Endocrinology is a rapidly advancing field. Chemical modification of the steroid nucleus for all classes of adrenal hormones has produced a whole array of molecules with varying degrees of agonist activity in vivo. The discovery of the receptor for steroid hormones gave a fresh start to the manner in which hormone action could be modulated in vitro and in vivo; latest developments are presented in this volume.

Antagonists for steroid hormones are important not only as tools to probe the molecular mechanisms of hormone action but also as important new adjuncts in the arsenal of modern medical therapy. Antihormones for androgens, estrogens and mineralocorticoids have already entered the clinic. Less well characterized are antiglucocorticoids and antiprogestagens.

The purpose of the workshop summarized in this volume was to discuss new advances in the field of adrenocorticoid antagonists. Pharmacological and clinical advances in the use of inhibitors of steroid hormone synthesis, and in the production of antibody to circulating hormones, have moved less rapidly than other approaches to antagonize hormone action.

Entirely new to the field of endocrinology, and reviewed here for the first time for the chemist, the clinician, and the general research worker, are multifaceted antihormones such as RU 38486. Other derivatives in this series are also reviewed and the material presented is so new that special precautions had to be taken to protect industrial rights. These suggest that specific antiglucocorticoids and antiprogestagens should be available to the clinician in the near future. This would also appear to be true of new derivatives of antagonists for mineralocorticoids.

Equally new to the literature of endocrinology is the so called glucocorticoid antagonizing factor, already well described in journals of microbiology. Further characterization of this endogenously produced material could have a far-reaching impact for both the fundamentalist and the clinician.

It is hoped that this volume will be useful to all workers interested in the area of endocrinology, including the clinician. The workshop would well have served its purpose by providing a forum where students from various fields could discuss their particular, and sometimes contradictory, data.

Thanks are due to Roussel-Uclaf, LKB, and Merck Sharp & Dohme-Chibret for their financial assistance. For local organization of the workshop, and for publicity, the Secretariat of the VII. International Congress of Endocrinology is to be commended. Finally, space does not permit individual mention of organizations, agencies, and persons who variously contributed to the success of the workshop and of this volume.

Paris, July 1984 The Editor

CONTENTS

EXPLOITABLE STRATEGIES IN ADRENAL STEROID ANTAGONISM

M. K. Agarwal and G. Lazar[+]
Centre National de la Recherche Scientifique
15 rue de l'Ecole de Médecine, 75270 Paris Cédex 06, France
[+] University Medical School, Szeged, Hungary

Introduction

A hormone has classically been defined as a substance, endogenous to the organism, which, in minute amounts, is capable of altering the rate of cellular processes innate to the target tissue. No level of physiological or pathophysiological hierarchy escapes the influence of one or several hormones. Thus, hormone antagonists are important not only as new tools to understand the organisation and expression of the complex mammalian genome but also as agents of much therapeutic potential.

An antihormone may be defined as a substance capable of attenuating, or antagonising to various degrees, a hormone induced response, regardless of their mechanism or site of action (Figure 1). This may involve a mere lowering of serum levels of the active hormone, as with specific antibodies directed either against the hormone in question or against the specific releasing factor, or the elaborating trophin. Other possibilities would include interference with the synthesis, the secretion, or the release of a naturally ocurring hormone. Modification of the bound vs the free ratio in the blood, as well as altered metabolism and/or elimination, are other potential possibilities. An agonist, natural or synthetic, would be a substance capable of mimicking the effects of the parent molecule whereas an antagonist would block the effect of an agonist. Evidently, various shades of agonist and antagonist activity can be en-

Adrenal Steroid Antagonism

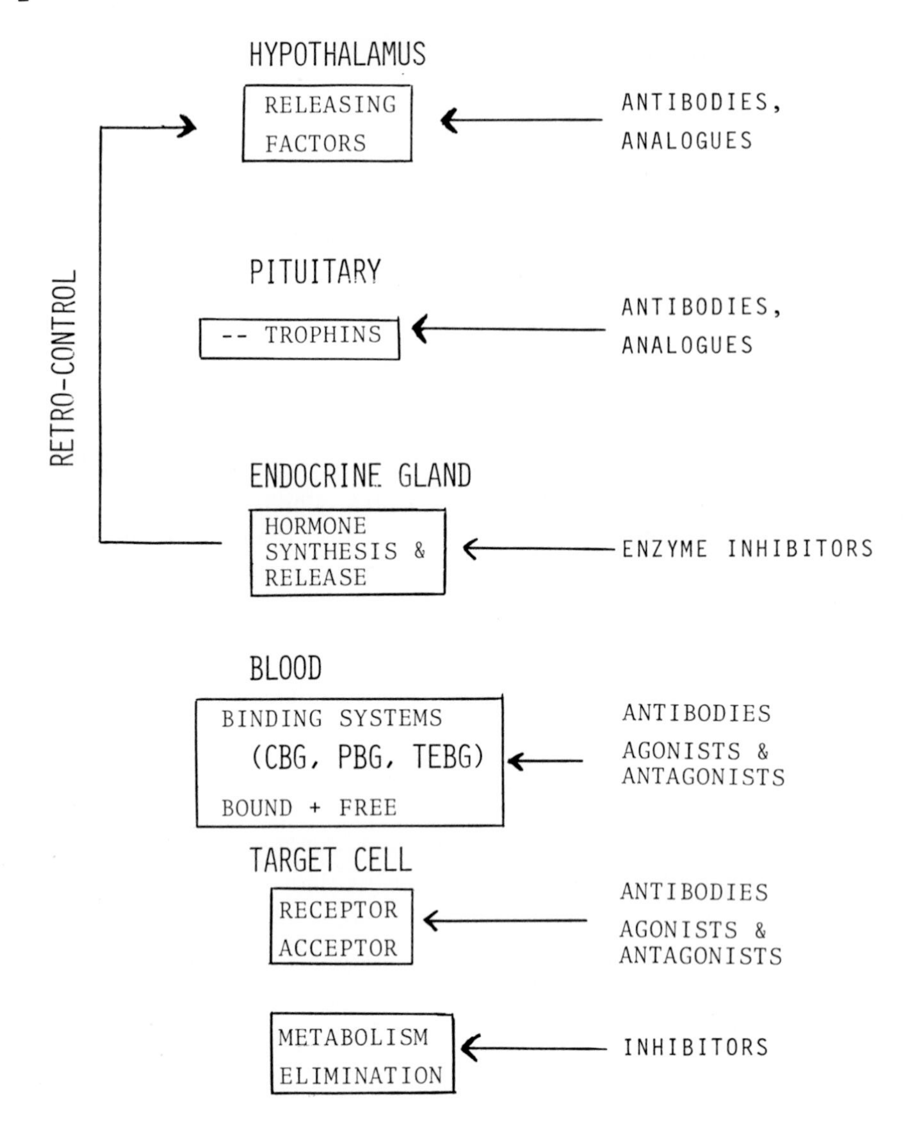

FIGURE I: Schematic representation of the endocrine hierarchy and the various exploitable levels where search for an active anti-hormone may be made. No indication is given of the relative importance of the various targets in this regard. For further details see text.

visioned depending upon the cell type and the hormone mediated process, as well as the chemical nature of hormones.

This chapter is limited to analysing the conceptual possibilities that can be exploited in the quest for a substance antagonising a defined process in a given cell type as this appears to be the only means fulfilling the definition of an ideal antihormone. Only two steroid hormones have been discussed in some detail with passing reference to other steroids. The reference citation refers freely to existing reviews in specific fields.

Antiglucocorticoids

Selected aspects of glucocorticoid action on individual target tissues can be blocked by hypoglycin (1), insulin (2), norepinephrine (3), cordicepin (4), protein synthesis inhibitors (5), vitamin A (6), aspirin (7), cystamine (8), adenosine (9), interferon (10), pyridoxine (11), RNA derivatives (12), and vitamin B_{12} (13) but this list is not exhaustive. The lack of specific antagonists is so disheartening that adrenalectomy has remained the major therapeutic arm in the treatment of Cushing's disease although adrenal insufficiency ensuing thereafter has given rise to Addison's syndrome, along with numerous other complications (14). Inhibitors of cortisol synthesis (eg. aminogluthemide, Op'DDD) in the human subject have received much attention but their use in the clinic has been limited to inoperable or recurring adrenal tumors that provoke hypercortisolemia (14). Another approach has been to switch off ACTH secretion via a retrocontrol following dexamethasone administration. Endogenous ACTH secretion can also be inhibited by reserpine, barbiturates, and morphine (14). Anti-ACTH antibodies, too, have potential application.

The preceding goes to demonstrate the near total lack of

products that may antagonise specific responses of glucocorticoids at given times in predetermined tissues. This, in part, is due to the very wide range of effects exerted by these hormones and the fact that almost all tissues tested respond to glucocorticoids. As far as catabolic response on lymphatic tissue (thymus) is concerned, cotexolone and progesterone derivatives oppose several aspects of cortisol action (inhibition of glucose uptake, uridine incorporation) but were inactive in this same tissue in vivo (15). In an anabolic response, testosterone and various other derivatives of progesterone inhibit tyrosine transaminase induction by either cortisol or dexamethasone in hepatoma culture cells in vitro. These same compounds were totally inactive in this regard in the intact liver in vivo (15, 16). Testosterone and oestradiol derivatives, however, did inhibit triamcinolone acetonide mediated liver gluconeogenesis in the adrenalectomised rats (17, 18). Thus, the most important conclusion from these studies is the fact that in vitro systems are poor models of the situation in vivo and that extrapolations need, indeed, be made with much caution.

In another series of studies, bacterial endotoxins were found to oppose the glucocorticoid mediated enzyme induction and gluconeogenesis in the intact liver (19, 20). Reversal of cortisol mediated responses on the cells of the reticuloendothelial system and on immunosupression was also observed after treatment with the toxin. The promising nature of these studies requires further elaboration, especially since they are active on both cellular types. Collectively, search for an antiglucocorticoid must account for both anabolic and the catabolic phases of the hormone and an ideal substance is conceptually difficult to envision.

More recently, a synthetic compound dubbed Ru 38486 appears to hold much promise as glucocorticoid antagonist.

Antimineralocorticoids

Aldosterone production in man is regulated by fluid K^+ concentration and by the renin angiotensin system (21-23). Analogues of angiotensin (Saralasin) have been tested for the control of aldosterone synthesis with good success. Aldosterone synthesis can also be inhibited by several chemicals (aminogluthemide, cyanoketone, amphenone, metapyrone) that lower adrenal hormone formation; more specific inhibitors (SU 9055, heparin) are also known. Landau et al., as early as 1954, revealed progesterone mediated antagonism of aldosterone secretion through an interference with the renin-angiotensin system but in more recent studies this appears to proceed by competitive inhibition of the mineralocorticoid receptor occupancy by aldosterone (24 25). Among other physiological antagonists one may note glucagon, calcitonin, thyroxine, and possibly prostaglandins (21-23); however, no studies are available to indicate either their specificity in this respect or their interaction with the cellular receptors specific for aldosterone. All these antagonists have little if any clinical use due primarily to the lack of specificity and the wide ranging physiological effects exhibited by them.

By far the greatest amount of research effort has been devoted to specific aldosterone antagonists over the past two decades. Spirolactones, canrenone, potassium canrenoate, are all competitive inhibitors of aldosterone for attachment to the specific cellular receptor. The physiological nature of this antagonism is evident from the fact that natriuretic diuresis induced by spirolactones is completely dependent on aldosterone levels (the antagonist has no intrinsic agonist or antagonist activity) as well as on the sodium load in the distal tubule which is why the effect is potentiated by thiazides. Spirolactones also inhibit aldosterone biosynthesis, possibly by a feedback action (21-23).

Clinical indications of spirolactones include primary hyperaldosteronism (Conn's syndrome), hypertension due to mineralocorticoids other than aldosterone (deoxycorticosterone, 18-hydroxydeoxycorticosterone), essential hypertension, fluid retention in chronic liver disease, nephrotic syndrome, and congestive heart failure. Undesirable side effects are related to hyperkalemia, hypernatremia, hyperuremia, diminished testosterone biosynthesis, increased plasma progesterone levels, and interference with cortisol assay (21-23). The most outstanding complication of these remains hyperkalemia. Long term therapy with spirolactones is also ill advised because of gynecomastia, impotence, menstrual abnormalities, all of which are possibly mediated through interference with testosterone biosynthesis and through the oestrogenic actions of spirolactones.

The Receptor

Some 15 years have elapsed since the earliest demonstrations of receptors for steroid hormones in the cytoplasm (for recent reviews see 30-32). In its simplest form, the genesis of a complex between the steroid and its specific, soluble receptor, with eventual activation and translocation into the nucleus, and attachment to nuclear acceptor sites, was said to trigger the physiological response by amplified transcription of the organ specific genes (28, 31, 33-35). It follows therefore that competition at the receptor level would constitute an efficient and specific means to antagonize the action of a given hormone.

A fundamental tenet of this initial receptor concept was the presence of a single protein in the cytoplasm capable of accepting all grades of agonists and antagonists for any one of the five classes of steroid hormones (28, 33-35 for reviews). The ease and rapidity of competition binding analyses, the inavailability of various steroid derivatives of high radiospecific activity, and the application of age old biochemical methods such as gradient centrifugation and gel filtration, all contributed to the protracted persistance of a fallacious notion that a single, unitary vector was the most likely explanation of steroid hormone mediated physiological action (29).

Over the past several years, an ever increasing body of evidence has been arguing against the 'one hit' type of mechanism. Thus, a number of model systems such as the one entry site, the induced fit, the allosteric (reviewed in 34) appear to be outdated.

An important milestone appeared to have been established in an earlier report where gluco- and mineralo- receptor systems were said to express polymorphism and heterogeneity depending upon the ligand, the tissue, and the animal species (29). A basic notion of this multipolar model stated that contrary to the viewpoint of a single peptide, various agonists and antagonists were bound to chromatographic peaks endowed with clearly distinct hydrodynamic properties. This was later extended to liver androgen and oestrogen receptors (36). A number of studies from other laboratories have confirmed the idea of an antagonist (tamoxifen) binding site that is clearly distinct from the agonist binding protein for the oestrogen receptor in rat uterine and liver cytoplasm (37). Furthermore, even nuclear acceptor sites have more recently been found to exhibit polymorphism and heterogeneity (38) in a manner akin to our initial propositions for the cytoplasmic vector (28, 29).

To date it has not been possible to establish whether these multiple peaks are indeed different proteins or modified versions of a more fundamental, basic unit. It is however inconceivable that the genetic repertory should possess sufficient heterogeneity to transcribe separate receptors for all possible synthetic agonists and antagonists that are being made available day after day. These considerations should be borne in mind during receptor purification.

Some of the confusion stems from the fact that current methods of analysis provide information on the nature of the receptor only after the latter has interacted with the ligand. It is currently unknown whether the native receptor inside the cell has much resemblance with isolated, ligand-bound component. A number of considerations indicate that the state of affairs inside the cytoplasm may be widely different.

First, vector stabilization by appropriate ligand is a well established phenomenon for all classes of receptors but is especially true of the mineralocorticoid receptor (28,33-35). Association with the ligand and subsequent activation result in the transfer of the receptor to the nuclear compartment (32). To account for the observed half life of the cytoplasmic receptor it follows therefore that the latter must be free of the ligand, and must have some sort of mechanism to assure equilibrium between the bound and the free receptor moieties.

Second, it has recently been shown that rat liver Glucoreceptor is associated with RNA, and RNAase A converts it from a heavy 7 - 8 S to a light 3 - 4 S form that is usually analyzed by conventional in vitro assay procedures (39). This calls to the mind some of the earliest experiments on enzyme induction where the amount of the enzyme, say liver tyrosine transaminase (TT), was believed to be regulated by the inducer by a change of equilibrium between the ribosome bound and free forms (40). This sort of mechanism assures a feedback economy to the protein synthetic capability and seems to be of wide occurrence.

It therefore seems very likely that a nascent, proreceptor is consistently present in the cytoplasmic fraction, it is not bound to the steroid, and its stability is assured by its association with the ribosome after translation of the receptor gene(s). This is depicted schematically in Figure 2. As soon as an appropriate ligand is presented to this nucleo-protein complex, the receptor protein may be freed of its ribosomal support either spontaneously or via the action of RNAses. Steroid mediated regulation of homologous and heterologous receptor concentrations is well established and fits in very nicely with the above scheme (32,37).

Transmission of conformational changes along long distances is already known for protein hormones (41). A basic tenet of Argusoid model presented here is the right conformation conferred upon the ribosome bound proreceptor by a physiologically active ligand thereby rendering it geometrically disposed for recognition by various factors for posttranslational modifications (Fig. 2). The name argusoid is derived from the greek myth of Argus whose chief role was the surveillance and coordination by virtue of 100 eyes, just as the receptor must be sensitive to a myriad cellular demands and processes, depending upon the physiological status.

A whole battery of ions, chelators, cofactors, receptor transformation factors, endogenous inhibitors and activators, have been documented by experimental analysis in vitro (reviews in 28,29,30). Since their role in vivo remains speculative, they may act either with the ribosome bound proreceptor, or the free, ligand-bound nascent receptor generated from the former.

The influence of proteases and inhibitors in recepterology has been widely documented (33,42,43). It is therefore entirely conceivable that the nascent, soluble receptor in Fig. 2 may really be a pre-receptor, much like the enzymes of the coagulation cascade, and a number of protein hormones, that require a proteolytic cleavage to be rendered biologically acttive (33,42). Protein phosphorylation is said to regulate protein synthesis (44) and this could also influence the free receptor. Once the mature holoreceptor has been formed, it may be fully active in the cytoplasm at the level of MTV translation (45), contrary to the classical viewpoint where the steroid-receptor complex could only modulate transcription. It

Figure 2.

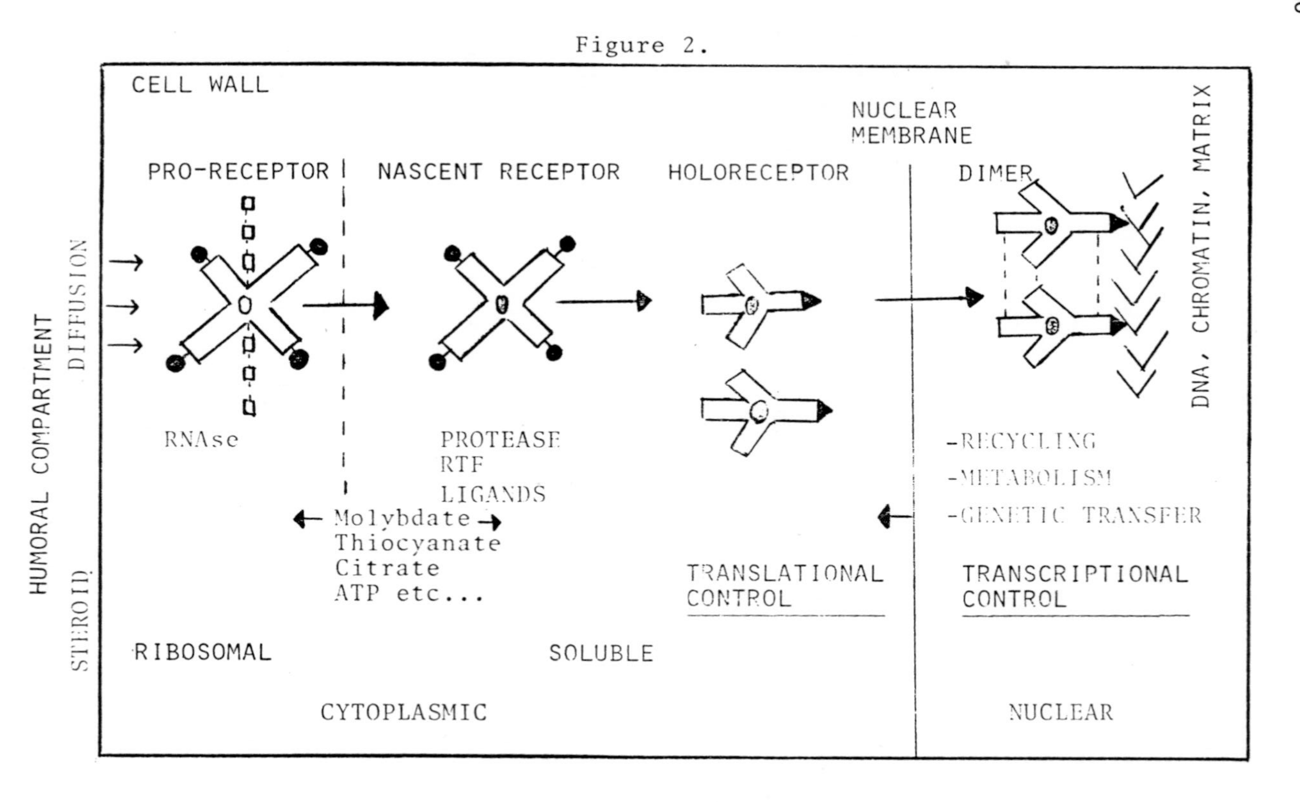

THE ARGUSOID MODEL

follows therefore that exploitable targets for hormone antagonism may be sought at any, and all, of the levels shown in Fig. 2, contrary to attempts directed only against receptor-ligand attachment.

Perspectives

Steroid hormone antagonism has not developed to a similar extent for various classes of corticoids. Steroid antiandrogens such as cyproterone acetate, and non steroid antiadrogens such as flutamide, are already in clinical use and act via the androgen receptor (24-26); retrosteroids, capable of inhibiting LH release, are still experimental (24-26). Antiprogestogen (RMI,12,936) was patented in the U.S. in 1974 and is known to inhibit progesterone synthesis (27-29); R 2323 may be more specific since it competes for the progesterone receptor (22, 46). Oestrogen antagonism is actually being exploited clinically both through non steroid (nafoxidine, clomiphen, tamoxifen) as well as steroid (testosterone, epithioandrostanol) derivatives (21,24,25,47,48).

Antimineralocorticoid action has already been described above. Glucocorticoid antagonism has been lagging so far but Ru 38486 seems to hold a promising future. Exploitable strategies in future should take into account various components of the receptor system presented here in the Argusoid model. These may turn out to be even more specific than the overall steroid-receptor antagonism attempted hitherto.

References

1. Kean, E.A., Walters P.: West. Ind. Med. 24, 206 (1975).
2. Smith, B.T., Giroud, C.J., Robert, M., Avery, M.E.: J. Pediatrics, 87, 953 (1975).
3. Sitaramam, V., Ramasarma, T.: Life Sci. 16 , 1387 (1975).
4. Young, D.A., Barnard, T., Mendelsohn, S., Giddings, S.: Endocr. Res. Comm. 1, 63 (1972).
5. Agarwal, M.K.: Sub-Cellular Biochem. 1, 207 (1972).
6. Griffin, M.J., Cama, H.R.: Nature 228, 762 (1970).
7. Sedlak, J., Demey, L., van Cauwenberge, H.: Arch. Int. Pharmacodyn. Ther. 200, 378 (1972).
8. Flemming, K., Geierhaas, B.: Experientia 28, 965 (1972).
9. Chagoya, D.E., Sanchez, R., Briones, R., Pina E.: Biochem. Pharmacol. 20, 2535 (1971).
10. Matsuno T., Shirasawa, N.: Biochem. Biophys. Acta 538, 188 (1978).
11. Milholland, R.J., Rosen, F., Nichol, C.A.: Ann. N. Y. Acad. Sci., 166, 126 (1969).
12. Laguna, J., Hamabata, A., Chagoya, V.: Gac. Med. Mex. 98, 634 (1968).
13. Hadnagy, C., Gyergy, F.: Inst. Z. Vitaminforsch. 37, 354 (1967).
14. Bricaire, H., Luton, J.P.: Therapie 29, 645 (1974).
15. Kaiser, N., Milholland, R.J., Turnell, R.W., Rosen, F.: Biochem. Biophys. Res. Comm. 49, 2516 (1972).
16. Samuels, H.H., Tomkins, G.M.: J. Mol. Biol. 52, 57 (1970).
17. Agarwal, M.K., Lazar, G., Sekiya, S.: Biochem. Biophys. Res. Comm. 79, 499 (1977).
18. Agarwal, M.K., Coupry, F. : FEBS Letters 82, 172 (1977).
19. Agarwal, M.K., Lazar, G.: Microbios 20, 183 (1977).
20. Agarwal, M.K.: Naturwissenschaften 62, 167 (1975).
21. Winter, I.C., ed. The clinical use of aldosterone antagonists., Charles Thomas, 1960.
22. Ochs, H.R., Greenblatt, D.J., Bodem, C., Smith, T.W.: Am. Heart J. 96, 389 (1978).
23. Ross, E.J.,, Aldosterone and Aldosteronism, Llyod-Duke, 1975.
24. Landau, R.L., Lugibihl, K.: Rec; Prgr. Horm. Res. 17, 249 (1961).

25. Wambach, G., Higgins, J.R.: Clin. Res. 25, 1 (1977).

26. Hubinot, P.O., Hendeles, S.M., Preumont, P., eds., Hormones and antagonists, S. Karger, 1972.

27. Clark, J.H., Klee, W., Levitzki, A., Wolff, J., eds., Hormone and antihormone action at the target cell, Dahlem Konf., 1976.

28. Agarwal, M.K., ed., Multiple Molecular Forms of Steroid Hormone Receptors, Elsevier/North Holland, 1977.

29. Agarwal, M.K.: FEBS Letters 85, 1 (1978).

30. Agarwal, M.K., ed., Principles of Receptology, Walter de Gruyter, Berlin, New York, 1983.

31. Pasqualinini, J.R.: Receptors and Mechanisms of Action of Steroid Hormones, Marcel Dekker, New York, 1977.

32. Jensen, E.V., Greene, G.L., Closs, L.E., De Sombre, E., Nadji, M.: Rec. Prgr. Horm. Res. 38, 1 (1982).

33. Agarwal, M.K., ed., Proteases and Hormones, Elsevier/North Holland, Amsterdam, N.Y., 1979.

34. Agarwal, M.K., ed., Antihormones, Elsevier/North Holland, Amsterdam, New York, 1979.

35. Agarwal, M.K., ed., Hormone Antagonists, Walter de Gruyter, Berlin, New York, 1982.

36. Agarwal, M.K.: Biochem. Biophys. Res. Comm. 109, 291 (1982).

37. Winnecker, R.C., Clark, J.H.: Endocrinol. 112, 1910 (1983).

38. Eckert, R.L., Katzenellenbogen, B.S.: J. Biol. Chem. 257, 8840 (1982).

39. Tymoczko, J.L., Phillips, M.M.: Endocrinol. 112, 142 (1983).

40. Know, W.E., Enzyme Patterns in Fetal, Adult and Neoplastic rat Tissues, S. Karger, Basel, 1972.

41. Chothia, C., Lesk, A.M., Dodson, G.G., Hodgkin, D.C.: Nature 302, 500 (1983).

42. Agarwal, M.K.: FEBS Letters 106, 1 (1979).

43. Agarwal, M.K., Philippe, M.: Biochem. Med. 26, 265 (1981).

44. Clemens, M.: Nature 302, 110 (1983).

45. Firestone, G.L., Payvar, F., Yamamoto, K.R.: Nature 300, 221 (1982).

46. Tamaya, T., Furuta, N., Motoyama, T., Biku, S., Ohono, Y., Oakoda, H.: Acta Endocrinol., 88, 190 (1978).

47. Danazol, J. Int. Med. Res. 5, 3 (1977).

48. Centchroman. Ind. J. Exp. Biol. 15, 12 (1977).

STRUCTURAL CHARACTERISTICS OF ANTAGONISTS FOR GLUCO AND MINERALOCORTICOIDS

William L. Duax and Jane F. Griffin

Medical Foundation of Buffalo, Inc.
73 High Street, Buffalo, NY 14203

Introduction

The characteristic responses of steroidal hormones require their binding to specific receptor proteins in target tissue (1). While response clearly depends upon the interaction of the receptor-steroid complex and nuclear chromatin, the precise details of this interaction and the role played by the steroid in this process remain undetermined (2,3,4). Structural details undoubtedly have a direct bearing upon receptor affinity and will directly or indirectly influence receptor activation, transport, and nuclear interaction. The existence of antagonists that compete for the steroid binding site of the receptor with high affinity demonstrates that the phenomena of binding and activity are at least partially independent. If agonists and antagonists compete for the same site on a receptor a comparison of their three-dimensional structures should make it possible to identify which structural features are responsible for binding and which control activity.

Crystallographic data on over 400 steroids (5,6) provide information concerning preferred conformations, relative stabilities and substituent influence on the interactive potential of steroid hormones. The analysis of subsets of these data (7,8,9) suggests strongly that the conformations (three dimensional shapes) observed in the solid state are at or near the global minimum energy position for the isolated molecules. In some cases two or more conformationally distinct isomers (molecules of identical composition but different three dimensional shape) co-crystallize in the same lattice (10,11). This co-crystallization indicates that these

Adrenal Steroid Antagonism

conformers may be of nearly equal energy, and that they were in equilibrium in the solvent from which the crystals were grown.

If the receptor-bound steroid is in its minimum energy conformation, then it should be possible to compare the crystallographically observed structures of a series of steroids that compete for a specific binding site and determine what structural features of the steroid are essential for binding, how tight a fit exists between the steroid and the receptor, and to what extent the binding site of the receptor protein is flexible. If a conformation of the steroid differing from that seen in the crystals is required for binding, this will place an additional energy requirement on the binding process. Certain steroids that exhibit exceptionally high affinity for the receptor might be expected to be in their minimum energy conformation when bound, thus eliminating the need for the additional energy of activation.

On the basis of an examination of the data on molecular structure, receptor binding, and biological activity for a series of estrogen and progestin agonists and antagonists, we have proposed a model for hormone action in which the A-ring end of the steroid is primarily responsible for initiating receptor binding, while the D-ring end controls the subsequent molecular interactions governing biological response (12,13).

This review will describe molecular conformational analysis of mineralo- and glucocorticoid agonists and antagonists and implications concerning the molecular basis for corticoid action based upon that analysis.

Mineralocorticoids

Of the naturally occurring steroids, aldosterone (11β,21-dihydroxy-3,20-dioxo-4-pregnen-18-al) is the most potent regulator of electrolyte excretion and is vital, therefore, to innumerable life processes. The isolation, structural elucidation, and total synthesis of aldosterone were accomplished in the early 1950's. From chemical studies Ham, _et al_. (14) postulated the existence of an equilibrium between the 18-aldehyde (I), the 11β,18-oxide (II), and the 18-acetal-20-hemiketal (III) isomers of Figure 1. A structural equilibrium is often described as existing between I and II only. On the basis of n.m.r. spectra there is little evidence for the presence of isomer I in solution (15). Simpson, et al.,(16) suggested that in solution the steroid reacts mainly as the 11β,18-oxide (II). The ease of preparation of 20,21-cyclic acetals of aldosterone cited by Gardi (17) suggests the ability of aldosterone to react equally well in solution as the 18-acetal-20-hemiketal (III). Because of the chiral centers at C(18) and C(20), there are two possible structural isomers of II and three possible isomers of III. The treatment of aldosterone 21-acetate with potassium carbonate in aqueous methanol leads to the formation of two additional isomers 17α-aldosterone (IV, Figure 1) and 11β,18: 18,21-diepoxy-20,21-dihydroxy-4-pregnen-3-one (V, Figure 1).

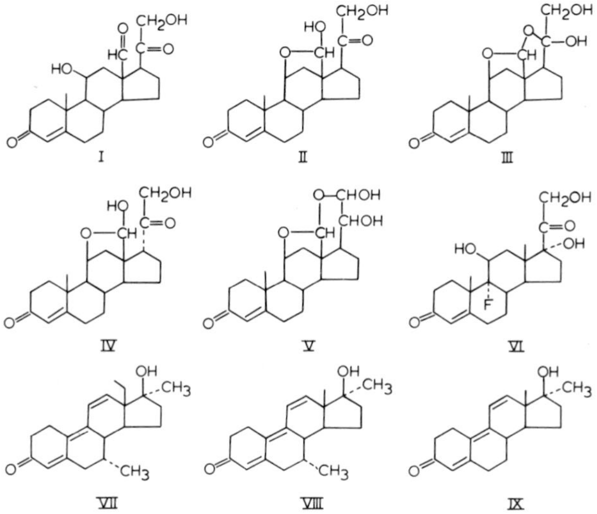

Figure 1. Chemical structures of natural and synthetic mineralocorticoids including five isomers of aldosterone (I-V).

On the basis of molecular models alone it is difficult to make reliable predictions about the relative energy and stability of these various isomers. The X-ray crystal structures of the 18R, 20R isomer of III (18), and a derivative of the 18R isomer of II (19) provide reliable information on the details of their conformation and tangible evidence of their stability. X-ray crystal structure analysis revealed the presence of both 18R, 20S, 21R and 18R, 20S, 21S isomers of V in a 3:1 ratio (unpublished data). The conformations of III, II and V observed in the solid state are illustrated in Figure 2. A comparison between II and III, and II and V illustrates that there is very little change in the conformation of the steroid backbone associated with the 18-,21 epoxide formation. There is however a significant change in the nature and location of substituents on the β face of the D ring that will alter the interactive potential of the molecule.

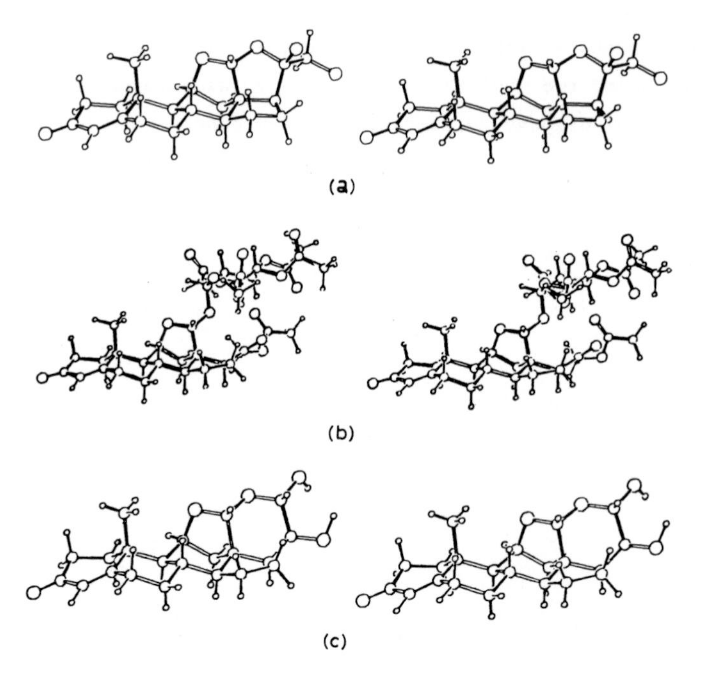

Figure 2. Stereo views of the observed conformations of (a) III, (b) the 18-glucopyronosiduronate of II, and (c) the 18R, 20S, 21S isomer of V.

In order to gain insight into the relative stability of their isomers, the crystallographically observed structures of III, and Va, and Vb and model structures of I and II were subjected to the energy minimization procedure using the program MM2p (20). The calculations indicate that conformers Va and Vb are of nearly equal energy in agreement with their observed cocrystallization. In the case of the aldosterone isomers the predicted greater stability of isomer I is in conflict with the data available from solid state and solution measurements. Nevertheless the conformations obtained from energy minimization of I and II provide a useful approximation of the overall shape of these isomers. The isomerism of aldosterone will have a direct bearing on its biological activity. Isomers will have different degrees of affinity for the various steroid receptors or metabolizing enzymes present in vivo. In the absence of other information it is impossible to determine which of the many possible isomers is (or are) responsible for mineralocorticoid receptor (Mr) binding and hormone action. However the examination of the structures of a number of steroids that compete effectively for the binding site of a specific Mr may provide a basis upon which to postulate the structural features most conducive to high affinity binding.

A recent study in which Raynaud and coworkers (21) compared the binding affinity of the same 104 steroids for estrogen receptor (Er) from mouse uterus, progestin receptor (Pr) from rabbit uterus, androgen receptor (Ar) from rat prostate, glucocorticoid receptor (Gr) from rat liver and mineralocorticoid receptor (Mr) from rat kidney provides a valuable survey of the structural basis for selectivity of binding to different receptors. Of the compounds studied aldosterone had the distinction of exhibiting high affinity for the Mr and very weak competitive binding to the other four receptors. Of the steroids in this sample only 9α-fluorocortisol (VI in Fig. 1) has higher affinity for the Mr. VI also has high affinity for the Gr, but not for the other three receptors. 7α,17α-dimethyl-18-methyl-17-hydroxy-estra-4,9,11-trien-3-one (VII) was found to have the same affinity for the Mr as aldosterone and even higher affinity for the Pr,Ar and Gr. In contrast 7α,17α-dimethyl-17-hydroxy-estra-4,9,11-trien-3 one (VIII) has less affinity for the Mr than aldosterone and reduced affinity for the Gr but enhanced affinity for the progestin and androgen receptors in comparison with VII. The relative affinities of these potent mineralocorticoids for the various receptors are compared in Table 1 using the graphical technique employed by Raynaud in his review (21). Examination of the chemical structures shows that the only common features are the steroid fused ring system and the 4-en-3-one A-ring. The X-ray

Table 1. Relative affinity of four potent mineralocorticoids for the estrogen (Er), progestin (Pr) androgen (Ar), mineralocorticoid (Mr) and glucocorticoid (Gr) receptors. The relative affinities are taken from the paper of Raynaud et al.(21) using their very effective graphical device.

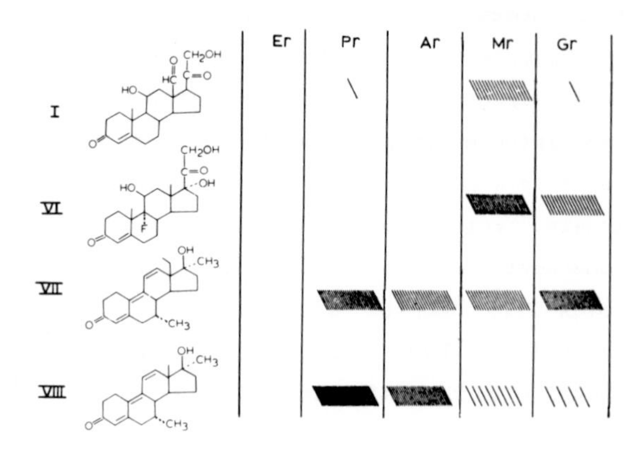

crystal structures of I and VI provide precise details concerning the overall shape of these molecules and the nature of their strongest intermolecular interactions. Although X-ray determinations of VII and VIII have not been reported, that of the parent structure 17α-methyl-17-hydroxy-estra-4,9,11-trien-3-one (IX) has been. The structure determination revealed the existence of the two conformational isomers of

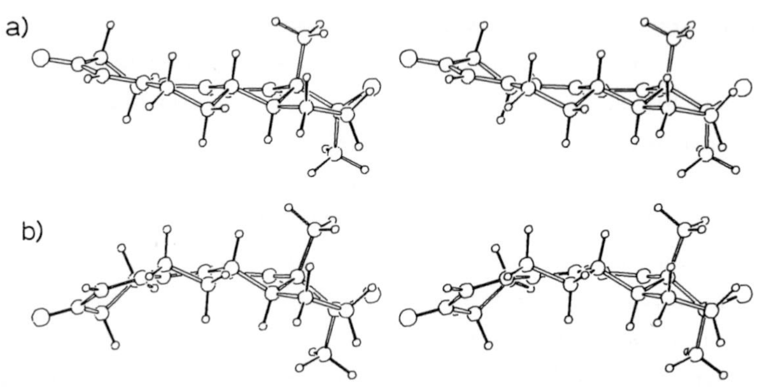

Figure 3. Stereo views of the conformational isomers of 17α-methyl-17 hydroxy-estra-4,9,11-trien-3-one that cocrystallize from methanol solution. Conformer A is flatter and bears resemblance to aldosterone in the A and B ring region, while conformer B is bowed toward the α face like 9α-fluorocortisol.

1X shown in Figure 3. The overall shapes of these isomers are significantly different. The shape of one of them has more in common with aldosterone (II) and the other resembles somewhat 9α-fluorocortisol (VI). The observed flexibility of structures with the 4,9,11-trien-3-one composition is a major contributing factor to their successful competition for four of the principal steroid hormone receptors. The structures of the two conformers were modified by methyl additions to create molecular models for VII and VIII. These models were then subjected to energy minimization using the MM2p program and the minimum energy conformations are illustrated in Figure 4 b,c,e and f.

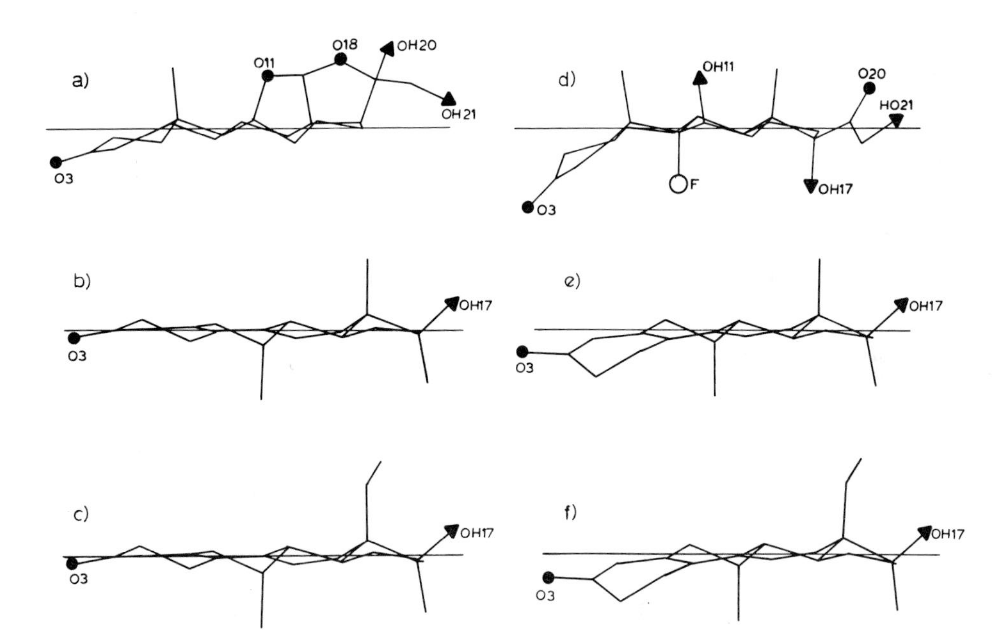

Figure 4. A projection parallel to the least-squares plane through atoms C(5)-C(17) in (a) III, (b) VIIIa, (c) VIIa, (d) VI, (e) VIIIb, and (f) VIIb.

The significant differences between the overall shapes of the four steroids that have high affinity for the mineralocorticoid receptor indicate that a high degree of complementarity between the entire surface of the substrate and the receptor cannot be required. The similarity in the shapes of the steroid backbone and the 4-en-3-one A rings of aldosterone III, and conformer VIIa and VIIIa suggest that this profile of the A and B rings might be conducive to close association with the receptor. Although conformations VIIb and VIIIb bear a superficial resemblance to 9α-fluorocortisol (VI), close examination reveals that the A ring conformations are entirely different. 9α-fluorocortisol has a normal 2β half chair conformation while the A-rings of VIIb and VIIIb have the inverted 1β,2α half chair conformation. The 2β hydrogen is axial in VI and equatorial in VIIb and VIIIb. A significant correlation between the inverted 1β,2α half chair conformation and high affinity binding to the Pr has been established (22). Conformers VIIb and VIIIb most probably account for the high binding affinity of VII and VIII for the Pr. When the A and B rings of the four structures in Table I are subjected to a least-squares process to maximize atomic overlap, the fit depicted in Figure 5 is achieved. The good agreement in overall shape of the A and B rings on the four structures suggest that a binding site might easily accommodate the A and B rings of any of these molecules with little distortion in binding site topography. The specific nature of the

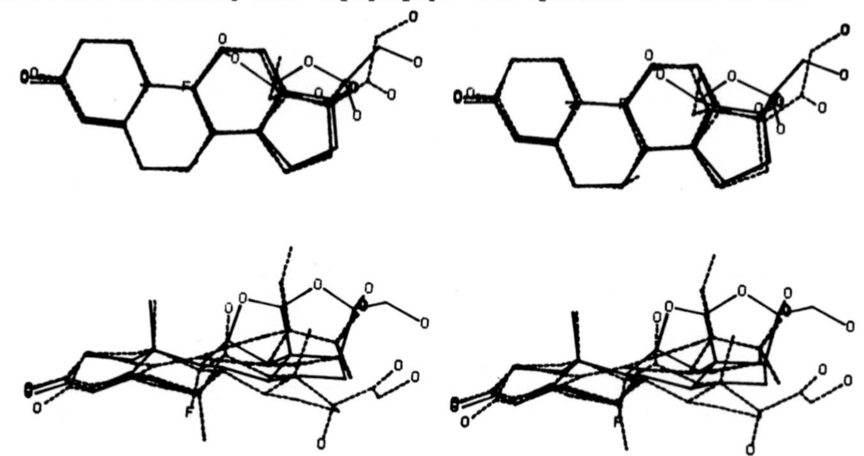

Figure 5. Comparison of the overall shapes of III, IV, VIIb and VIIIb when the A and B rings are superimposed by means of a least-squares fit of the atoms of those rings.

substrate-receptor interactions (hydrogen bond formation, hydrophobic contact and charge transfer interactions) constitute equally important factors in stabilizing the binding. When the projections perpendicular to the least-squares planes through atoms C(5) to C(17) of aldosterone and 9α-fluorocortisol are superimposed (Fig. 6a), a significant difference in the orientation of the A rings relative to the reference planes in these two active compounds is revealed.

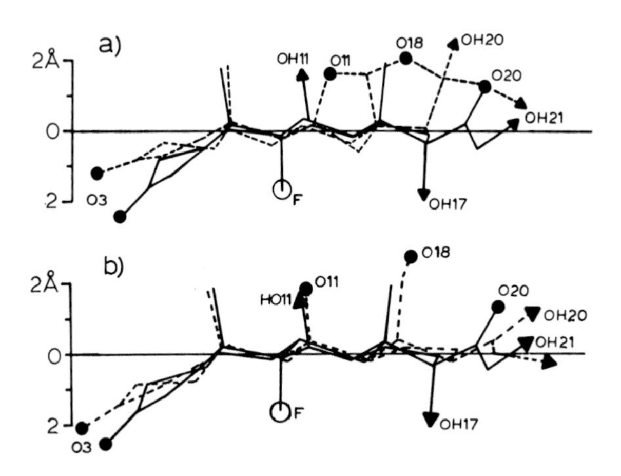

Figure 6. Comparison between the structures of 9α-fluorocortisol and the crystallographic observed conformation of aldosterone III (a) and the theoretical conformation of the 18-aldehyde form of aldosterone (b).

If the 18-aldehyde form of aldosterone were present in vivo and in vitro, it is very likely that β-face crowding in the molecule would produce a conformation bent toward the α-face similar to that observed in 9α-fluorocortisol (Fig. 5b), suggesting that this is the form relevant for comparative analysis. The initial interaction between receptor and substrate may require only the flat Δ^4-3-one A ring. If a bowing toward the α-side and a 21-hydroxyl group are required for activity, a shift of the configurational equilibrium of aldosterone toward form (I) would produce the desired conformation.

Antagonists of mineralocorticoid activity include compounds that compete for the aldosterone binding site of the Mr but fail to elicit the characteristic hormonal response. By comparing the structures of these antagonists with those of the potent mineralocorticoid agonists, additional information on the structural features that are responsible for initiating receptor binding and differentiating between agonist and antagonist response can be obtained. Spironolactone (23) its metabolites (24) and analogues (25,26) (Figure 7) are among the most potent antagonists for the Mr. Analysis of binding and activity data indicate that the intact 17-spironolactone ring is essential for binding to the Mr (25,27,28) and is most conducive to the generation of an antagonist response.

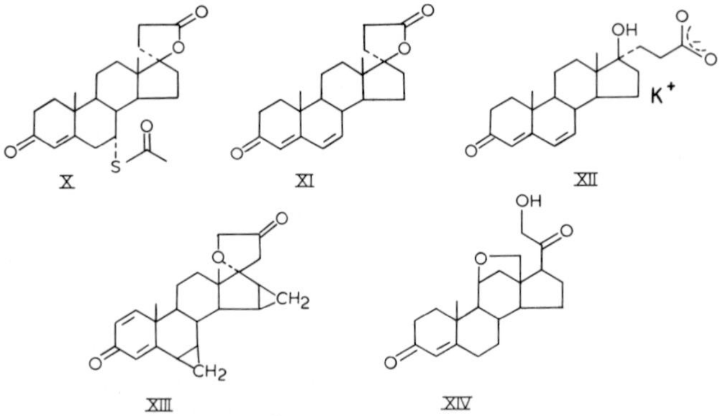

Figure 7. The chemical structures of the mineralocorticoid antagonists (X) spironolactone, (XI) canrenone, (XII) potassium canrenoate, (XIII) spirorenone and (XIV) 18-deoxyaldosterone.

When the A and B rings of canrenone, a principal metabolite of spironolactone, and those of aldosterone are superimposed in Figure 8, the spiro ring is seen to be well below the plane of both molecules. The antagonist behavior of spironolactone and canrenone may be due in part to steric interference caused by the spiro ring placement but it is more likely to be due to the absence of hydrogen bond donating groups capable of mimicking the function of the aldosterone O(20) and O(21) hydroxyl groups. These groups on aldosterone can act as hydrogen bond donors and/or acceptors while those on canrenone and spironolactone could only act as hydrogen bond acceptors.

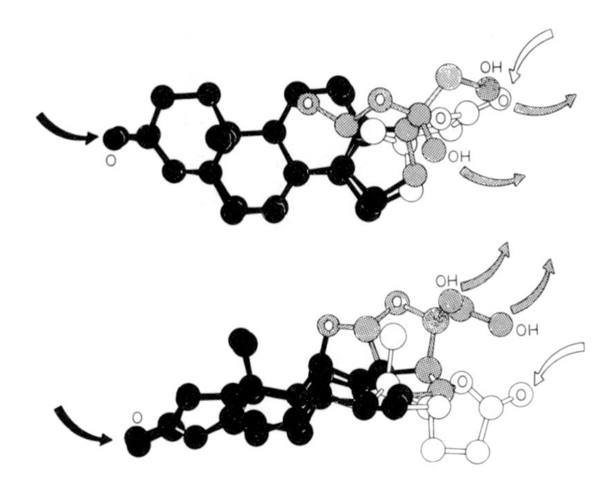

Figure 8. Comparison of the conformation and hydrogen bonding potential of aldosterone and the antimineralocorticoid, canrenone. Common structure dark, agonist shaded, antagonist unshaded.

18-Deoxyaldosterone (21-hydroxy-11β,18-oxido-4-pregnene-3,20-dione) is a derivative of aldosterone in which the aldehyde hemiacetal is replaced by an 11β,18-oxide ring (XIV, Figure 7). The 18-deoxy derivative possesses one-third of the binding affinity of aldosterone for the cytoplasmic aldosterone receptor, but exhibits a 2:1 antagonist to agonist ratio (29). Crystals of 18-deoxyaldosterone contain two crystallographically independent molecules that differ significantly from one another in the orientation of the 17β-side chain (Figure 9). These two conformers are of approximately equal energy and in equilibrium in solution. Both 18-deoxyaldosterone molecules resemble aldosterone in the overall shape of the A, B, C and E rings, and thus would be expected to compete successfully for the aldosterone receptor. In the aldosterone crystal, the two F-ring hydroxyls, O(20) and O(21), each donate and accept a hydrogen bond. Similarly, the O(21) hydroxyl of 18-deoxyaldosterone (both conformations) donates and accepts a hydrogen bond. Although the O(20) carbonyl of 18-deoxyaldosterone is a potential hydrogen bond acceptor, no hydrogen bonding involving this oxygen is observed in the solid state. The similarity between the O(21) hydrogen bond orientations of aldosterone and molecule I of 18-deoxyaldosterone (Fig. 9a) suggests that this conformer

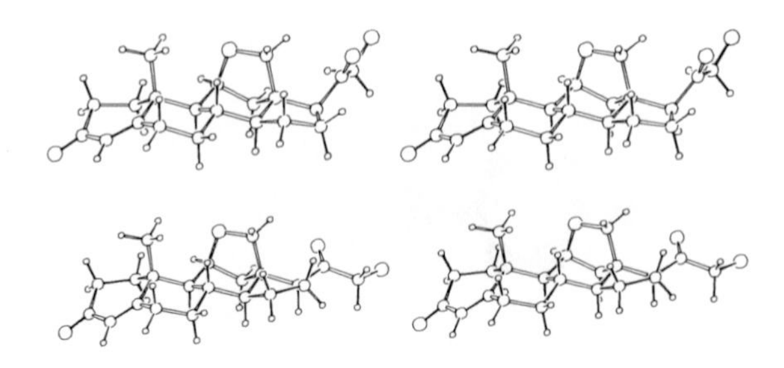

Figure 9. Stereo ORTEP views of the two molecules of 18-deoxyaldosterone illustrating difference in 17β-side chain orientation.

of 18-deoxyaldosterone is responsible for the partial agonism exhibited by the molecule. On the other hand, molecule II appears to have a side-chain orientation that would elicit little or no subsequent activity (Figure 10). Thus, this conformer may account for the antagonist properties of the molecule.

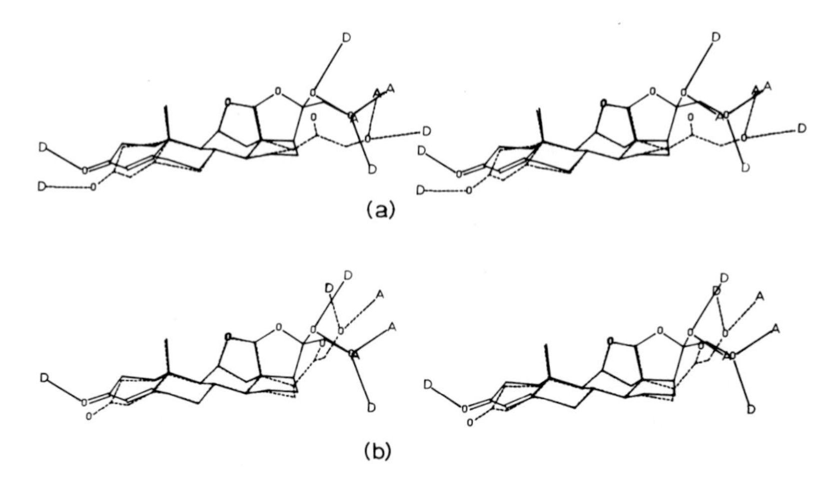

Figure 10. Comparison of crystallographically observed conformations of 18-deoxyaldosterone and aldosterone (solid line) after least-squares fitting of the C and E rings of (a) molecule I and aldosterone, and (b) molecule II and aldesterone. The observed direction of hydrogen bond donors (D) and acceptors (A) in the crystal are indicated.

Glucocorticoids

The glucocorticoid activity of cortisol as measured by the glycogen deposition and anti-inflammatory assays is considerably enhanced by 9α-fluorination. 9α-Chlorination or dehydrogenation of atoms C(1) and C(2) - results in a smaller increase, and 9α-bromination decreases activity (30). The relative activities associated with these changes in the cortisol structure are summarized in Table 2. The conformations of cortisol, its 9α-halogen derivatives, and 6α-methylprednisolone (6α-methyldehydrocortisol), as revealed by single crystal X-ray analysis, have been compared in order to determine whether there are any conformational differences that might be correlated with the differences in biological activity.

Table 2

Enhancement Factors for Various Modifications of Cortisol (30)

Functional group	Glycogen deposition, Rat	Anti-inflammatory, Rat
9α-Fluoro	10	7-10
9α-Chloro	3-5	3
9α-Bromo	0.4	-
1-Dehydro	3-4	3-4

The areas of the glucocorticoid molecule that might be presumed _a priori_ to exhibit the greatest conformational variations are the unsaturated A ring and the C(17) side chain, where some degree of free rotation about the C(17)-C(20) and C(20)-C(21) bonds might be expected. X-ray analysis of over 90 pregnanes has shown that the C(17) side-chain orientation is largely invariant (8). On the other hand, comparative analysis of cortisol and 9α-fluorocortisol suggested that the 9α-substituent influence upon A-ring orientation plays a significant role in altering activity (31). Studies of 9α-bromo- and 9α-chlorocortisol suggested modifications of the original hypothesis.

Since the biological effects of 9α-substitution are proportional to the electronegativity of the 9α–substituent, it was originally proposed that the biochemical differences resulted from an inductive effect of the 9α–substituent on the ionization constant of the 11β–hydroxy group. Consequently, it might be expected that there would be differences in the interatomic distances and valency angles in the C(9) region of cortisol and its halogenated derivatives. Inspection of these parameters reveals only a small amount of variation, and there are no trends that can be correlated with the pharmacological data. The C(9)–halogen distances are consistently longer than expected, but this characteristic of the 9α–halogenated structures is also not correlated with their activities. The bond lengthening may be related to repulsion between the 9α–substituent and the surrounding α–axial hydrogens, and the carbon-bromine distance in 9α–bromo–cortisol agrees closely with the value of 2.028Å observed in another 9α–substitued steroid, 9α–bromo–17β–hydroxy–17α–methylandrost–4–ene–3,11–dione (32).

Cortisol and its fluorinated derivative differ in the orientation of the A ring with respect to the plane defined by the B–, C–, and D–rings. This difference is seen most clearly in Figure 11a, where the two molecules are superimposed in a projection parallel to the least–squares plane through atoms C(5) to C(17) inclusive. This view shows that the A ring in 9α–fluorocortisol is bent underneath the molecule to a much greater extent

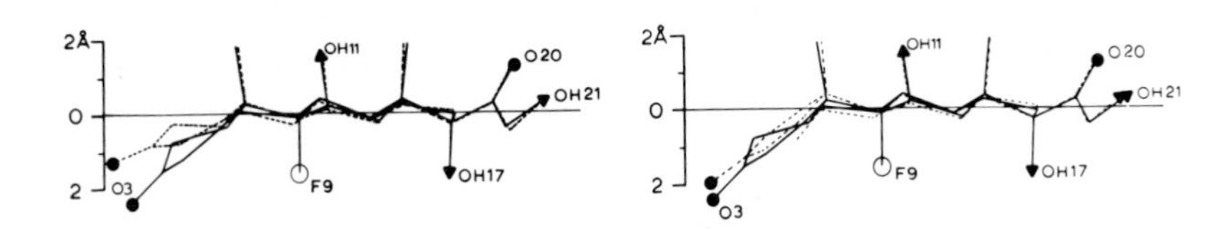

Figure 11. Superposition of the projections perpendicular to the least–squares plane through C(5)–C(17) of (a) cortisol and (b) 6α–methylprednisolone on 9α–fluorocortisol illustrate the similarity in A ring orientation in the two potent glucocorticoids.

than it is in cortisol, and in this respect the conformation of 9α–fluorocortisol resembles the conformation of 6α–methylprednisolone (Figure 11b). This similarity may be significant from a pharmacological viewpoint since both 9α-fluorination and 1,2-unsaturation increase glycogen deposition and anti-inflammatory activity. From inspection of a Dreiding model of 6α-methylprednisolone, it can be seen that the presence of the 1-2 double bond will force the A ring to adopt a conformation similar to what is actually observed in the crystal structure. However, the reason for the sharp bowing of the A ring toward the α-face in 9α-fluorocortisol is not so obvious, and the bowing was not anticipated prior to the crystallographic investigation. Nevertheless, numerous crystallographic observations (33) reveal that this bowing of the A ring toward the α-face is characteristic of steroids having 9α-fluoro, 11β-hydroxy-4-en-3-one composition (Figure 12).

Figure 12. Superposition of three pregnanes having 9α-fluoro substituents showing nearly identical bowing of the A ring toward the α-face.

A second crystallographic determination of the conformation of cortisol is provided by the recently reported pyridine complex (34). The A ring of the cortisol molecule is bent toward the α-face to a higher degree in the pyridine complex than in the methanol complex (Fig. 12). Although the bending is not as great as that observed in 9α-fluorocortisol, it is consistent with the suggested ability of the cortisol to adopt such a bent conformation. Since 9α-fluorination and dehydrogenation of atoms C(1) and C(2) both increase glucocorticoid activity, it is tempting to speculate that the 9α-fluorocortisol conformation is an optimum conformation for binding to a receptor protein. Two conformers of cortisol may exist in equilibrium in vivo with the larger percentage of molecules in a conformation similar to those observed in the pyridine and methanol complexes and a small fraction assuming the 9α-fluorocortisol conformation. Although it may be relatively easy for cortisol to make the

the transition to the 9α-fluorocortisol conformation, the presence of the larger halogens and more severe nonbonded contacts in 9α-chlorocortisol and 9α-bromocortisol may prevent these molecules from assuming the bent conformation.

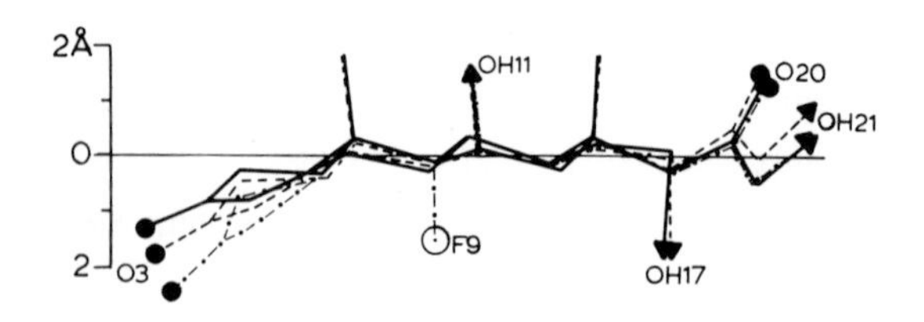

Figure 13. Superposition diagram of the two crystallographically observed conformations of cortisol, solid; cortisol methanol, ---; 9α-fluorocortisol, --•--•-.

If a relationship exists between A-ring orientation and glucocorticoid activity, it must be possible to find a physical basis for the correlation. It is highly probable that steroid hormones interact with receptor proteins in vivo through the formation of hydrogen bonds involving the oxygenated functional groups. Any significant change in A-ring orientation will affect the distances between O(3) and the other oxygen atoms; these distances in the crystallographically observed structures of the corticoids, are given in Table 3. There is little difference in the O(3)-O(11) distances, but the O(3)-O(17) and O(3)-O(21)

Table 3. Distances (Å) between O(3) and Other Oxygens

	Cortisol	9α-Fluoro-cortisol	9α-Chloro-cortisol	9α-Bromo-cortisol	6α-Methyl-prednisolone
O(3)-O(11)	6.847	6.827	6.71	6.81	6.593
O(3)-O(17)	9.683	9.032	9.774	9.668	9.196
O(3)-O(20)	11.739	11.453	11.691	11.815	11.423
O(3)-O(21)	12.522	12.091	12.487	12.479	11.905

distances average about 0.5Å less in 9α-fluorocortisol and 6α-methylprednisolone. Consequently, a protein with functional groups at an optimum separation for simultaneous binding to the O(3) and O(17) or O(21) atoms of these molecules might not bind as readily to cortisol.
The relative distances between the positions of the atoms acting as hydrogen bond donors and acceptors in the crystal structures of cortisol-methanol and 9α-fluorocortisol are illustrated in Fig. 14. It is conceivable that analogous hydrogen bond donors and acceptors are present on the corticoid receptor in approximately these positions.

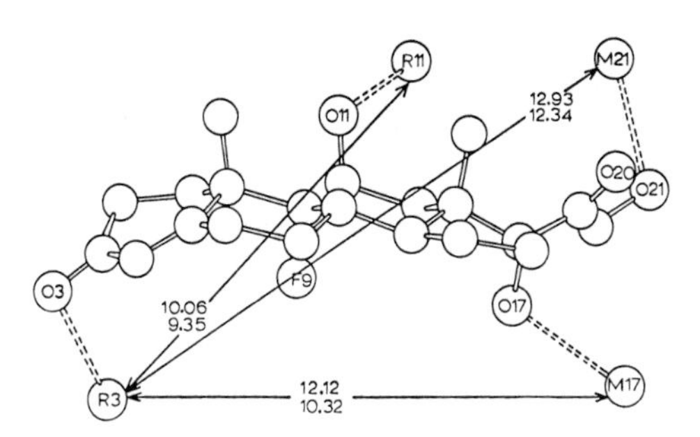

Figure 14. Distances between projected points of attachment to cortisol (top) and 9α-fluorocortisol, R = receptor, M = macromolecule.

A number of steroids that have been found to have varying degrees of antiglucocorticoid activity (35-45) are illustrated in Figure 15. The only common structural feature of these compounds is their 4-en-3-one composition on a steroid frame. Apparently this structural feature which they also share with the most potent glucocorticoid agonists is responsible for their ability to compete successfully for binding to the glucocorticoid receptor. The variation in hydrophobic and hydrophilic character and steric bulk of the substituents on the B, C, and D rings suggests that outside of the 4-en-3-one ring the structural requirements for binding are not very stringent. The crystallographically observed conformation of the glucocorticoid antagonists XV, XVI, XVII, XXI) and XXIII (Figure 15) are compared with that of the potent agonist 9α-fluorocortisol in figure 16. The antagonist properties of testosterone (35), progesterone (36), cortexolone (37,38,39), Δ1,9(11)-deoxycortisol (40), and Δ'1,4-11-oxa-11-deoxycortisol (41) appear to stem from the

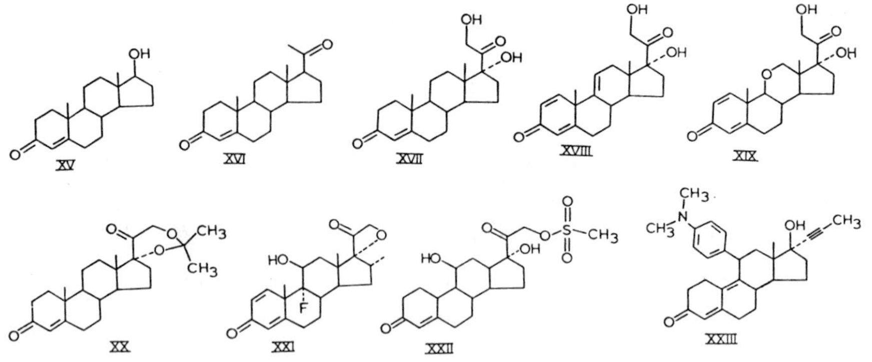

Figure 15. Structures exhibiting varying degrees of antiglucocorticoid activity include (XV) testosterone, (XVI) progesterone, (XVII) cortexolone, (XVIII) Δ1,9(11)-deoxycortisol, (XIX) Δ 1,4-11-oxa-11-deoxycortisol, (XX) cortexolone 17,21-acetonide, (XXI) dexamethasone oxetanone, (XXII) cortisol 21-mesylate, and (XXIII) 11β-(4-dimethyl aminophenyl)-17β-hydroxy-17α-(prop-1-ynyl)-estra-4,9-dien-3-one, RU38486.

absence of the hydroxyl groups at C(11), C(17) and C(21). While these hydroxyl groups are not required for binding, one or more of them must perform a function subsequent to binding that is crucial to hormonal response.

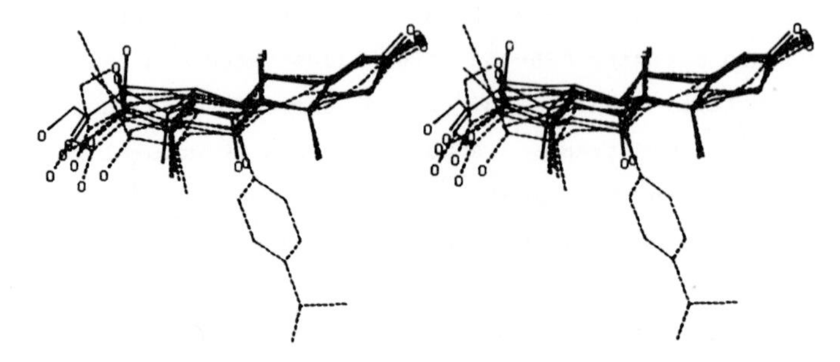

Figure 16. A comparison of the overall conformations of the glucocorticoid antagonists XV, XVI, XVII, XXI and XXIII (dashed) with that of the potent agonist 9α-fluorocortisol (solid line).

The antagonist properties of cortexolone 17,21-acetonide (42), can be due not only to the absence of the 11-hydroxyl substituent but to the fact that the 17α and 21 hydroxyls have been substituted by the acetonide, removing the possibility of hydrogen bond donation. The replacement of the 11-hydroxyl with a 4-dimethyl aminophenyl group in RU38486 indicates that not only is the 11-hydroxy not required for receptor binding (43) but that the receptor can tolerate the presence of a very bulky substituent in the 11β-position. Since no X-ray crystal structure has been reported for RU38486, the conformation shown in Figure 17, was obtained using the MM2p program. The antagonist behavior of dexamethasone oxetanone (44)

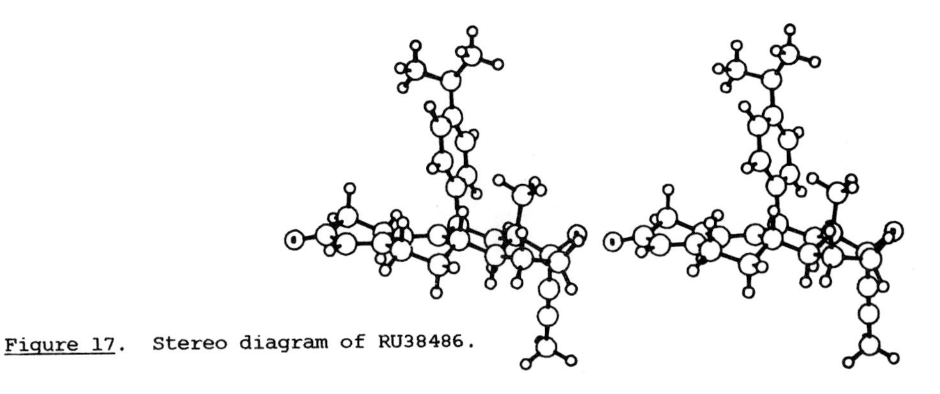

Figure 17. Stereo diagram of RU38486.

and cortisol mesylate demonstrates that antagonism is not simply due to removal of the 11β-hydroxy group. In dexamethasone oxetanone the hydrogen bonding capabilities of O(17α) and O(21) have both been removed by the oxetanone formation. In the case of cortisol 21-mesylate (45) it would appear that blocking the 21 hydroxyl group alone is sufficient to produce antagonist behavior.

When the X-ray crystal structures of dexamethasone and dexamethasone oxetanone are compared, the A, B, and C rings of the agonist and antagonist are found to have nearly identical conformations but the differences in the D-ring are appreciable (Figure 18). In this case there does not appear to be a steric impediment to receptor interaction since the bulk fit in the D-ring region is good. The primary difference between the structures is in the chemical character of the D-ring region of the molecule. The most obvious difference is in the hydrogen bonding capabilities of the D-ring substituents. While both agonists and antagonists can accept a hydrogen bond to O(20) only the agonist can also donate two hydrogen bonds to effect or stabilize a receptor or macromolecular interaction.

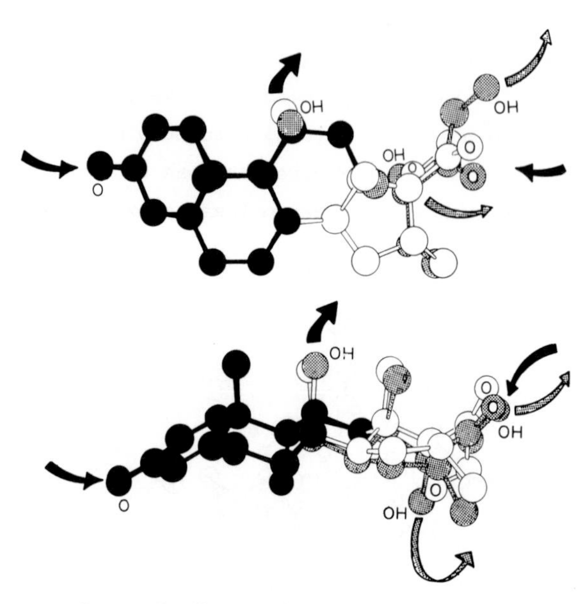

Figure 18. Comparison of the conformation and hydrogen bonding potential of dexamethasone and the antiglucocorticoid dexamethasone oxetanone. Common structure dark, agonist shaded, antagonist unshaded.

It has been suggested that the pregn-4-en-11-ol-3,20-one structure contains the minimal structural requirements for glucocorticoid receptor binding and hormonal expression (46,47,48). Examination of the structure and conformation of potent glucocorticoid agonists and antagonists indicates that their competition for the same binding site is primarily due to the similarity in composition of their A-rings. The 11-hydroxy group appears to be more important for activity than for binding. Many of the best antagonists (Figure 15) do not have the 11-hydroxyl. Cortexolone, which lacks the 11-OH, forms a complex with the receptor which translocates to the nucleus, but does not bind to chromatin or induce the characteristic hormonal response (49,50). The 11β-hydroxy substituent slows down the rate of dissociation from the receptor (51) and is essential to the expression of glucocorticoid action (51,52). The glucocorticoid activity of 11β,17β-dihydroxy-17α-pregna-1,4,6-trien-20-yn-21-methyl-3-one (RU26988) (Figure 19) demonstrates that neither a 17α-hydroxyl nor a 21 hydroxyl is essential for binding or activity (53).

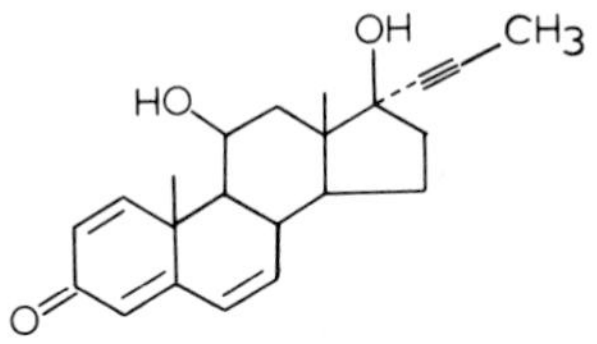

Figure 19. 11β,17β-dihydroxy-17α-pregna-1,4,6-trien-20-yne-21-methyl-3-one, a potent competitor for the glucocorticoid receptor having high local antiinflammatory activity.

RU26988 has over twice the ability of unlabeled dexamethasone to compete with dexamethasone receptor (53) and exhibits high local antiinflammatory effects (54). Nevertheless blocking of both the 17α and 21 hydroxyls by an acetonide or of the O21 hydroxyl only with a mesylate is sufficient to convert a potent glucocorticoid into an anticorticoid. It would appear that the 17β-hydroxy group of RU26988 is serving as a surrogate for one or both of the missing 17α and 21 hydroxyls.

D-ring control of activity

The structural features that are required for binding to the corticoid receptors, and the structural features that differentiate agonism from antagonism, suggest that the A ring plays the primary role in binding, whereas the substituents on the B, C and D rings play the primary role in controlling activity. The possible means by which these other substituents might control activity include (i) inducing or stabilizing an essential conformational state in the receptor (allostery), (ii) influencing the aggregation state of the receptor, or (iii) participating in a direct interaction with DNA or chromatin (see Figure 20).

In model (i), contact between the bound steroid and the receptor (Figure 20) such as hydrogen bonding involving O11 induces and/or stabilizes a change in the receptor conformation. If the required substituent is missing, the steroid will bind, but the conformational change in the receptor will not be stabilized and the hormone will be more readily released. Such a mechanism would be compatible with the slower dissociation rate of glucocorticoids that have an 11β-hydroxy substituent (51).

Alternatively (ii), the functional groups on the D-ring might stabilize or destabilize the formation of different multimeric forms of the receptor. The nuclear processing steps essential for the expression of corticoid activity, which are partially or completely impaired in corticoid antagonists, could be the stabilization either of a conformational state (i) or of an aggregate state of the receptor (ii) by the substituents on the B, C, and D rings. Finally, when the receptor-steroid complex interacts with DNA, the steroid D ring may be sufficiently exposed to contact the DNA directly. A possible model for such an interaction is provided by the crystal complex of deoxycorticosterone and adenine (55), in which the carbonyl and hydroxyl substituents on the corticoid D-ring form hydrogen bonds to the two nitrogens of adenine that would normally be involved in Watson-Crick base-pairing. Such contacts might be critically involved, either in DNA sequence recognition, or in the activation of transcription by the steroid-receptor complex.

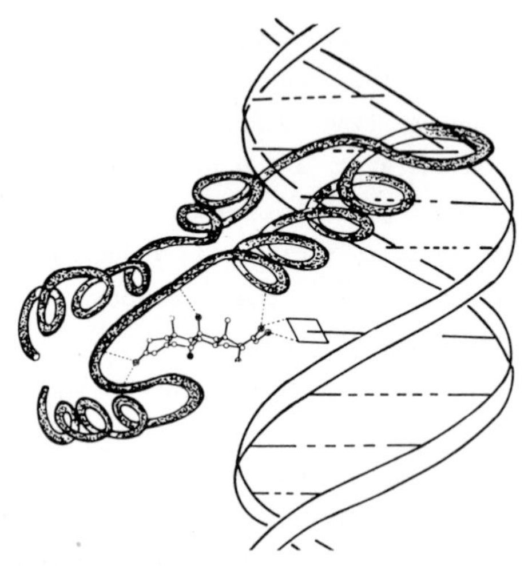

Figure 20. The mechanism of action of glucocorticoid hormones appears to be initiated by A-ring association with the receptors. The 11β-hydroxyl may stabilize a conformation in the receptor slowing the hormone's off rate, and the D ring substituents may participate in the interaction of the steroid receptor complex with DNA.

Summary

Examination of the structures of compounds having high affinity for mineralocorticoid and glucocorticoid receptors strongly suggests that receptor binding is primarily the result of a tight association between the receptor and the steroidal A ring. High-affinity binding to the mineralocorticoid receptor appears to be correlated to a complementary fit between amino acids of the receptor site and a flat 4-en-3-one A ring similar to that imposed upon aldosterone by the 11-18-epoxide formation. The glucocorticoid receptor appears to prefer a 4-en-3-one A ring that is bowed toward the α-face as is the case in structures having a 9α-fluoro substituent or additional unsaturation at C(1)-C(2). The data suggest that specific interactions between the steroid B, C, and D rings and the receptor play at best a minor role in receptor binding but are the most important factor in determining agonist versus antagonist behavior subsequent to binding.

Antagonists that compete for a steroid receptor site may be expected to have the A-ring composition and conformation necessary for receptor binding but lack the 11β-OH and the D-ring conformational features and functional groups that induce or stabilize subsequent receptor functions. Antagonists might also be compounds with A-ring conformations appropriate for binding but other structural features that interfere with susequent receptor functions essential to activity. The possible means by which D rings might control activity include: (i) inducing or stabilizing a specific conformational state in the receptor (allostery), (ii) influencing the aggregation state of the receptor, and/or (iii) participating in a direct interaction with DNA or chromatin. It should be emphasized that these three mechanisms are not mutually exclusive.

Acknowledgements

Research supported in part by NIAMDD Grant No. AM-26546 and DDR Grant No. RR-05716. The organization and analysis of the data base associated with this investigation and several of the illustrations were carried out using the PROPHET system, a unique national computer resource sponsored by the NIH. The authors wish to express their appreciation to Q. Bright, G. Del Bel, C. DeVine, J. Gallmeyer, D. Hefner, K. McCormick, P. Strong and M. Tugac for assistance in the organization and preparation of this manuscript.

References

1. Jensen, E. V ., Jacobson, H. I.: in "Recent Progress in Hormone Research", R. O. Greep (Ed.), Academic Press, New York pp. 387-414 (1962).

2. King, R. J. B., Mainwaring, W. I. P.: "Steroid-Cell Interaction", University Park Press, Baltimore (1974).

3. O'Malley, B. W., Birnbaumer, L.: "Receptors and Hormone Action", Vol. 1, Academic Press, New York (1978).

4. Milgrom, E.: in "Biochemical Actions of Hormones", G. Litwack, (Ed.), Vol. 8, Academic Press, New York, pp. 465-492 (1981).

5. Duax, W. L., Norton, D. A.: "Atlas of Steroid Structure", Vol. 1, Plenum Press, New York (1975).

6. Griffin, J. F., Duax, W. L., Weeks, C. M.: "Atlas of Steroid Structure", Vol. 2, Plenum Press, New York (1984).

7. Duax, W. L., Weeks, C. M., Rohrer, D. C.: in "Topics in Sterochemistry", E. L. Eliel and N. L. Allinger (Eds.), Wiley Interscience, New York, Vol. 9, pp. 271-383 (1976).

8. Duax, W. L., Griffin, J. F., Rohrer, D. C.: J. Amer. Chem. Soc. **103**, 6705-6712 (1981).

9. Duax, W. L., Fronckowiak, M. D., Griffin, J. F., Rohrer, D. C.: in "Intramolecular Dynamics", J. Jortner and B. Pullman (Eds.), D. Reidel Publishing Co., Dordrecht, pp. 505-524 (1982).

10. Campsteyn, H., Dideberg, O., Dupont, L., Lamotte, J.: Acta Cryst. **B35**, 2971-2975 (1979).

11. Duax, W. L., Griffin, J. F., Strong, P. D., Funder, J. W., Ulick, S.: J. Amer. Chem. Soc. **104**, 7291-7293 (1982).

12. Duax, W. L., Cody, V., Griffin, J. F., Rohrer, D. C., Weeks, C. M.: J. Tox. Environ. Health **4**, 205-227 (1978).

13. Duax, W. L., Weeks, C. M.: in "Estrogens in the Environment", J. A. McLachlan (Ed.), Elsevier/North Holland, New York pp. 11-31 (1980).

14. Ham, E. A., Harmon, R. E., Brink, N. G., Sarett, L. H.: J. Amer. Chem. Soc. **77**, 1637-1640 (1955).

15. Genard, P., Palem-Vliers, M., Denoel, J., Cauwenberg, H. V., Eechaute, W.: J. Steroid Biochem. **6**, 201-210 (1975).

16. Simpson, S. A., Tait, J. F., Wettstein, A., Neher, R., Euw, J. V., Schindler, O., Reichstein, T.: Experientia **10**, 132-133 (1954).

17. Gardi, R.: "Hormonal Steroids", L. Martin and A. Pecile (Ed.), Academic Press, New York, NY, p. 107 (1965).

18. Duax, W. L., Hauptman, H.: J. Amer. Chem. Soc. **94**, 5467-5471 (1972).

19. Weeks, C. M., Rohrer, D. C., Duax, W. L.: J. Steroid Biochem. **7**, 545-551 (1976).

20. Allinger, N. L., Yuh, Y. H.: Q.C.P.E. **13**, 395 (1981).

21. Raynaud, J. P., Ojasoo, T., Bouton, M. M., Philibert, D: in "Drug Design", E. J. Ariens (Ed.), Academic Press, New York, Vol. 8, pp. 169-214 (1978).

22. Duax, W. L., Cody, V., Griffin, J. F., Hazel, J., Weeks, C. M.: J. Steroid Biochem. **9**, 901-907 (1978).

23. Cella, J. A., Kagawa, C.: J. Amer. Chem. Soc. **79**, 4808-4809 (1957).

24. Karmin, A., Brown, E. A.: Steroids **20**, 41-62 (1972).

25. Funder, J. W., Feldman, D., Highland, E., Edelman, I. S.: Biochem. Pharmacol. **23**, 1493-1501 (1974).

26. Krause, W., Kühne, G.: Steroids **40**, 81-90 (1982).

27. Wambach, G., Casals-Stengel, J.: Biochem. Pharmacol. **32**, 1479-1485 (1983).

28. Peterfalvi, M., Torelli, V., Fournex, R., Rousseau, G., Claire, M., Michaud, A., Corvol, P.: Biochem. Pharmacol. **29**, 353-357 (1980).

29. Ulick, S., Marver, D., Adam, W. R., Funder, J., W.: Endocrinology (Baltimore) **104**, 1352-1356 (1979).

30. Fried, J.: in "Biological Activities of Steroids in Relation to Cancer", G. Pincus and E. P. Vollmer (Eds.), Academic Press, New York, N.Y., pp. 9-24 (1960).

31. Weeks, C. M., Duax, W. L., Wolff, M. D.: J. Amer. Chem. Soc.**95**, 2865-2868 (1973).

32. Cooper, A., Lu, C. T., Norton, D. A.: J. Chem. Soc., Section B, 1228-1237 (1968).

33. Weeks, C. M., Duax, W. L.: Acta Cryst. **B32**, 2819-2825 (1976).

34. Campsteyn, H., Dupont, L., Dideberg, O.: Acta Cryst. **B30**, 90-94 (1974).

35. Sasson, S., Mayer, M.: Endocrinology **108**, 760-766 (1981).

36. Turnell, R. W., Kaiser, N., Milholland, R. J., Rosen, F.: J. Biol. Chem. **249**, 1133-1138 (1974).

37. Kaiser, N., Mayer, M., Milholland, R. J., Rosen, F.: J. Steroid Biochem. **10**, 379-386 (1979).

38. Kaiser, N., Mayer, M.: J. Steroid Biochem. **13**, 729-732 (1980).

39. Naylor, P. H., Gilani, S. S., Milholland, R. J., Ip, M., Rosen, F.: J. Steroid Biochem. **14**, 1303-1309 (1981).

40. Chrousos, G. P., Barnes, K. M., Sauer, M. A., Loriaux, D. L., Cutler, G. B., Jr.: Endocrinology **107**, 472-477 (1980).

41. Chrousos, G. P., Sauer, M. A., Loriaux, D. L., Cutler, G. B., Jr.: Steroids **40**, 425-431 (1982).

42. Rousseau, G. G., Cambron, P., Brasseur, N., Marcotte, L., Matton, P., Schmit, J.-P.: J. Steroid Biochem. **18**, 237-244 (1983).

43. Chobert, M.-N., Barouki, R., Finidori, J., Aggerbeck, M., Hanoune, J., Philibert, D., Deraedt, R.: Biochem. Pharmacol. **32**, 3481-3483 (1983).

44. Duax, W. L., Griffin, J. F., Simons, S. S.: Abst #690, 63rd Annual Meeting Endocr. Soc., Cincinnati, OH (1981).

45. Simons, S. S., Jr., Thompson, E. B., Johnson, D. F.: Proc. Natl. Acad. Sci. **77**, 5167-5171 (1980).

46. Harmon, J. M., Schmidt, T. J., Thompson, E. B.: J. Steroid Biochem. **14**, 273-279 (1981).

47. Goldstein, A., Aronow, L., Kalman, S. M: "Principles of Drug Action: The Basis of Pharmacology" 2nd Edn., Wiley, New York, 36-39 (1974).

48. Liddle, G. W.: in "Textbook of Endocrinology: Part 1: The Adrenal Cortex", Williams, R. H. (Ed.), W. B. Saunders, Lippincott, PA, 246 (1974).

49. Turnell, R. W., Kaiser, N. Milholland, R. J., Rosen, F: J. Biol. Chem. **249**, 1133-1138 (1974).

50. Wira, C. R., Munck A.: J. Biol. Chem. **249**, 5328-5336 (1974).

51. Aranyi, P.: Eur. J. Biochem. **138**, 89-92 (1984).

52. Naylor, P. H., Gilani, S. S. H., Milholland, R. J., Rosen, F.: Endocrinology **107**, 117-121 (1980).

53. Gomez-Sanchez, C. E., Gomez-Sanchez, E. P.: Endocrinology **113**, 1004-1009 (1983).

54. Teutsch, G., Costerousse, G., Deraedt, R., Benzoni, J., Forten, M., Philibert, D.: Steroids **38**, 651-660 (1981).

55. Weeks, C. M., Rohrer, D. C., Duax, W. L.: Science **109**, 1096-1097 (1975).

11 BETA-SUBSTITUTED 19-NORSTEROIDS : AT THE CROSSROADS BETWEEN HORMONE AGONISTS AND ANTAGONISTS

Georges Teutsch
Centre de Recherche Roussel-Uclaf
93230 Romainville (France)

Introduction

The long and largely unsuccessful search for potent anti-glucocorticoids [1] has recently experienced a major breakthrough with the report on the *in vitro* and *in vivo* activity of RU 38 486, a novel 11β-substituted-19-norsteroid [2,3,4,5]. This compound also proved to be a potent anti-progestin [6,7,8] and has already shown effectiveness as an abortifacient in humans [9]. Numerous analogues of this compound have now been synthesized and tested for their hormonal and (or) anti-hormonal activities [10]. The results obtained so far have allowed us to gain new insights into hormone-receptor interaction. It is the purpose of this report to stress the usefulness of 11β-substituted-19-norsteroids for receptor mapping and their mechanistically relevant importance in the expression of agonistic versus antagonistic activities. Although the main emphasis will rest on the glucocorticoid receptor, consequences for the progestin receptor will also be considered. Since the past lack of convenient ways of access to this class of compounds may be responsible for the relatively late unveiling of their fascinating characteristics, a review on available synthetic pathways has been included.

Syntheses of 11β-Substituted-19-Norsteroids

The 11-keto pathway :

The first documented synthesis of an 11β-alkyl-19-norsteroid is the report in 1970 by Baran et al. on a variety of 11β-methyl-estra-1,3,5(10)-trienes which were further transformed to the estr-4-ene series [11]. The synthetic scheme was based on a methodology independently devised in the tigogenin series by Elks [12] and Kirk and Petrow [13], involving the reaction of an 11-keto function with an organometallic reagent followed by dehydration and catalytic hydrogenation. This approach (scheme 1) has been subsequently applied to the synthesis of 11β-ethyl estrone by Baran et al [14], and to that of 11β-ethyl, n-propyl and n-butyl estrenes by van den Broek et al [15].

Scheme 1

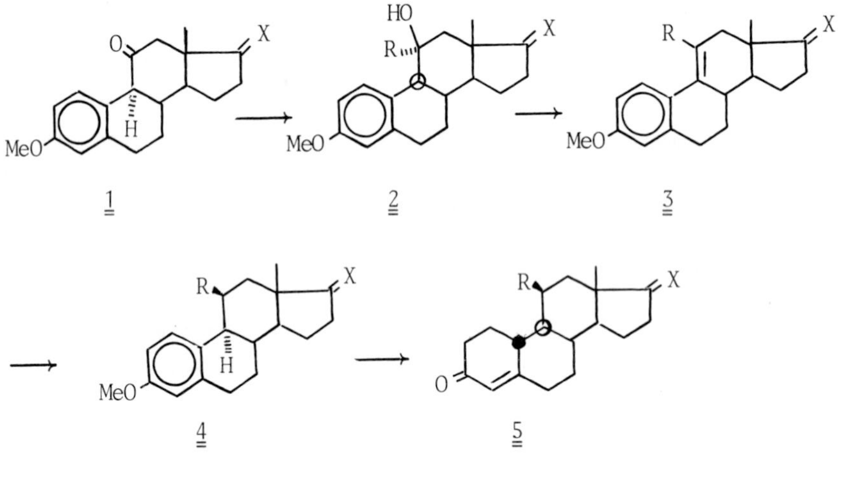

Although this scheme appears to be relatively simple, it is limited by a number of drawbacks which make it impracticable as a convenient and general route to 11β-substituted steroids. Indeed, the starting material, 11-oxo-9β-estradiol-3-methyl ether (1, X =α-H, β-OH) is a readily epimerizable, oxidizable compound which has to be prepared from estradiol in five steps (Liang et al [16]) and handled with great care. Poor yields of the adduct 2 are obtained already for R = ethyl due to competing reduction of the ketone by the Grignard reagent [17]. Furthermore, hydrogenation of the Δ9(11) double bond of intermediate 3 proceeds with poor stereoselectivity, yielding non negligeable amounts of the unwanted 9β-H, 11α-R isomer of 4 [15]. It is also obvious that substituents which would be readily hydrogenated cannot be introduced by this method. Finally it is also of interest to mention that this synthetic scheme does not succeed in the 13-ethyl gonane series [17].

As a close variation, van den Broek [15] used the selective hydrogenation of 11-methylene-5-estrene-3,17-dione-3,17-diethylene acetal 6 to produce the 11β-methyl-estrene 7 (Scheme 2).

Scheme 2

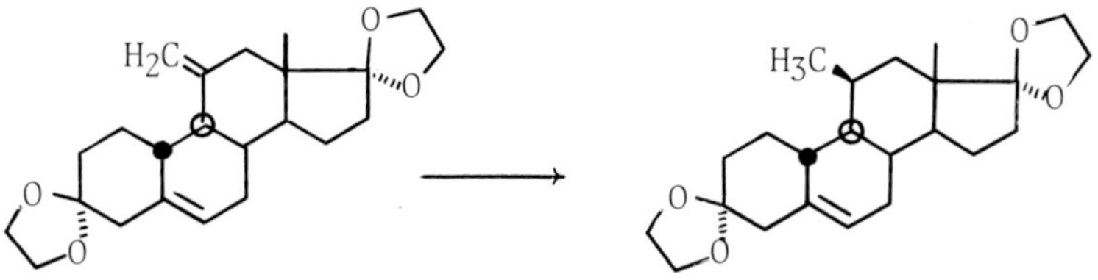

H_2C 6 → H_3C 7

However, the relative difficulty in obtaining the homologous 11-alkylidene-19-norsteroids [18] precludes any generalization of the method. Nevertheless the 11-methylene derivative was used successfully for the generation of the hydroxymethyl, methoxymethyl and chloromethyl substituents [15]. Yet another variantof the 11-keto pathway has been reported by Baran et al [14] who further transformed the previously described 11β-methyl-tigogenin acetate 8 [12][13] to a 11β-methyl-19-norsteroid via reductive aromatization [19] of the corresponding 1,4-dien-3-one 9 (Scheme 3).

Scheme 3

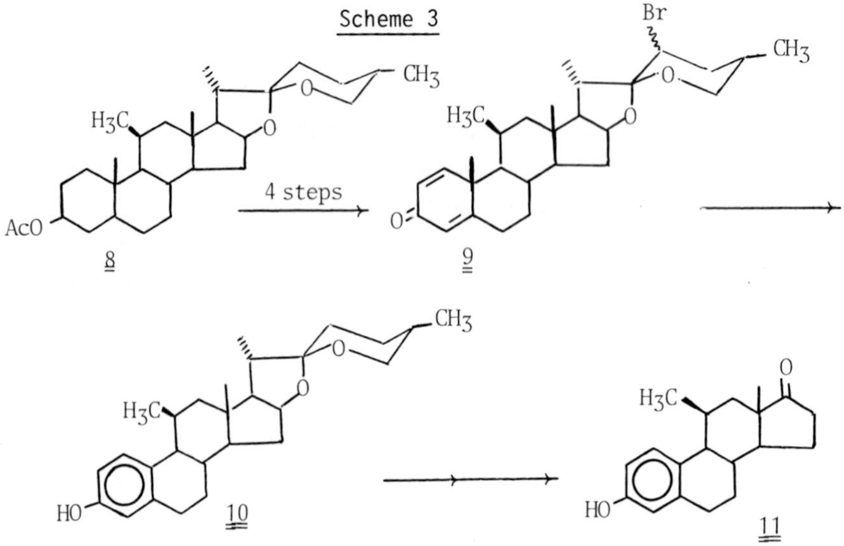

Br
CH_3
O
O
H_3C
AcO
4 steps
CH_3
O
O
H_3C
O
8
9

CH_3
O
O
H_3C
HO
10
O
H_3C
HO
11

Coombs and coworkers also used the 11-keto pathway for the synthesis of 9α, 11β-dimethyl-19-norsteroids [20] starting with 9α-methyl-11-oxo-estrone.

Finally the so far unreported route to a 11β-methyl-estra-4,9-dien-3-one devised at Roussel Uclaf by R. Bardoneschi [21] (Scheme 4) should be mentioned.

Scheme 4

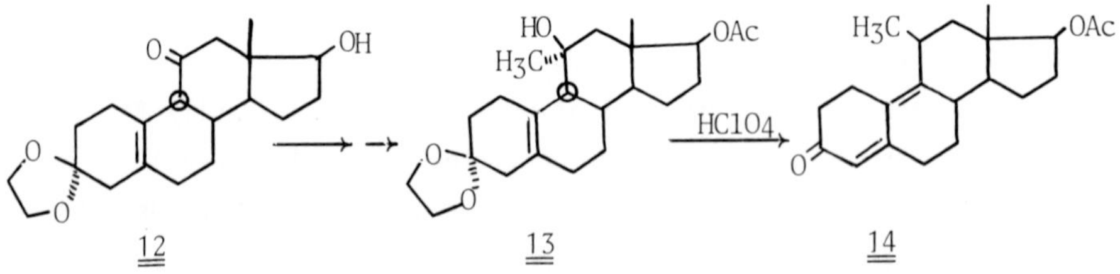

O
OH
O
O
12
HO
H_3C
OAc
O
O
13
$HClO_4$
H_3C
OAc
O
14

The outcome of the dehydration-deketalization step suggests that compound 14, which has an axial 11β-methyl substituent is thermodynamically more stable than the 11α-methyl analogue 15.

This result is confirmed by calculation of the respective minimized energies of the two molecules using the SCRIPT program [22], which shows that 14 is more stable than 15 by approximately 3 Kcal/mole. The minimum energy conformations of the two compounds are depicted in Fig. 1.

Fig. 1

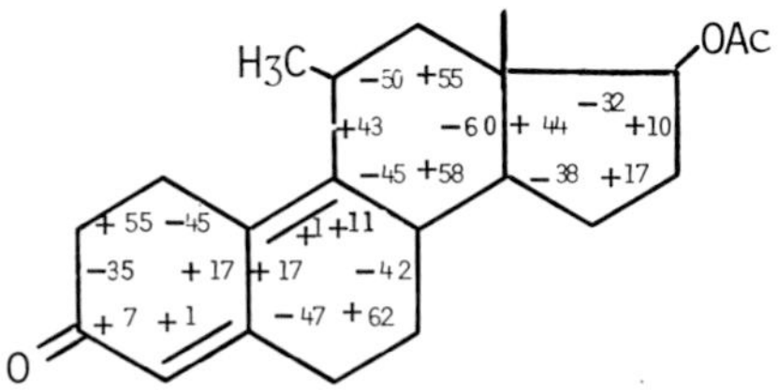

14 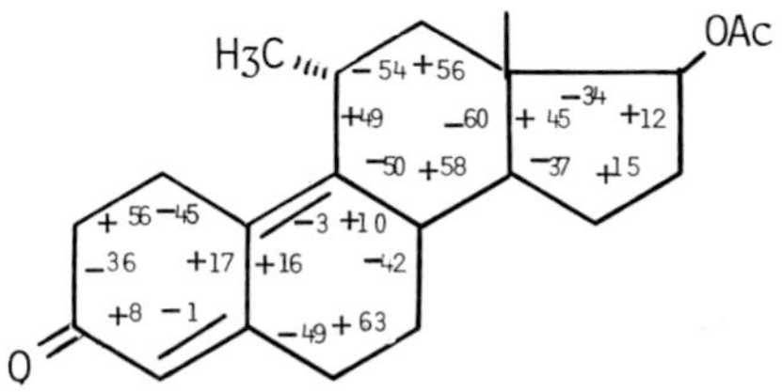 15

Total synthesis :

The inherent difficulties associated with the 11-keto pathway led Garland and Pappo [17] to seek a total synthesis of 11β-methyl-19-norsteroids based on the Torgov estrone synthesis [23]. Although the original scheme did not succeed, probably because the pro-C-11 methyl substituent introduces too much strain in the transition state of the planned cyclisation, the modification depicted in Scheme 5 afforded the desired 11β-methyl derivatives both in the estrane and 13-ethyl-gonane series. However for obvious reasons it could not be expected to constitute a ready and general access to 11β-substituted-19-norsteroids.

Scheme 5

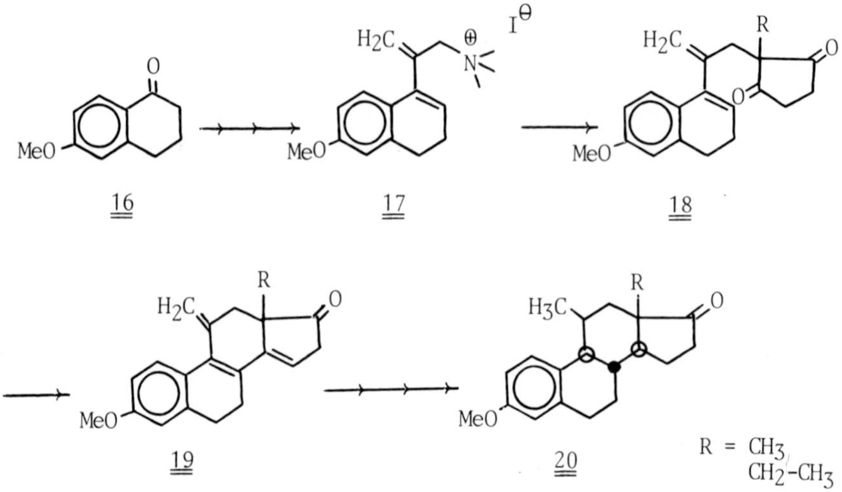

16 17 18

19 20

R = CH_3, CH_2-CH_3

Alternatively Groen and Zeelen investigated the possibility of achieving a synthesis of 11β-methyl estrone via the biomimetic polyene cyclization [24] developed by Johnson [25]. However they found that the major isomer formed had the unwanted 11α configuration.

The epoxide pathway :

Epoxidation of 5(10), 9(11)-estradienes by meta-chloroperbenzoic acid had been shown to yield mainly the 5α,10α epoxide along with smaller amounts of the 5β, 10β and 9α, 11α isomers [26]. The selectivity, which is also dependant to some extent on ring D substitution could be improved by replacing meta-chloroperbenzoic acid with hexafluoroacetone hydroperoxide [27][28] or its catalytic equivalent - 85% hydrogen peroxide in the presence of 0.2 equivalents of hexafluoroacetone sesquihydrate [29][30]. The combination 30% hydrogen peroxide-hexachloroacetone proved to be equally effective [31], (Table 1).

Table 1 : Epoxidation of a steroidal 5(10), 9(11)-diene by various reagents

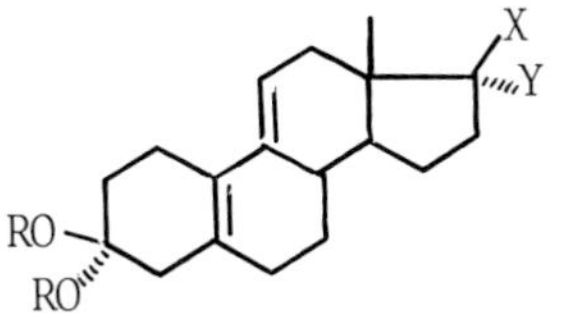

21

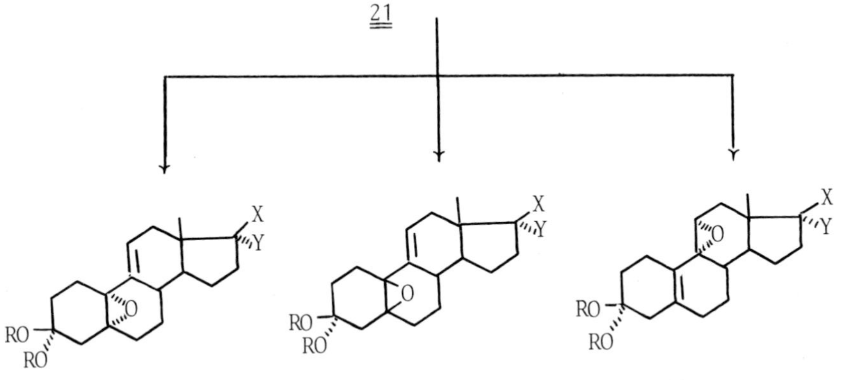

22 23 24

Reagent	Product(s) obtained	References
m-Cl perbenzoic acid	22 + 23 + 24	[26]
$(CF_3)_2C{<}^{OH}_{OOH}$	22 + 23	[28]
$(CF_3)_2C{=}O$, 1.5 H_2O + H_2O_2 or $(CCl_3)_2C{=}O$ + H_2O_2	22 + 23	[31]
H_2O_2, PhCN	24	French Patent 2 213 272 (1973)

The 5α, 10α epoxides of type 22 are prone to nucleophilic opening with Grignard reagents to afford 10β-substituted steroids [32]. This property has been used to advantage in the formal total synthesis of cortisone [33] and of the 10-ethynyl analogue of hydrocortisone [34] (Scheme 6).

Scheme 6

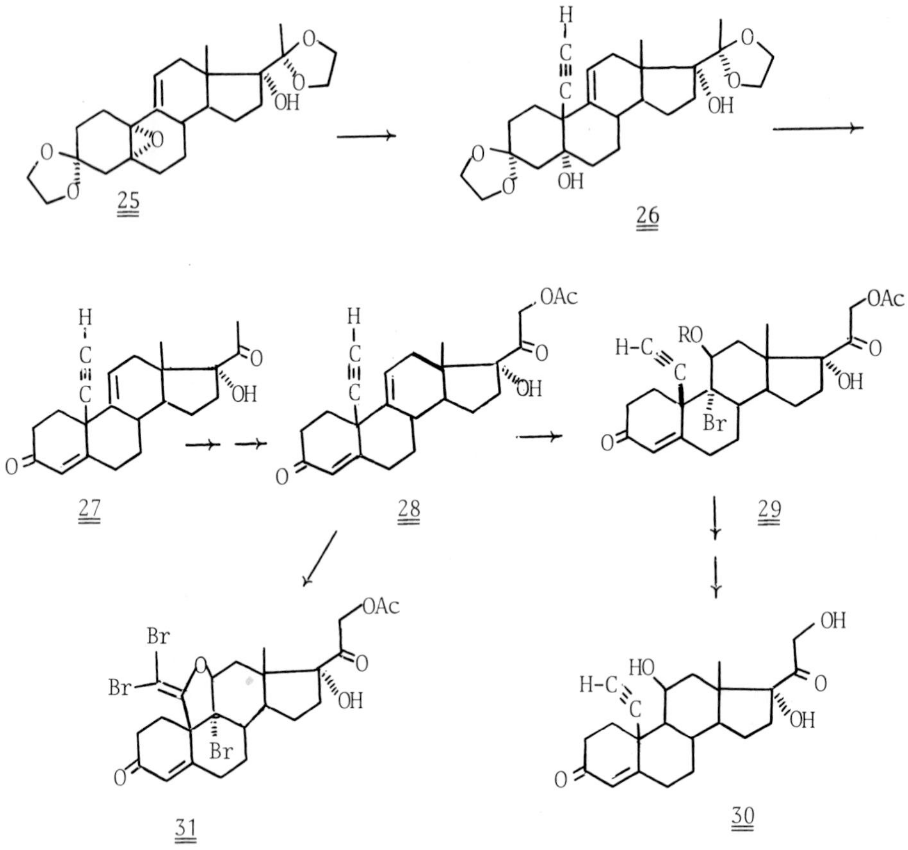

But yields of the product decrease rapidly with increasing carbon chain length of the aliphatic organo magnesium halides, due to a reductive side reaction. Thus, the scope of 10β-substitution seems to be quite limited.

When the Grignard reagent was replaced by a lithium cuprate, we were delighted to find that the epoxide had been opened in a conjugate manner, to yield the 11β-substituted compound as the sole reaction product. Copper chloride catalyzed Grignard reagents led to the same result [35]. This synthetic scheme was subsequently used to prepare a number of 11β-substituted 4,9-estradienes [36,37] and 1,3,5(10) estratrienes [36] (Scheme 7).

Scheme 7

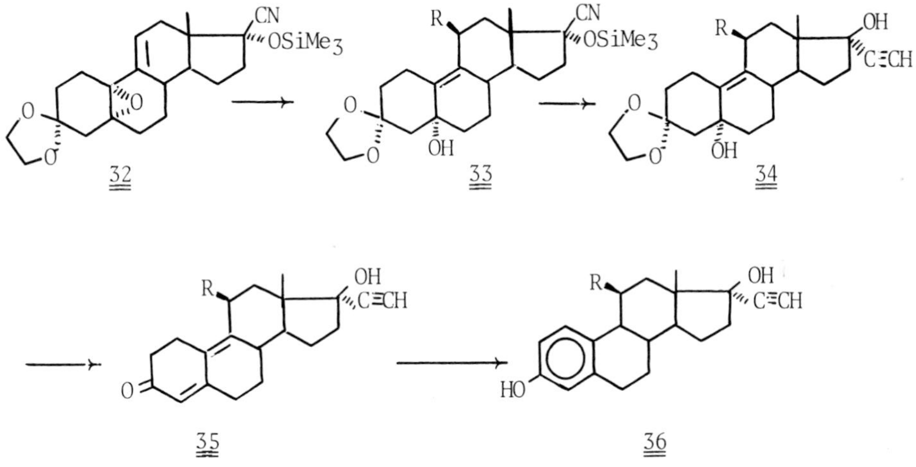

This pathway proved to be extremely efficient, and with a few exceptions almost any R group can be introduced, provided that the corresponding lithium (RLi) or Grignard (RMgX) reagent is accessible. Thus for instance some difficulties which arose with the pyridyl group were easily solved by using the magnesium cuprate [38].

Steric hindrance, is in general not a problem as shown by the high yield introduction of an 11β-tertiary butyl group [35]. Table 2 shows a few representative examples of organic groups which can be introduced in the 11β-position of 19-norsteroids.

Table 2 : Representative examples of organic groups which have been introduced in the 11β-position of 4,9-estradienes

Saturated	Cyclic	Unsaturated	Aromatic	Heteroaromatic
Methyl	Cyclopropyl	Vinyl	Phenyl	2,3 or 4-pyridyl
Ethyl	Cyclopentyl	Allyl	o,m,p-OMe-phenyl	2 or 3-thienyl
n-Propyl		1-Methoxyvinyl	m,p-halophenyl	2-furyl
i-Propyl		Allenyl	etc.	
n-Butyl				
t-butyl				
n-decyl				
CH_2SiMe_3				

In spite of the fact that conjugate opening of allylic epoxides with organocuprates had been known in the butadiene [39,40] and cyclohexadiene [41,42] series the strict regio- and stereospecificity we observed came somewhat as a surprise. Indeed, a one step SN_2' pathway, by which the incoming nucleophile would develop an increasing 1,3-diaxial interaction with the angular methyl group appears highly unlikely. The fact that good nucleophiles like mercaptide, azide, thiocyanate [43] or acetylide [32,34] give only 10β-substitution of the epoxide (Scheme 8) led us to propose a two step mechanism (Scheme 9) which could account for the experimental facts [35,44].

Scheme 8

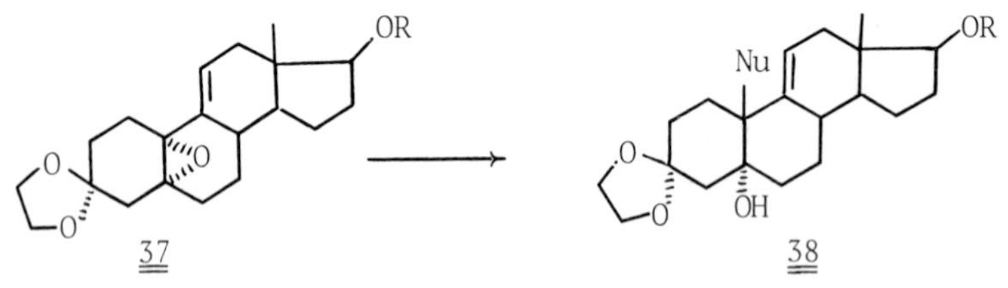

37 → 38

Nu = SPh, N_3, NCS, C≡CH

Scheme 9

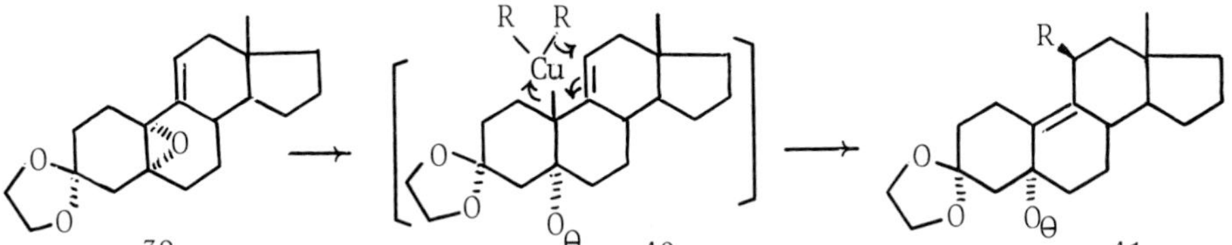

39 40 41

As a conclusion to this section, it should be stressed that conjugate epoxide opening is by far the most convenient method to synthesize the 11β-substituted 4,9-estradienes and 1,3,5(10) estratrienes.

Entry into the 4-estrene series involves a reduction followed by reconjugation of the 5(10) double bond (Scheme 10).

Scheme 10

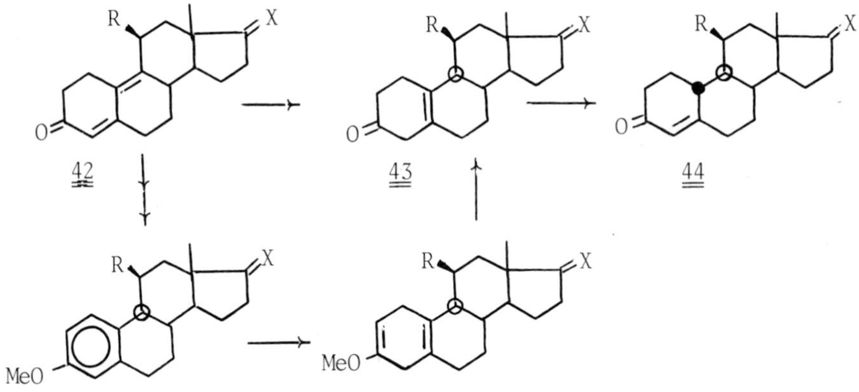

42 43 44

45 46

The direct reduction to the non conjugated ketone 43 with the correct 9α-stereochemistry is generally possible [45,46] except for very bulky 11β-substituents like t-butyl. In this latter case, three main components were isolated after lithium-ammonia reduction, resulting respectively from 5α, 5β and 9β protonation (Scheme 11)[47].

Scheme 11

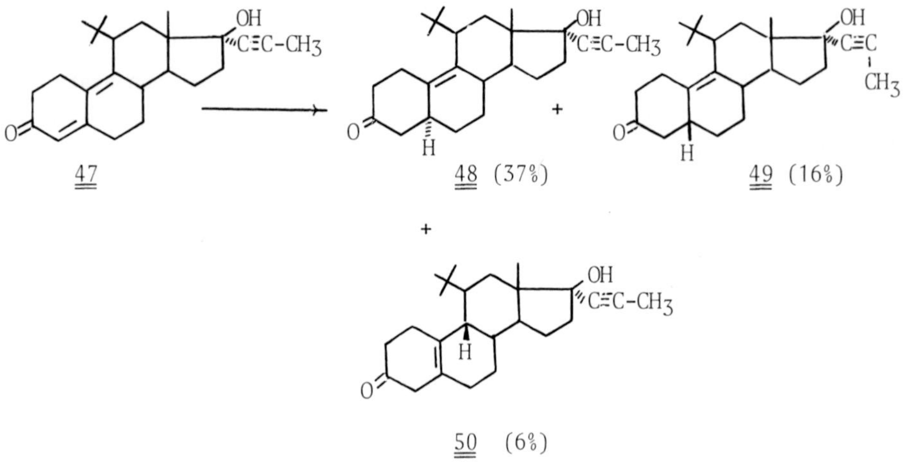

OH
C≡C-CH3
47
48 (37%)
49 (16%)
50 (6%)

A less direct access to 43 is also possible via Birch reduction of the A-ring aromatic steroid 45 (Scheme 10) and has been shown to work with 11β-substituents such as methyl [11][15], ethyl, n-propyl, i-propyl, n-butyl [15], vinyl [48] and isopropoxymethyl [49]. Reconjugation of the double bond by protonation at carbon 10 is possible for a variety of 11β-substituents (methyl, ethyl, n-propyl, i-propyl, n-butyl, vinyl, benzyl and isopropoxymethyl)[11,15,46,48,49] but has so far failed for aromatic substituents [46,48]. This result suggests that an 11β-phenyl ring could be more of a steric hindrance to 10β-protonation than an 11β-isopropyl substituent, a hypothesis which does not appear to be obvious on examination of molecular models. Another, so far undetected reason, may be responsible for this unusual behaviour.

Structural Features :

As we are going to discuss interactions of this class of compounds with receptor proteins it is important to know their three dimensional structure, and most importantly establish the beta-configuration of the C-11 substituent.

Early determinations were mainly based on ^{1}H-NMR although the pairs of compounds to be compared differed both by the configurations at C-9 and at C-11 [14][15]. The major argument to distinguish between the 11β and the 11α-methyl derivatives in the estra-1,3,5(10) triene series was that the 11α-methyl group would be close to the plane of the aromatic A-ring, whatever the C-9 configuration and would thus be considerably deshielded. This prediction turned out to be correct and further evidence was provided in the case of 11β-hydroxymethyl 4-estrene-3,17-dione by using the INDOR technique [50] to assign the coupling constants of the ABX system demonstrating the equatorial nature of the 11α-proton [15]. Still more convincing proof came with the advent of the epoxide route which allows access to both the 11β and the 11α isomers [35][44]. We found that for the pair of compounds <u>51</u> and <u>52</u>, the difference in chemical shift of the C-13 methyl group in the ^{1}H-NMR spectrum was small but significant (0.07-0.1 ppm) when R is an aliphatic group, the β substituent leading to the more deshielded (low field) signal as predicted by intramolecular Van der Waals interaction [51]. If however, R is an aromatic group, β substitution will induce strong shielding ($\sim$ 0.5 ppm) of the C-13 methyl group relative to α substitution. This effect was expected as the methyl group is located within the shielding cone [52] of the aromatic ring. This anisotropic effect of the phenyl ring has also been recently used for the structural assignement in the bicyclo [3.2.1] octane series [53].

The same conclusion can be drawn by using the dienones represented by the general formula <u>53</u> and the chemical shift of the angular methyl protons for a few selected compounds are summarized in Table 3.

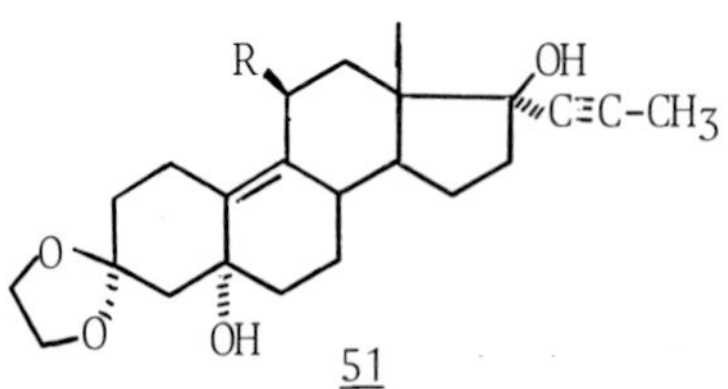

Table 3 : Chemical shifts of the C-13 Me protons in the 60 MHz spectrum of 11β-R-17β-hydroxy-17α-propynyl-4,9-estradiene-3-ones (53)

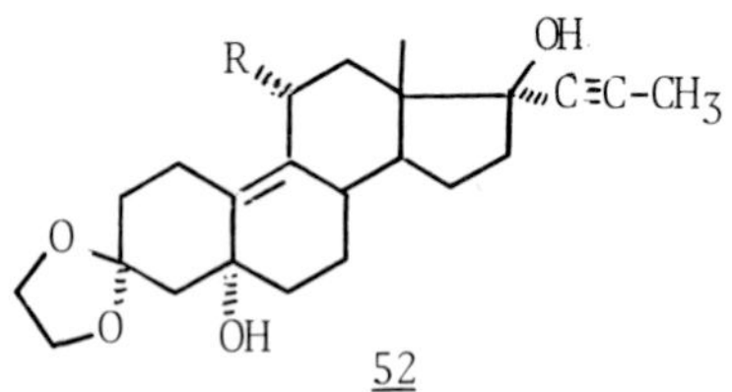

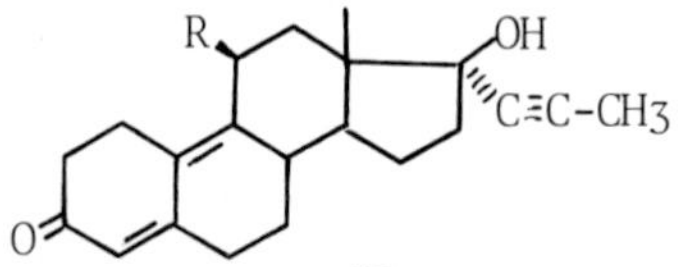

R	δ (Hz)	R	δ	R	δ	R	δ
H	60.5	Cyclo Pr	70	Ph	31	2-thienyl	40.5
Me	64	Cyclo Pen	68.5	p-OMe C_6H_4	32	3-thienyl	34
Et	63	Vinyl	58.5	p-Cl C_6H_4	31.5	2-furyl	36
n-Pr	64	Allenyl	62.5	p-F C_6H_4	30.5	2-pyridyl	27
i-Pr	65			p-Br C_6H_4	30.5	3-pyridyl	31
t-Bu	66.5			p-CF_3 C_6H_4	29.5	4-pyridyl	31
				p-NMe_2 C_6H_4			

Fig. 2 : Eclipsing of the plane of the phenyl ring in an 11β -phenyl-4,9-estradiene with de C-9-11 bond

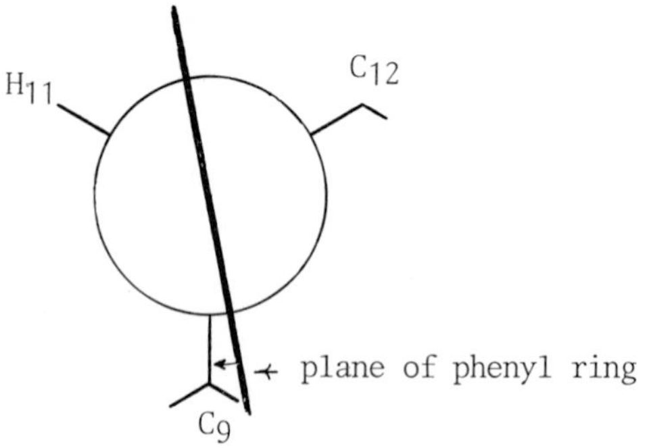

Finally direct proof of structure was obtained by X-ray diffraction measurements on certain 11β-substituted dienones [54][55]. These studies also gave some idea concerning the conformational aspects related to 11β-substitution. It was found, for instance, that the planes of aromatic or vinylic substituents closely eclipse the C_9-C_{11} single bond. A very similar situation is believed to occur in solution as can be infered from the strong shielding of the angular methyl group in the NMR. Indeed, using the correlations developed by Bovey for the anisotropic effect of the benzene ring [52] it is found that the experimental 0.5 ppm shielding can be related to an approximately zero degree dihedral angle between the plane of a 11β phenyl ring and the C_9-C_{11} single bond (Fig. 2).

Energy minimisation, using the SCRIPT program [22] led to similar though not identical results, as the value found for the dihedral angle was + 20° (compared to +9° from X-ray data).

However for practical purposes it can be assumed that the quasi-eclipsing corresponds very closely to reality. This knowledge was essential for the subsequent mapping of the receptor.

Structure-Activity Relationships :

The much improved access to 11β-substituted 19-norsteroids prompted us to investigate the effect of various C-11 substituents on relative binding affinity (RBA) for the steroid hormone receptors. The first results showed a completely unexpected enhancement of the RBA's for the progestin receptor (PR), when the substituent was vinylic or aromatic [36]. This property was subsequently used to design RU 25 253 (54) a very potent progestin [37]. Even more surprising, at the time, was the observation that the same substituents also induced an unusually high RBA for the glucocorticoid receptor (GR)[56][57]. The 11β-thienyl derivative RU 25 055 (60a) was shown to be devoid of any glucocorticoid type activity [58] but proved to antagonize dexamethasone in vitro for TAT induction in HTC [57,59,60], however the detection of an antiglucocorticoid activity in vivo remained elusive [58]. These results were encouraging enough to start the search for more potent anti-glucocorticoids in this series of compounds. As a first approach we decided to make large variations in the nature of the 11β-substituent, keeping the C-17 substitution unchanged. To that effect, we chose to use the 17α-propynyl side chain instead of ethynyl. Indeed, we had previously shown that it is possible to mimic the characteristic dihydroxy-acetone side chain of glucocorticoids like hydrocortisone or prednisolone by the 17α-propynyl-17β-hydroxy moiety [61,62,63] but not by the 17α-ethynyl-17β-hydroxy moiety [61](Table 4).

When transposed to the estradiene series, this concept appeared to be partly justified as can be deduced from the results on Table 5. Generally, for the same 11β-substituent both RBAs for GR and PR increased when replacing the 17α-ethynyl group by a propynyl group.

One of the first compounds synthesized in this research, RU 38 486 (61d), displayed a remarkable anti-glucocorticoid activity both in vitro and in vivo [2,3,4,5,64]. The high RBA for the PR, on the other hand, turned out to be linked with potent anti-progestin activity [6,7,8,64]. Thus one single compound provided us with the long sought breakthrough, both in the anti-glucocorticoid [1] and in the anti-progestin [65] fields. Numerous compounds have been synthesized since, with the objective of dissociating the anti-hormonal activities. It was assumed that if we could achieve selectivity in binding to either the PR or the GR, the specificity in biological activity would follow automatically. Table 6 displays some of the results which were obtained, showing that it is possible to a limited extent, to reduce the RBA for the PR without affecting too drastically the binding to the GR.

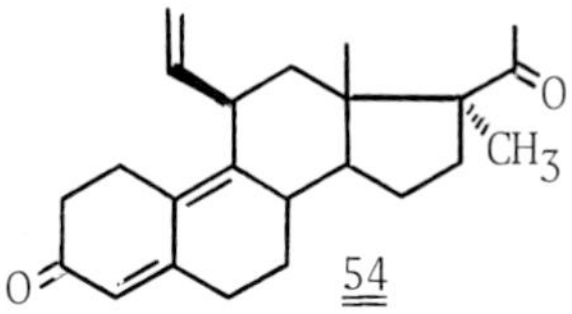

54

Table 4 : Comparative biological evaluation of 17α-alkynyl steroids and cortisol (58) and prednisolone (59)

		RBA GR(a)	RBA MR(b)	Ear Edema EC_{50} (mg/ml)	Compd.
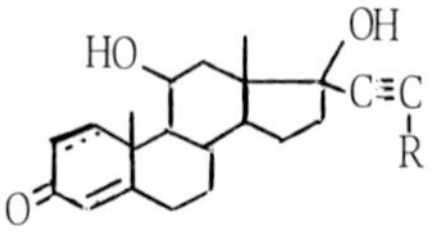 HO, OH, C≡C-R	R=H Δ4	5.1	0	IN 1	55
	R=Me Δ4	68	0.3	0.6	56
	R=Me Δ1,4	100	0.4	0.07	57
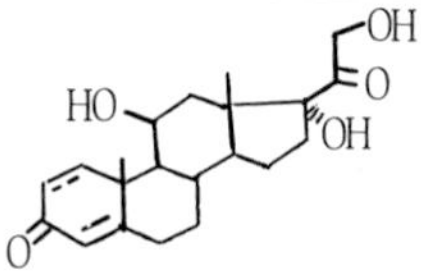 HO, OH, O, OH	Δ4	31	19	2.5	58
	Δ1,4	47	18	0.5	59

a) Rat thymus glucocorticoid receptor, 4 hrs, 0°C (Dexamethasone = 100)
b) Rat kidney mineralocorticoid receptor, 30 min., 0°C (Aldosterone = 100)

Table 5 : Receptor binding of 11β-substituted 19-norsteroids

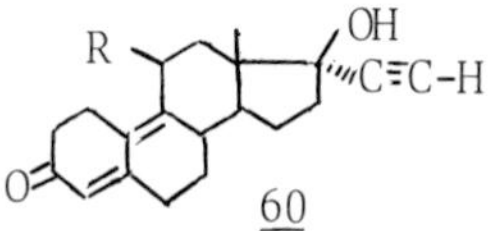

60 (R, OH, C≡C-H) 61 (R, OH, C≡C-CH_3)

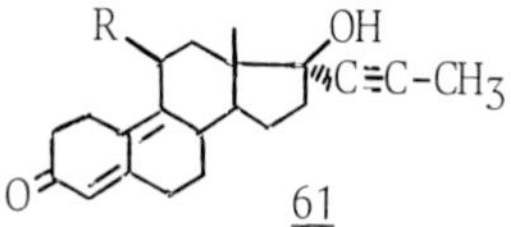

R	GR[a] (60)	PR[b] (60)	GR[a] (61)	PR[b] (61)
a) 2-thienyl	136-54	70-85	268-299	230-438
b) p-MeO-C_6H_4	237-87	130-335	279-299	136-506
c) p-F-C_6H_4	45-23	38-36	216-283	46-85
d) p-Me_2N-C_6H_4	279-235	81-350	283-302	78-530

a) Rat thymus, 4 hrs and 24 hrs at 0°C (Dexamethasone = 100)
b) Rabbit uterus, 2 hrs and 24 hrs at 0°C (Progesterone = 100)

Table 6: Search for selective binding of 11β-R-17β-hydroxy-17α-(prop-1-ynyl))-4,9-estradien-3-ones (53) to the glucorticoid receptor

R	GR*	PR**	URID***	Compound
n-Pr	83	11	13	a
t-Bu	59	0.9	49	b
p Me-C_6H_4	296	295	77	c
m Me-C_6H_4	203	112	62	d
p OMe-C_6H_4	299	506	87	e
m OMe-C_6H_4	246	14	51	f
p SMe-C_6H_4	181	605	64	g
m SMe-C_6H_4	163	13	36	h
p NMe_2-C_6H_4	302	530	77	i
m NMe_2-C_6H_4	51	4	25	j

* Rat thymus, 24 hrs, 0°C (Dexamethasone = 100)
** Rabbit uterus, 24 hrs, 0°C (Progesterone = 100)
*** % inhibition of 5.10^{-8}M Dexamethasone effect on uridine incorporation into thymocytes.

For instance it was found that aliphatic substituents are compatible with the GR binding while decreasing considerably PR binding. Similarly, meta substituted phenyl rings led to somewhat improved dissociation as compared to their para analogues. Modifications of the D-ring substitution had very little effect on selectivity, except by inversion of the ethynyl group from α to β. Unfortunately, none of these compounds possessed an anti-glucocorticoid activity as potent as that of RU 38 486. It has been hypothesized that both binding to GR and PR would be necessary for good anti-glucocorticoid activity based on the report that progestins accelerate the dissociation of glucocorticoids from the GR [66,67,68] although no evidence to support this mechanism has been provided so far in our series. A similar hypothesis had been put forward by Raynaud when he assumed that anti-glucocorticoid activity could be related to progestational activity [57][69]. This assumption may only be erroneous in a matter of semantics, as until recently, all compounds with high RBAs for the PR were progestin agonists. However the basic reason was the observation that progestins interact with the GR with high association and dissociation rates. This property, which was believed to be typical of all antihormones [69], had been applied successfully to the detection of anti-estrogens, anti-androgens and anti-mineralocorticoids [70,71]. It does clearly not apply to the most potent anti-hormones of the 11β-substituted-19-nor-steroid series [64], suggesting a different and new mechanism of action which will be discussed in a later section of this chapter. Efforts to dissociate RBAs in the reverse direction, that is, high RBA for PR and low RBA for GR, met with even less success as the most selective compound with respect to receptor binding, had poor anti-progestional activity in the Clauberg test against progesterone (Table 7).

Table 7 : Search for selective binding of 11β-substituted-19-nor-4,9-pregnadiene-3,20-diones to the rabbit progestin receptor

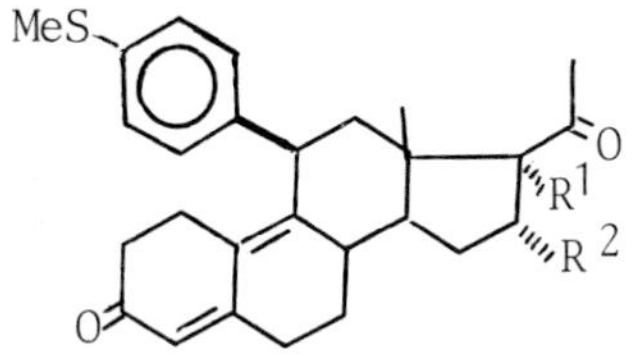

MeS R1 R2 O O

		PR[a]	GR[b]	compound
R^1=H	R^2=Me	176	43	62
R^1=Me	R^2=H	230	27	63

a) Rabbit uterus, 24 hrs, 0°C (Progesterone = 100)
b) Rat thymus, 24 hrs, 0°C (Progesterone = 100)

Though chemical modifications had not yielded so far compounds with highly specific anti-hormonal activities, they provided us with a wealth of RBA data on a coherent series of molecules. We selected a few representative examples which enable us to gain closer insight into the hormone-receptor interaction in the vicinity of the steroidal C-ring.

In the absence of any crystallographic data on the not yet available pure receptor proteins, or, better, receptor-steroid complexes, the only way to make deductions about the receptor site is by trying to fit various steroidal ligands to the receptor. The routinely measured RBAs are thought to represent a good tool to evaluate the degree of fitting of the steroid with its receptor [72,73,74]. The use of this technique has led to the emergence of an, albeit vague, picture of the progesterone receptor, in which binding interactions are provided by hydrogen bonds at the C-3 and C-17 (C-20) oxygenated extremities of the steroid, and hydrophobic forces which can be optimized by interaction of various steroid substituents with lipophilic pockets included in the receptor site [72,73,74,75,76]. Other receptors, and in particular the glucocorticoid receptor also comply with this simplified model [77,94]. From the RBAs presented above (Tables 5, 6 and 7) it is obvious that both the PR and the GR possess a large hydrophobic pocket able to accomodate substituents as large as a phenyl ring. For the PR, hydrophobic pockets were suspected to be present in the close vicinity of positions 6α, 11β, 16α and 17α [76] but their size was believed not to exceed the bulk of a methyl group. Concerning 11β-substituted 19-norsteroids, it had been shown that substituents like methyl [11] and chloro [78] increased significantly progestational activity and (or) RBA for the progesterone receptor [79]. Much less information is available for the glucocorticoid receptor as relatively few RBA data have been published. This is probably the reason why it is generally implicitely believed that one of the most important pharmacophores in this class of compounds is the 11β-hydroxy group which is considered as a hydrogen bond donor [77]. How then to explain the activity and (or) RBA's of compounds in which this hydroxy group has been replaced by a chlorine atom as in Dichlorisone [80] or 64 [81][88] (Table 8). Similarly, some trienones in the 19-nor series also possess significant affinities for the glucocorticoid receptor [83] even though a related in vivo activity has not been demonstrated so far. These examples show that the presence of a hydrogen bond donor is clearly not needed for good binding. The 11β-chlorine atom could be suspected to act to the contrary, as a hydrogen bond acceptor, a function which can also be performed by the hydroxyl group, but our results with 11β-phenyl and 11β-t-butyl-19-nor-dienones (vide infra) tend to reject this hypothesis, provided that all these compounds bind to the same receptor site. Of this we can be quite sure as in a first modification step it had been possible to replace in the androstane series the dihydroxy-acetone side chain by

Table 8 : RBA of 9,11-dichloro-pregnanes for the rat-liver glucocorticoid receptor (0°C, 24 hrs, Dex = 100)

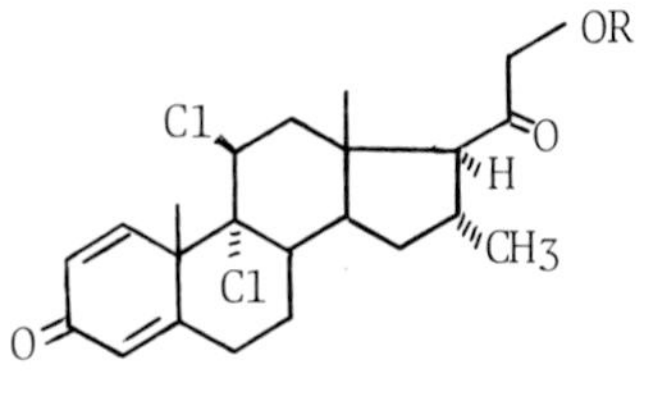

	R	RBA
a)	R=H	120
b)	R=Ac	41

a 17α-propynyl-17β-hydroxy substitution without loss of either affinity or *in vivo* activity [61]. Further modification by removal of the angular C-10 methyl group and concomitant introduction of a lipophilic 11β-substituent also maintained good affinity for the glucocorticoid receptor and we now discovered with satisfaction that the compound in which R is a vinyl group (scheme 12) possesses significant glucocorticoid activity [84]. It is thus highly likely that the binding occurs with one and the same fully functional receptor site.

Scheme 12 : Structural modifications of Prednisolone which led to retention of glucocorticoid activity

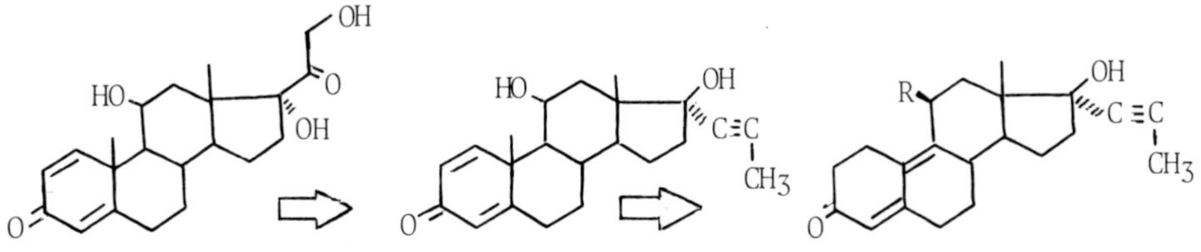

R = vinyl

In consequence, the mapping of the glucocorticoid receptor using 11β-substituted 19-nor-dienones seems to be quite justified. To that end we selected the largest substituents compatible with receptor binding (RBA⩾20) and represented them with their Van der Waals radii on a single steroid backbone using the SCRIPT programm [22]. In this way one obtains a fake steroid probe in which the overlapping 11β-substituents can be approximated to a vague representation of the hydrophobic pocket into which they fit. As a first approach we used only groups which possess axial symetry in order to reduce the uncertainty about the location of the individual atoms : t-Bu and p-tBu-C_6H_4 for the GR and p-tBu-C_6H_4 for the PR (see Table 9). The result, shown in Figure 3 represents the minimum size of the hydrophobic pocket above C-11 of the steroid, both in the GR and the PR (left and right respectively).

Table 9 : RBAs of 11β-substituted 19-norsteroids used for receptor mapping

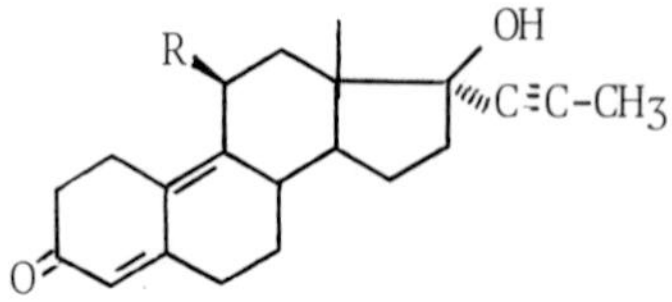

R	GR[a]	PR[b]	Compd.
tBu	(see Table 6)		53b
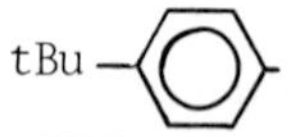 tBu-C6H4- (p-tBu-phenyl)	38	48	65
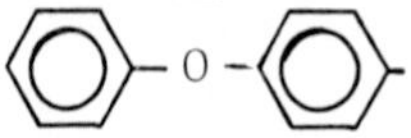 C6H5-O-C6H4- (p-phenoxyphenyl)	100	177	66
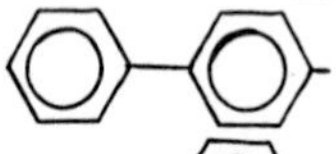 C6H5-C6H4- (biphenyl)	160	278	67
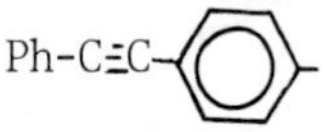 Ph-C≡C-C6H4-	7	8	68

a) Rat thymus, 0°C, 24 hrs (Dexamethasone = 100)
b) Rabbit uterus, 0°C, 24 hrs (Progesterone = 100)

In fact larger substituents are tolerated in the GR as well as in the PR. This is the case namely with the p-phenoxyphenyl group which retains excellent binding affinity (Table 9). If we include it in the visualization program allowing free rotation of the distant phenyl ring around the C-O bond, a much wider pocket is obtained as seen in Fig. 4. Of course it is not obvious and more probably unlikely that all the rotamers would be accepted by the pocket and more work is needed to

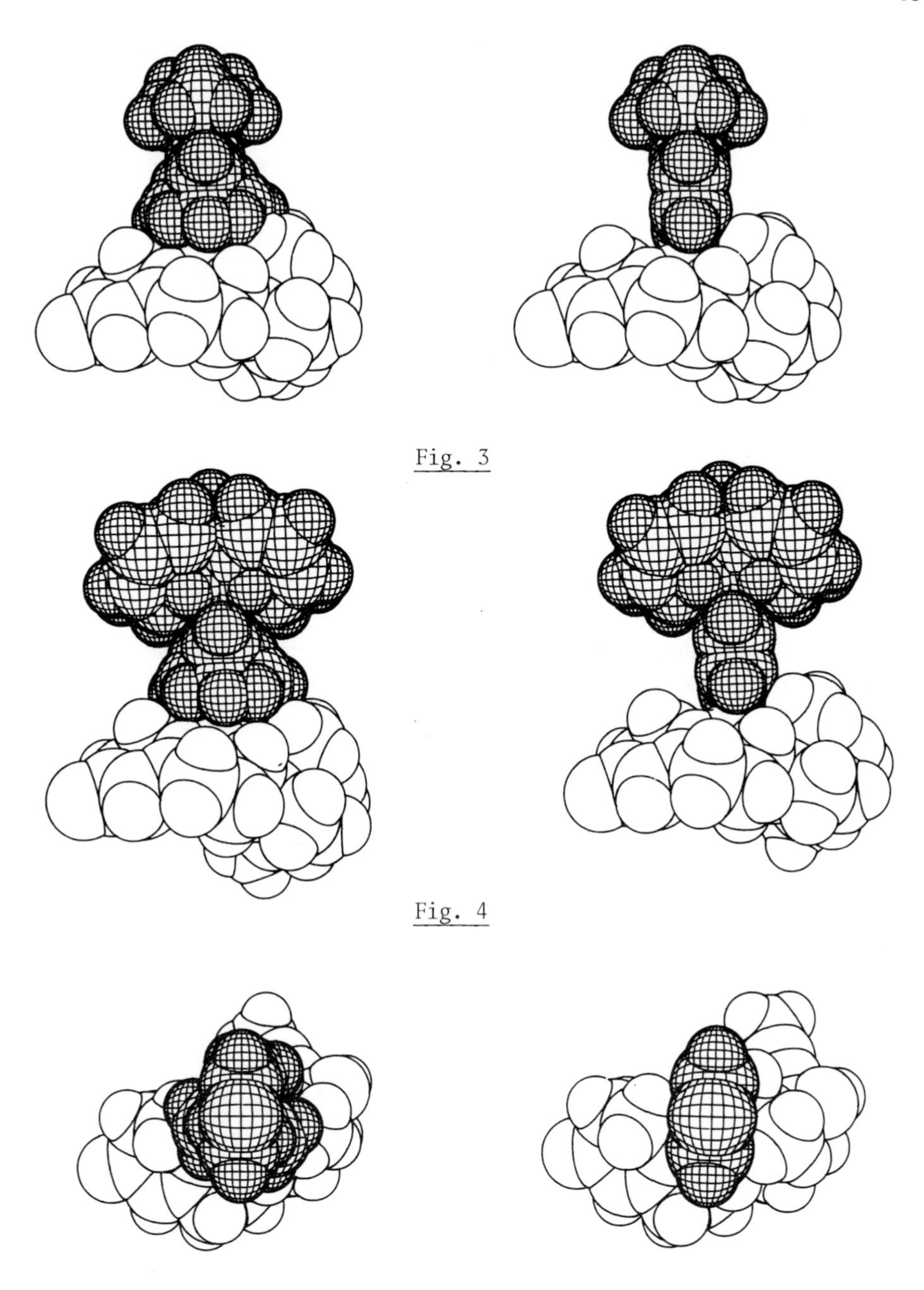

Fig. 3

Fig. 4

Fig. 5

achieve further refinement. Concerning the "depth" of the pocket it can be roughly estimated at 10-12Å based on the RBAs of compounds 67 and 68 for both receptors under discussion. The shapes are quite different in the close proximity of the steroid, the GR having a much wider opening than the PR which is more selective and will not accept a substituent with a diameter superior to the thickness of a phenyl ring (3.4Å). This is best seen in Fig. 5 which represents the estimated section of the pocket at a distance of 3Å from C-11 along the axis of the 11β-bond.

Furthermore by superimposing the glucocorticoid RU 26 559 (compound 57 in Table 4) with the fake steroid used in Fig. 3 to visualize the hydrophobic pocket of the GR, it can be seen (Fig.6) that the angular methyl group (dotted spheres) overlaps in part with the pocket, suggesting that the latter most probably extends over C-10 in the GR.

Fig. 6 : Computer drawn fitting of a steroid of the androstane series (RU 26 559) with a fake steroid probe, showing that the C-10 angular methyl group fits into the hydrophobic pocket of the GR which also harbours the 11β-substituents.

Even though more work is needed to improve this very crude picture it already gives us a satisfactory answer to the question about the effect of the C-10 methyl on GR and PR binding : the wider section of the lipophilic pocket in the GR at close range to the steroid allows the methyl group to be included leading to increased affinity as compared to the 19-nor analogues, while the situation is reversed for the PR where the methyl group would experience an out of pocket interaction reducing the affinity.

Table 10 = Qualitative evaluation of agonistic-versus antagonistic activities of a few 11β -substituted-19-nordienones

RU CODE	R	GR[a]	PR[b]	GLU Ago[c]	GLU Antag.[d]	PRO Ago[e]	PRO Antag.[f]
38 275	CH_3	197-62	137-234	+	-	n.d.	n.d.
42 764	$CH=CH_2$	478-342	140-390	++	-	++	-
39 300	$(CH2)_2CH_3$	207-83	14-11	±	±	n.d.	n.d.
38 502	$C(CH_3)_3$	106-59	0.9-0.9	±	+	n.d.	n.d.
38 486	>N-(phenyl)	283-302	78-539	-	++	-	++

a) Rat thymus glucocorticoid receptor 0°C, 4 hrs-24 hrs (Dex=100)
b) Rabbit uterus progestin receptor 0°C, 2 hrs-24 hrs (Progest=100)
c) Glucocorticoid agonistic activity as evaluated on inhibition of uridine incorporation in thymocytes and thymolytic activity in vivo : ++ = more than 50% thymus weight loss after 50 mg/kg, s.c. administration in mice + = 20 to 49%; + = 5 to 19%; - = less than 5%.
d) % Inhibition of 5.10^{-8}M dexamethasone effect on uridine incorporation in thymocytes by 10^{-7}M of the compound to be tested. The -, ±, +, ++ code is the same as in c.
e) Progestational activity evaluated by the Clauberg test
f) Anti-progestional activity against progesterone on the Clauberg test.
n.d = not determined

Having now established the existence of a large hydrophobic pocket both in the GR ad the PR and in view of the fact that the mere modification of the C-11 substituent is able to transform an agonist to an antagonist (Table 10) we feel that this pocket must be implicated in the mechanism of action of the hormones under scrutinity. Indeed from the results on the Table, it can be seen that the quality of the biological response (agonist-antagonist) is independent of RBAs as well as of the RBA time-course, but is solely related to the size of the 11β-substituent. Thus the 11β-vinyl derivative RU 42 764 is active as a full glucocorticoid and progestional agonist whereas RU 38 486 is a pure antagonist in both types of activity. Intermediate sized substituents can lead to partial agonists-antagonists, a good evidence that all compounds bind to the same receptor site. As the larger substituents do not seem to interfere with the first step of steroid-receptor interaction, that is "binding of the receptor to the steroid" it is likely that the interference would have to occur in the subsequent activation step. This step is believed to consist of a profound conformational change of the receptor protein. If a large hydrophobic hole exists within the steroid-receptor complex, it can be assumed that it will also be involved in the transformation, perhaps even as a driving force as its collapse could provide some of the energy needed for the conformational change. Thus, excessive occupancy of this vital empty space by a bulky steroid substituent could lead to incomplete or modified activation. Whether the final result will be instability of the activated complex or incomplete activation with its possible consequences (impeded chromatin binding or inoperative effector) is not known with certainty so far. It has however been shown recently that in rat-thymus cytosolic GR preparations, RU 38 486 unlike agonists dissociated much faster from the "activated" receptor complex than from the "native" receptor [85]. The same authors also showed that the activated complex bound to DNA-cellulose with a much lower affinity than did the activated GR-Dexamethasone complex. This would suggest that two of the three proposed mechanisms are operating unless the steroid receptors are not truly cytosolic as has been implied several times [86,87]. In that case activation and reinforced nuclear binding would be one and the same thing. Whatever the true nature of steroid receptors, the basic mechanism of action remains valid, that is, binding of the steroid to the receptor, followed by activation of the complex involving a modification in the DNA binding domain of the receptor protein as proposed by Gustafsson for the GR [88]. Taking into account our finding on the existence of a large hydrophobic pocket within the GR (and the PR) and its likely involvement in the final expression of biological activity, i.e. agonistic versus antagonistic response, the simplified representation of the mechanisms of action of GR binding steroids [89,90] would have to be modified as shown in scheme 13.

Scheme 13

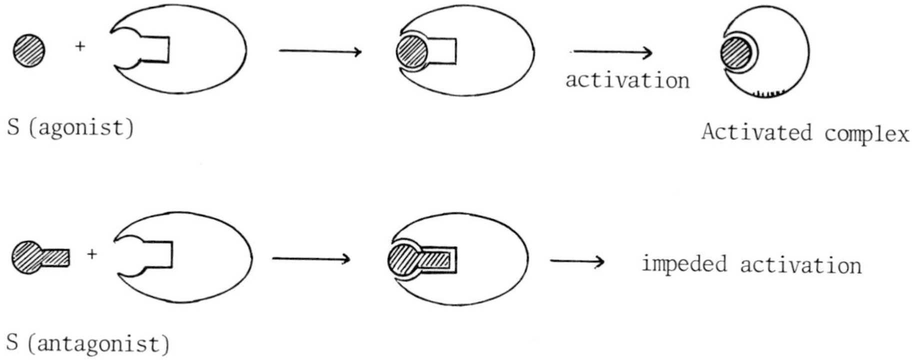

activation

S (agonist)

Activated complex

impeded activation

S (antagonist)

Thus depending on the size of the substituent which protrudes into the secondary site the interaction can lead either to an agonistic or to an antagonistic response. This model is not without some similarity with the one once proposed by Bell and Jones [91] but our results absolutely contradict the prediction of these authors that "the development of an anti-glucocorticoid with a high affinity for the receptor is not possible".

Conclusion and Perspectives :

An original synthetic pathway has allowed us to prepare a large selection of 11β-substituted 4,9-estradien-3-ones, and subsequently to explore extensively the effects of 11β-substitution on GR binding affinity and biological activity. The results obtained so far demonstrate the existence of a very large hydrophobic pocket linked to the main receptor site in the GR (and the PR). This secondary site has an estimated depth of about 12Å and is believed to be involved in the activation step of the primary steroid-receptor complex. Its occupancy by a bulky steroid substituent is thought to interfere with activation and leads to an antagonistic effect, that is to say, no biological response. It thus becomes possible to devise at will, either a glucocorticoid agonist or an antagonist, in the 19-norsteroid series. In addition, the lipophilic pocket could be used to harbour a variety of probes such as fluorescent or Raman active groups and affinity labels. Further mapping of this site using the Van der Waals signature technique [91] and (or) real time computer graphics molecular modelling [93] should enable us to answer such basic questions as the choice between the "lock and key" model and the "induced fit" model. It should also help us to design more selective agonists and antagonists both in the glucocorticoid and in the progestin series.

Aknowledgments :

I would like to thank all my friends and collaborators who participated at different times in this work :

1. Chemistry : A. Bélanger (Université Laval, Québec), G. Cahiez (Université Pierre et Marie Curie, Paris), L. Carbonaro (ENSC, Mulhouse), G. Costerousse, S. Didierlaurent, F. Goubet, A.M. Guerin, G. Millot (Roussel Uclaf).

2. Biology : M. Moguilewsky, D. Philibert and C. Tournemine (Roussel Uclaf).

3. Computing with the SCRIPT Program : G. Lemoine (Roussel Uclaf).

4. X Ray Studies : J.P. Mornon, E. Surcouf (Université Pierre et Marie Curie, Paris).

5. Typing of the manuscript : P. Acqueberge and D. Salaün (Roussel Uclaf)

References

[1] G.P. Chrousos, G.B. Cutler Jr, S.S. Simons Jr, M. Pons, D. Lynn-Loriaux, L.S. John, R.M. Moriarty : in The Proceeding of the VIth Annual Clinical Pharmacy Symposium, Heyden and Son Inc. Phila, PA, p. 152 (1982).

[2] D. Philibert, R. Deraedt and G. Teutsch : 8th International Congress of Pharmacology, Tokyo, July 1981, Abst. No 1463.

[3] M. Moguilewsky, R. Deraedt, G. Teutsch and D. Philibert : J. Steroid Biochem, 6th International Congress on Hormonal Steroids, Jerusalem, Abst. 203 (1982).

[4] M.N. Chobert, R. Barouki, J. Finidori, M. Aggerbeck, J. Hanoune, D. Philibert and R. Deraedt : Biochem. Pharmac., 32, 3481 (1983).

[5] C.I. Phillips, K. Green, S.M. Gore, P.M. Cullen and M. Campbell: The Lancet, April 7, (1984) 767.

[6] D. Philibert, R. Deraedt, G. Teutsch, C. Tournemine and E. Sakiz : 64th Annual Meeting, Endocrine Society, San Francisco, Abst. 668 (1982).

[7] D. Philibert, R. Deraedt, C. Tournemine, J. Mary and G. Teutsch : J. Steroid Biochem., 6th International Congress on Hormonal Steroids, Jerusalem, Abst. 204 (1982).

[8] D.L. Healy, E.E. Baulieu and G.D. Hodgen : Fertility and Sterility 40, 253 (1983).

[9] W. Herrmann, R. Wyss, A. Riondel, D. Philibert, G. Teutsch, E. Sakiz and E.E. Baulieu : Comptes Rendus Acad. Sciences 294 933 (1982).

[10] G. Teutsch, G. Costerousse, D. Philibert and R. Deraedt : Eur. Pat. 0057115 A2 (1982).

[11] J.S. Baran, H.D. Lennon, S.E. Mares and E.F. Nutting : Experientia 26, 762 (1970).

[12] J. Elks : J. Chem. Soc. (1960) 3333.

[13] D.N. Kirk and V. Petrow : J. Chem. Soc. (1961) 2091.

[14] J.S. Baran, D.D. Langford, I. Laos and C.D. Liang : Tetrahedron 33, 609 (1977).

[15] A.J. Van den Broek, A.I.A. Broess, M.J.v.d. Heuvel, H.P. De Jongh, J. Leemhuis, K.H. Schönemann, J. Smits, J. de Visser, N.P. van Vliet and F.J. Zeelen : Steroids 30, 481 (1977).

[16] C.D. Liang, J.S. Baran, N. Allinger and Y. Yuh : Tetrahedron 32, 2067 (1976).

[17] R.B. Garland, J.R. Palmer and R. Pappo : J. Org. Chem. 41, 531 (1976).

[18] A.J. van den Broek, C. van Bokhoven, P.M.J. Hobbelen and J. Leemhuis : Rec. J. Royal Neth. Chem. Soc. 94, 35 (1975).

[19] H.L. Dryden, G.M. Webber and J.J. Wieczorek : J. Amer. Chem. Soc. 86, 742 (1964).

[20] R.V. Coombs, J. Koletar, R.P. Danna and H. Mah : J. Chem. Soc. Perk I 792 (1975).

[21] R. Bardoneschi : unpublished results.

[22] N.C. Cohen, P. Colin and G. Lemoine : Tetrahedron 37, 1711 (1981).

[23] S.N. Ananchenko and I.V. Torgov : Tetrahedron Letters 1553 (1963).

[24] M.B. Groen and F.J. Zeelen : Recueil 98, 32 (1979).

[25] W.S. Johnson : Bioorg. Chem. 5, 51(1976).

[26] L. Nedelec : Bull. Soc. Chim. France 2548 (1970).

[27] C.T. Ratcliffe, C.V. Hardin, L.R. Anderson and W.B. Fox : J. Chem. Soc. Chem. Commun. 784 (1971).

[28] J.C. Gasc : Fr. Demande 2201287 (1974).

[29] G. Costerousse and G. Teutsch : (1975) unpublished results.

[30] R.P. Heggs and B. Ganem : J. Amer. Chem. Soc. 101, 2484 (1979).

[31] G. Costerousse and G. Teutsch : Eur. Pat. Appl. 5100 (1979), CA 93, 186675 n.

[32] L. Nedelec and J.C. Gasc : Bull. Soc. Chim. France 2556 (1970).

[33] J.C. Gasc and L. Nedelec : Tetrahedron Letters 2005 (1971).

[34] G. Teutsch and C. Richard : J. Chem. Res. (S) 87 (1981).

[35] G. Teutsch and A. Bélanger : Tetrahedron Letters 2051 (1979).

[36] A. Bélanger, D. Philibert and G. Teutsch : Steroids 37, 361 (1981).

[37] G. Teutsch, A. Bélanger, D. Philibert and C. Tournemine : Steroids 39, 607 (1982).

[38] G. Teutsch and G. Costerousse : J. Chem. Res. (S) 294 (1983).

[39] R.J. Anderson : J. Amer. Chem. Soc. 92, 4978 (1970).

[40] R.W. Herr and C.R. Johnson : J. Amer. Chem. Soc. 92, 4979 (1970).

[41] B. Rickborn and J. Staroscik : J. Amer. Chem. Soc. 93, 3046 (1971).

[42] C.R. Johnson and D.M. Wieland : J. Amer. Chem. Soc. 93, 3047 (1971).

[43] G. Teutsch : Abst. of the 2nd French-Japanese Symposium on Medicinal and Fine Chemistry (1982).

[44] G. Teutsch : Tetrahedron Letters 23, 4697 (1982).

[45] G. Teutsch, V. Torelli, R. Deraedt and D. Philibert : European Patent Applic. 834011785 (1983).

[46] G. Neef, G. Sauer and R. Wiechert : Tetrahedron Lett. 24, 5205 (1983).

[47] F. Goubet and G. Teutsch : unpublished results.

[48] G. Teutsch : unpublished results.

[49] H.J.J. Loozen and M.S. de Winter : Recueil 99, 311 (1980).

[50] D. Shaw : Fourier Transform NMR Spectroscopy, Elsevier (1976) pp. 279-282.

[51] T. Schaefer, W.F. Reynolds and T. Yonemoto : Can. J. Chem. 41, 2969 (1963).

[52] C.E. Johnson, Jr. and F.A. Bovey : J. Chem. Phys. 29, 1012 (1958).

[53] T.R. Kasturi, S.M. Reddy and P.S. Murthy : Org. Magn. Reson. 20, 42 (1982).

[54] J.P. Mornon : unpublished results.

[55] J.P. Mornon and E. Surcouf : Thèse de doctorat d' Etat, E. Surcouf, Université Paris VI (1982).

[56] D. Philibert, M. Fortin and G. Teutsch : unpublished results.

[57] J.P. Raynaud, T. Ojasoo and F. Labrie; in Mechanisms of steroid action, G.P. Lewis and M. Ginsburg, eds., MacMillan Publishers LTD, England (1982), pp. 145-158.

[58] R. Deraedt : unpublished results.

[59] E.M. Giesen and G. Beck : Horm. Metab. Res. 14, 252 (1982).

[60] E.M. Giesen, C. Bollack and G. Beck : Mol. Cell. Endocrinol. 22, 153 (1981).

[61] G. Teutsch, G. Costerousse, R. Deraedt, J. Benzoni, M. Fortin and D. Philibert : Steroids 38, 651 (1981).

[62] J.P. Raynaud, M.M. Bouton, M. Moguilewsky, T. Ojasoo, D. Philibert, G. Beck, F. Labrie and J.P. Mornon : J. Steroid Biochem. 12, 143 (1980).

[63] E.P. Gomez-Sanchez, C.E. Gomez-Sanchez and M.W. Ferris: J. Steroid Biochem. 19, 1819 (1983).

[64] D. Philibert : (this book).

[65] K.E. Kendle : in Hormone Antagonists, M.K. Agarwal, ed., Walter de Gruyter and Co. (1982) pp. 233-246.

[66] T.R. Jones and P.A. Bell : Biochem. J. 188, 237 (1980).

[67] M. Moguilewsky and R. Deraedt : J. Steroid Biochem. 15, 329 (1981).

[68] P.A. Bell and T.R. Jones : in Hormone Antagonists, M.K. Agarwal, ed., Walter de Gruyter and Co. (1982) pp. 391-405.

[69] J.P. Raynaud, M.M. Bouton and T. Ojasoo : Trends Pharma. Sci. 1 324 (1980).

[70] J.P. Raynaud : in Advances in Pharmacology and Therapeutics vol 1, J. Jacob ed., Pergamon Press (1979) pp. 259-278.

[71] J.P. Raynaud, T. Ojasoo and V. Vaché; in Medecine de la reproduction-gynécologie endocrinienne, P. Mauvais-Jarvis, R. Sitruk-Ware, F. Labrie, ed., Flammarion Medecine-Sciences (1982) pp. 461-475.

[72] J. Delettré, J.P. Mornon, T. Ojasoo and J.P. Raynaud; in Perspectives in Steroid Receptor Research, F. Bresciani, ed., Raven Press, New York (1981) pp. 1-21.

[73] J.P. Raynaud, J. Delettré, T. Ojasoo, G. Lepicard and J.P. Mornon; in Physiopathology of Endrocrine Diseases and Mechanisms of Hormone Action, A.R. Liss, Inc. New York (1981) pp. 461-476.

[74] J.P. Raynaud and T. Ojasso : in Steroid Hormone Receptors : Structure and Function, H. Eriksson and J.A. Gustafsson, eds. Elsevier Science Publishers B.V. (1983) pp. 141-170.

[75] W.L. Duax, J.F. Griffin, D.C. Rohrer and C.M. Weeks : in Hormone Antagonists, M.K. Agarwal, ed., Walter de Gruyter & Co. (1982) pp. 3-24.

[76] D.L. Lee, P.A. Kollman, F.J. Marsh and M.E. Wolff : J. Med. Chem. 20, 1139 (1977).

[77] M.E. Wolff, J.D. Baxter, P.A. Kollman, D.L. Lec, I.D. Kuntz, E. Bloom, D.T. Matulich and J. Morris : Biochem. 17, 3201 (1978).

[78] H.G. Gilbert, G.H. Phillipps, A.F. English, L. Stephenson, E.A. Woollett, C.E. Newall and K.J. Child : Steroids 23, 585 (1974).

[79] K. Kontula, O. Jänne, R. Vihko, E. de Jager, J. de Visser and F. Zeelen : Acta Endocrinol. 78, 574 (1975).

[80] C.H. Robinson, L. Finckenor, E. Oliveto and D. Gould : J. Amer. Chem. Soc. 81, 2191 (1959).

[81] G. Teutsch, G. Costerousse and R. Deraedt : French Patent 2342 738 (1976).

[82] J.P. Raynaud, T. Ojasoo, M.M. Bouton and D. Philibert : in Drug Design Vol. VIII, Academic Press, Inc. (1979) p. 196.

[83] Ibid p. 193.

[84] G. Teutsch, C. Tournemine and D. Philibert : Manuscript in preparation.

[85] M. Moguilewsky and D. Philibert : J. Steroid Biochem. 20, 271 (1984).

[86] W.J. King and G.L. Greene : Nature 307 745 (1984).

[87] W.V. Welshons, M.E. Lieverman and J. Gorski : Nature 307, 747 (1984).

[88] J.A. Gustafsson, J. Carlstedt-Duke, S. Okret, A.C. Wikström, O. Wrange, F. Payvar and K. Yamamoto : J. Steroid Biochem. 20, 1 (1984).

[89] A. Munck and A. Foley : Nature 278, 752 (1979).

[90] N.C. Lan, M. Karin, T. Nguyen, A. Weisz, M.J. Birnbaum, N.L. Eberhardt and J.D. Baxter : J. Steroid Biochem. 20, 77 (1984).

[91] P.A. Bell and T.R. Jones : in Antihormones, M.K. Agarwal, ed., Elsevier/North-Holland Biomedical Press (1979) pp. 35-50.

[92] E. Surcouf and J.P. Mornon : C.R. Acad. Sc. Paris, 295, série II, 923 (1982).

[93] P.A. Bash, N. Pattabiraman, C. Huang, T.E. Ferrin and R. Langridge : Science 222, 1325 (1983).

[94] P.H. Eliard and G.G. Rousseau : Biochem. J. 218, 395 (1984).

RU 38486 : AN ORIGINAL MULTIFACETED ANTIHORMONE IN VIVO

D. Philibert
Centre de Recherches Roussel Uclaf,
93230, Romainville, FRANCE

Introduction

Molecules antagonizing the action of steroid hormones, at the level of their specific receptor, have been sought for many years. Mineralocorticoid, estrogen and androgen antagonists which exhibit this property are actually used therapeutically (1,2,3).

No antiglucocorticoid however has reached the clinical stage, yet (4,5). Although several compounds show antagonist activity in cellular models *in vitro* (6-12), unfortunately, most of them are either poorly effective, or act as partial antagonist *in vivo* only in selected tests, and at very high doses (13-21). Similarly, antiprogestins, too, until now, are partial antagonist/agonist in animal models or in clinical trials (22-25).

In 1980 we decided to concentrate our efforts mainly on the last 2 types of antihormones. Six months later we discovered RU 38486 which displayed, both potent antiprogesterone and antiglucocorticoid activity *in vitro* as well as *in vivo*, and a moderate antiandrogenic effect without exhibiting any agonist properties (26-35). The two principal activities of this compound have been confirmed during the ongoing intensive clinical trials (36-39).

The aim of this paper is to present the biochemical and pharmacological activities of RU 38486 in experimental animals.

Adrenal Steroid Antagonism

Results

A) Biochemical studies on RU 38486

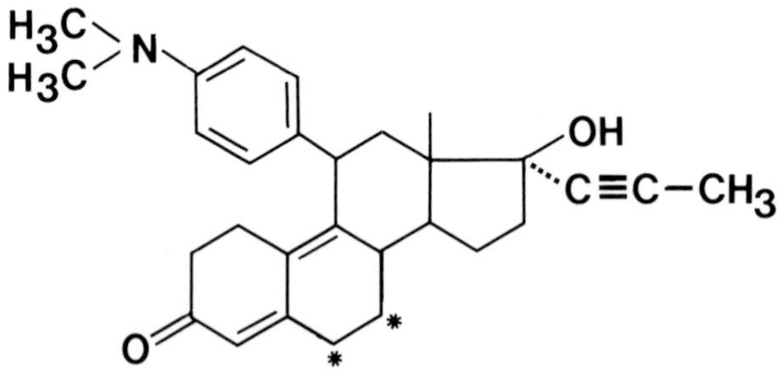

RU 38486

Fig 1 Structure of RU 38486, the asterisk indicates the position of the tritium on [6,7 ^{3}H]-RU 38486 (37 Ci/mmole)

RU 38486 (11β-(4-dimethylaminophenyl)-17β-hydroxy,-17α-(prop-1-ynyl)-estra-4,9-dien-3-one) (fig 1) was first studied in competition experiments to determine its relative binding affinity (RBA) for the five classes of cytoplasmic steroid receptors.

Relative binding affinity for steroid receptors

As shown in table 1, it displays a very strong RBA for both rat thymus glucocorticoid, and rabbit uterine progestin-receptor, respectively, 3 times that of dexamethasone (dexa) and 5 times that of progesterone (progest.) after 24 hours of incubation. Its RBA for the rat prostate androgen receptor is about 1/4 that of testosterone (testo.) whereas it has negligible affinity for rat kidney mineralocorticoid and mouse

Table 1 - Relative binding affinity (RBA) of RU 38486 for the cytoplasmic steroid receptors

RECEPTOR	GLUCOCORTICOID		PROGESTIN		ANDROGEN		MINERALOCORTICOID		ESTROGEN	
TISSUE	ADX RAT THYMUS		RABBIT UTERUS		CX RAT PROSTATE		ADX RAT KIDNEY		MOUSE UTERUS	
RADIOLIGAND	^{3}H-DEXAMETHASONE		^{3}H-R5020		^{3}H-R1881		^{3}H-ALDOSTERONE +RU 28362$^+$		^{3}H-ESTRADIOL	
RBA REFERENCE STEROID	DEXA=100		PROGEST.=100		TESTO.=100		ALDOSTERONE=100		ESTRADIOL=100	
INCUBATION TIME AT 0-4°C (HOURS)	4H	24H	2H	24H	0.5H	24H	1H	24H	2H	5H*
RBA-RU 38486	280	300	78	530	10	23	0.4	0.1	0.1	0.1

* AT 25°C
$^+$FOR FURTHER DETAILS SEE (40)

The tissues were homogenized with a teflon glass-potter in 0.01M Tris-HCl (pH 7.4) 0.25M sucrose buffer (+2mM DTT for thymus). The homogenates were centrifuged at 105000 g x 60 min at 0°-4°C to obtain cytosols which were incubated with the suitable radioligand in presence of increasing concentration of cold reference compound or of cold RU 38486. Bound radioactivity was measured by dextran coated charcoal (DCC) adsorption technique. The ratio of the concentration of reference compound over the concentration of test compound required to displace radioligand bound specifically by 50% (RBA) was determined. RBA's of reference compound were taken arbitrarily equal to 100%. Each value is the mean of 3 determinations. For further details see (41-43).

uterine estrogen receptor. Furthermore its RBA is stable and even increases during the incubation time.

Scatchard plot analysis

With tritiated RU 38486, we have compared in rats its binding parameters for thymus glucocorticoid receptor with those of ^{3}H-dexamethasone and for uterine progestin receptor with those of ^{3}H-R5020, a potent progestin (44). As seen in Fig 2a

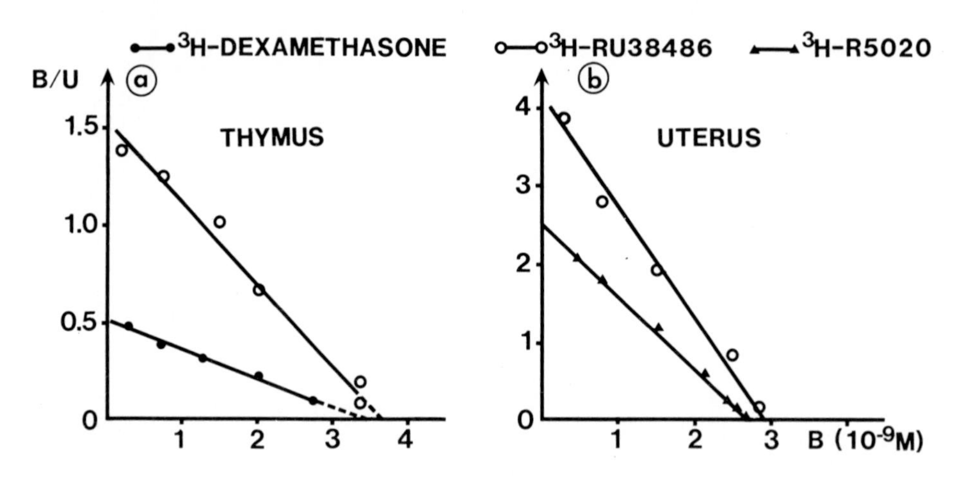

Fig 2 Scatchard plot analysis of ^{3}H-RU 38486 binding to thymus glucocorticoid and uterine progestin receptors in rats

a) thymus cytosol diluted 1/10 (w/v) in buffer (see Table 1) from Adx rats was incubated at 0°-4°C for 4 hours in presence of increasing concentrations of ^{3}H-RU 38486 or [2-^{3}H]dexamethasone (25 Ci/mmole) with or without a 500 fold excess of the corresponding radioinert steroid.
b) Ovx and Adx female animals received a single subcutaneous injection of 10 µg estradiol dissolved in 5% ethanol normal saline solution, to enhance the progestin receptor level. The animals were killed 40 hours later and the uterus removed. Uterine cytosol diluted 1/10 (w/v) was prepared (in presence of 100 nM of cold dexa to mask the glucocorticoid binding sites) and incubated with ^{3}H-RU 38486 and (6-7 ^{3}H)-R 5020 (51 Ci/mmole) as described above. Bound radioactivity was measured by DCC adsorption technique.

^{3}H-RU 38486 revealed the same number of glucocorticoid binding sites as ^{3}H-dexamethasone, whereas its association constant ka ($\simeq 4 \times 10^8 M^{-1}$) was about 3 times that of ^{3}H-dexamethasone ($\simeq 1.4 \times 10^8 M^{-1}$). For the progestin receptor (fig 2b), its association constant ka ($1.8 \times 10^9 M^{-1}$) was 2 times stronger than that of R5020 ($\simeq 0.9 \times 10^9 M^{-1}$). In previous publications it was shown that the ka of R 5020 was at least 7 times higher than that of progesterone in various species (45-48).

Dissociation rate from receptors

As illustrated in fig 3 the dissociation rate at 0°-4°C of ^{3}H-RU 38486 from the glucocorticoid receptor was very slow compared to that of dexamethasone and corticosterone. In fact, the ^{3}H-RU 38486-receptor complex dissociates with a half life time (t1/2) of about 100 hours compared to 16 hours for ^{3}H-dexamethasone, and 150 min for corticosterone. Similar results were obtained with the progestin receptor where the ^{3}H-RU 38486 half life time ($t1/2 \simeq 16$ hours) is much higher than that of R 5020 ($t1/2 \simeq 70$ min) and progesterone ($t1/2 \simeq$ 10 min), respectively.

Sucrose gradient analysis

Sucrose gradient analysis of thymus cytosol, incubated with ^{3}H-RU 38486 or with ^{3}H-dexamethasone showed (fig 4a) that the two radioligands form a complex with the glucocorticoid receptor which sediments in exactly the same 9S region and that 100 nM of cold dexamethasone nearly totally displaces the ^{3}H-RU 38486-9S peak. When similar studies were performed with uterine cytosol there was no difference in sedimentation profile for ^{3}H-RU 38486 and ^{3}H-R 5020 ; ^{3}H-RU 38486-8S peak was also completely inhibited by 100 nM of radioinert R 5020.

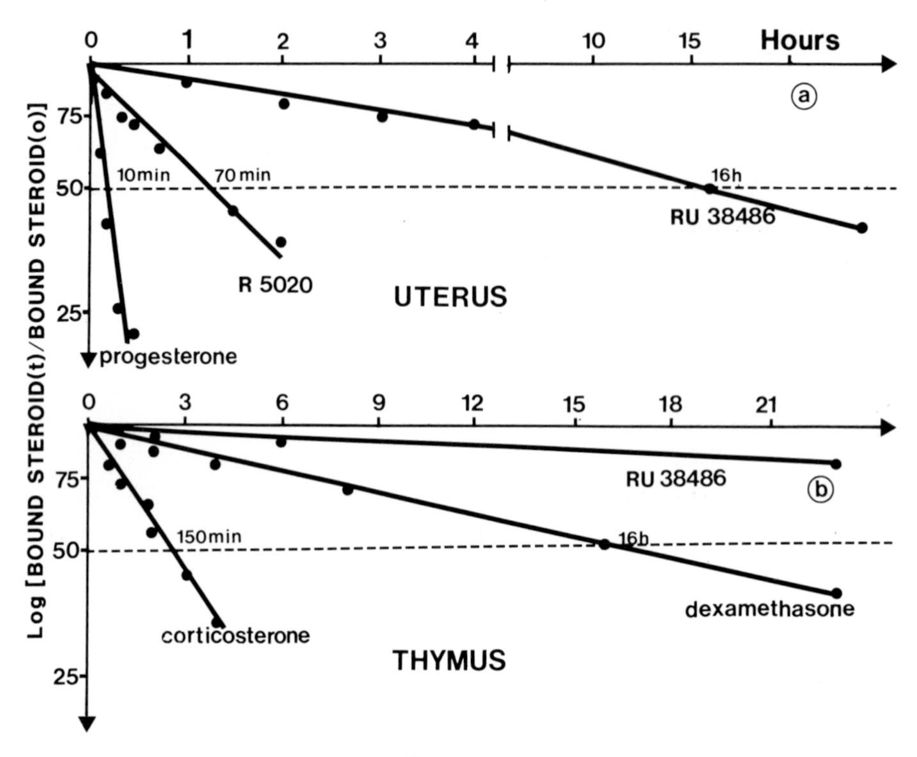

Fig 3 Dissociation rate of ^{3}H-RU 38486 from thymus glucocorticoid and uterine progestin receptors in rats

The same uterine cytosol preparations described in fig 2 were used.

a) Cytosol was incubated with 10^{-8}M ^{3}H-RU 38486 or ^{3}H-R 5020 or [1-^{3}H]-progesterone (27 Ci/mmole) at 0°-4°C for 2 hours. At the end of the incubation, a 1000 fold excess of the corresponding non radioactive steroid was added (Bo) and the amount of specifically bound radioligand (Bt) determined at different time intervals thereafter using the DCC adsorption technique. Log Bt/Bo x100 was plotted versus time. t1/2 was the time required for 50% of the receptor-steroid complex to dissociate.

b) The procedure described above was also used to study the dissociation rate of ^{3}H-RU 38486, ^{3}H-dexamethasone and |1-2^{3}H|-corticosterone (58 Ci/mmole) from the thymus glucocorticoid receptor.

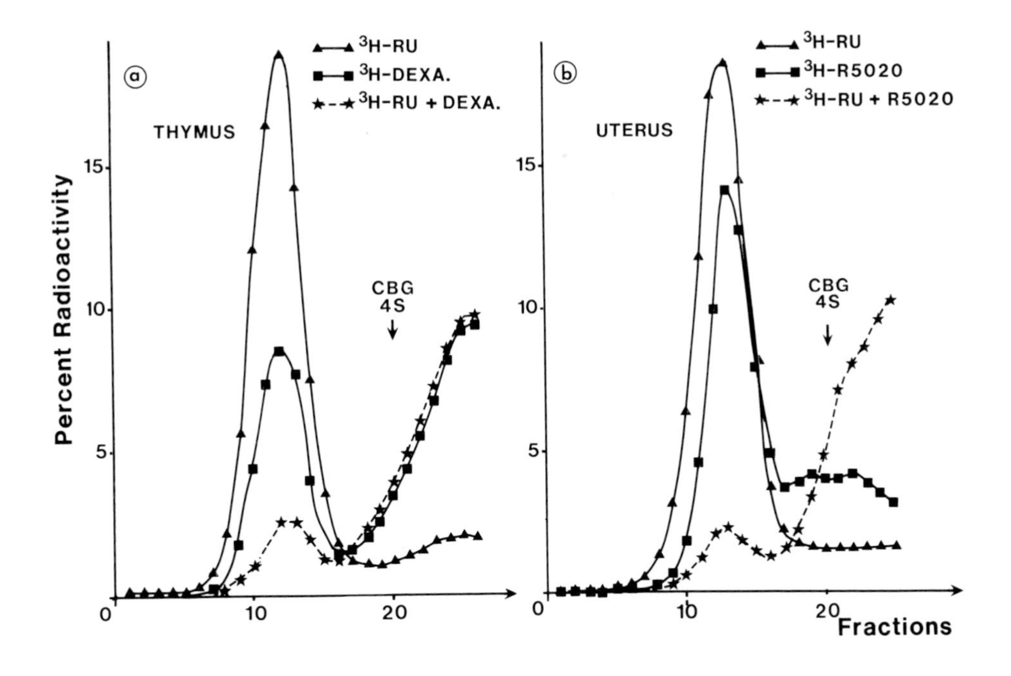

Fig 4 Sucrose gradient analysis of ^{3}H-RU 38486 (^{3}H-RU) binding to rat thymus and uterine cytosols

Tissues were homogenized in 10 mM Tris-HCl (pH 7.4) 1.5 mM EDTA, 1.5 mM 2-mercaptoethanol, 10 mM sodium molybdate buffer. Linear 5-20% sucrose gradients were prepared in the homogenization buffer.

a) Thymus cytosol samples (0.2ml) preincubated for 2 hours at 0°-4°C with ^{3}H-dexamethasone or ^{3}H-RU 38486, with or without 100 nM of cold dexamethasone, were centrifuged in a VTI65 rotor for 1 h at 65000 rev/min (L8-70 Beckman ultracentrifuge). The gradients were then fractionated into 2 drop samples by needle puncture at the bottom of the tubes. Rat corticosteroid binding globulin (CBG) was run on separate gradients as a sedimentation standard (4S).

b) Aliquots of 0.2 ml of uterine cytosol were incubated with ^{3}H-R 5020, or ^{3}H-RU 38486 with or without 100 nM R 5020 and analyzed as described above.

Thus, RU 38486 presents at the cytoplasmic receptor level a biochemical profile similar to that of potent agonists. However, as will be shown later, its binding to the receptor does not trigger biological responses observed with natural hormones or with synthetic agonists : on the contrary RU 38486 fully blocks the action of these steroids.

B) Biological studies on RU 38486

SPF rats of Sprague Dawley strain weighing 160-180 g and immature New-Zealand rabbits weighing about 1 kg were used. Rats were adrenalectomized (Adx) or/and ovariectomized (Ovx) 4 to 5 days before use.
RU 38486 was administered by oral route in suspension in aqueous solution containing 0.25% of carboxymethylcellulose and 0.2% of Polysorbate 80 (CMCP), in a volume of 5 ml/kg of body weight in rats and 1 ml/kg in rabbits. Estradiol, progesterone and testosterone were injected by subcutaneous route in sesame oil containing 5% benzylic alcohol. Statistical calculations were carried out using the Dunnett test (49).

I) Antiglucocorticoid activity

It is well known that glucocorticoids induce biological responses in many target cells and tissues. Therefore glucocorticoid antagonist activity of RU 38486 was studied *in vitro* as well as *in vivo* using various bioassays in intact and Adx male rats since the majority of antiglucocorticoids have been effective only in some tests (4,5,20). Thymocytes (6,8,10) and HTC cells (7,9,50) are the two principal cellular models commonly used to study the mechanism of glucocorticoid action. *In vitro* we used the former to assess the activity of RU 38486.

Thymocytes

As illustrated in fig 5, dexamethasone inhibited uridine incorporation into RNA in thymocytes, maximally at 10^{-7}M but RU 38486 showed no glucocorticoid effect even at 10^{-6}M. When $5x10^{-8}$M of dexamethasone was incubated with increasing concentrations of RU 38486 or cortexolone (8), RU 38486 fully antagonized at $5x10^{-6}$ the effect of dexamethasone and its 50% effective concentration EC_{50} was lower than the concentration of dexamethasone used. Cortexolone demonstrated a much weaker antiglucocorticoid effect which did not exceed 50% inhibition at 10^{-6}M.

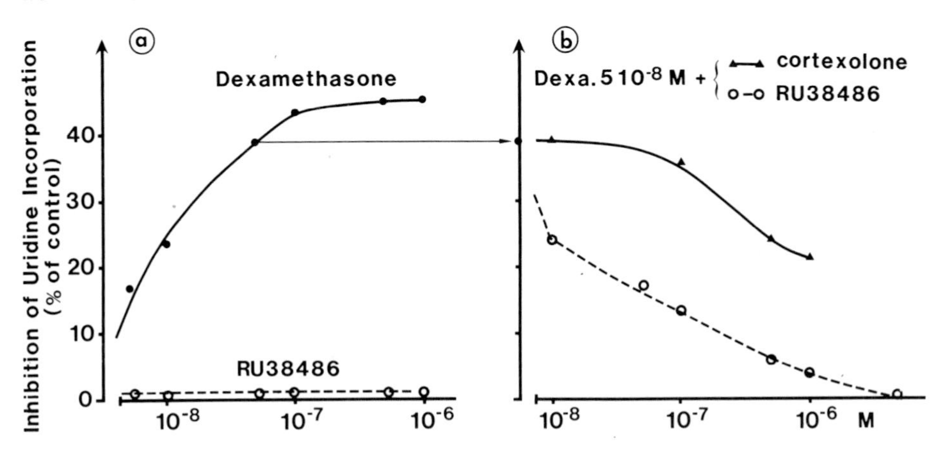

Fig 5 Agonist and antagonist activities of RU 38486 on uridine incorporation in rat thymocytes

Thymocytes were prepared by mincing pooled thymi of Adx rats in Hank's buffer, washed and suspended at a final concentration of 10^7 cells/ml in MEM (GIBCO) supplemented with sodium pyruvate (1mM), L glutamine (2mM) and non essential amino-acid solutions (1% v/v). Aliquots of 300µl were then incubated under carbogen for 3 h at 37°C with increasing concentrations of dexa or RU 38486 (5a), or with $5x10^{-8}$M of Dexa in presence of increasing concentrations or RU 38486 or cortexolone (5b). 1µCi of ^{3}H-uridine was added and the incubation was continued for 1 h. Radioactivity incorporated into trichloroacetic precipitable material was determined. For further details see (10).

It is well known that glucocorticoids elicit thymocyte death after overnight incubation and one of the first signs of lethal effects is the appearance of pycnotic cells (p.c.) (cells with condensed chromatin) (51).
After incubation of thymocytes for 6 hours at 37°C, in the experimental conditions described in legend to fig 5, 10^{-7}M dexamethasone induced the appearance of 40% of pycnotic cells compared to 8% obtained with the control. RU 38486, alone, up to 10^{-6}M had no effect (5% of p.c.), whereas at this concentration it fully prevented the action of 10^{-7}M dexamethasone (6% of p.c.).
In vivo, the effect of RU 38486 was studied with respect to many biological responses induced by dexamethasone in rats. "Acute" bioassays such as : liver glycogen, tyrosine aminotransferase (TAT) and tryptophan oxygenase (TPO) induction, increase of urinary volume and potassium excretion (52), inhibition of ACTH secretion ; chronic tests such as : decrease of body, thymus, and adrenal weights, anti-inflammatory effect on cotton-induced granuloma (53).

Liver, TPO, TAT and glycogen

As illustrated in fig 6, a single intraperitoneal injection of 0.01 mg/kg of dexamethasone provoked a sharp increase of hepatic TAT and TPO in Adx rats. These effects were inhibited in a dose dependent manner by an oral administration of RU 38486. The 50% effective dose (ED_{50}) of this compound was approximately 5 mg/kg whereas the full antagonist effect was obtained between 10 and 25 mg/kg. Under the same experimental conditions, RU 38486 was totally effective at 5 mg/kg on liver glycogen (results not shown). RU 38486, alone, up to 50 mg/kg did not exhibit any agonist activity.

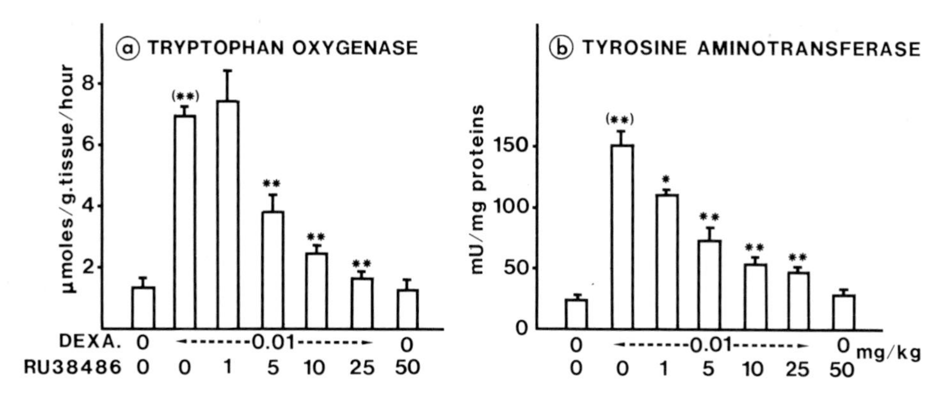

Fig 6 Effect of RU 38486 on hepatic TPO and TAT in Adx rats

Adx male rats, fasted for 17 hours, received an oral administration of RU 38486 1 hour before the i.p. injection of dexamethasone (CMCP). 4 hours after the dexa injection, the animals were killed and the livers removed. About 1 g of tissue was immediately dissolved in hot KOH. After precipitation with ethanol and hydrochloric hydrolysis, glycogen was assayed in the form of glucose (54). The other part of liver was homogenized in 10 volumes of 0.14M KCl, 0.0024M NaOH iced buffer. The homogenate was then centrifuged at 105000 g for 60 min. The assays of TPO and TAT were performed (55,56). All figures are means of 9 values (3 rats per dose and 3 assays per rat).

*$p < 0.05$	(**)	compared to control group
**$p < 0.01$	**	compared to dexa-treated group.

Diuresis

As illustrated in fig 7, a single subcutaneous injection of 0.05 mg/kg of dexamethasone provoked in Adx rats, a significant increase of urinary volume and potassium excretion. No effect on sodium excretion was observed. RU 38486 administered, orally 1 hour before, at a dose of 10 mg/kg, fully reversed the effect of dexamethasone on these two parameters.

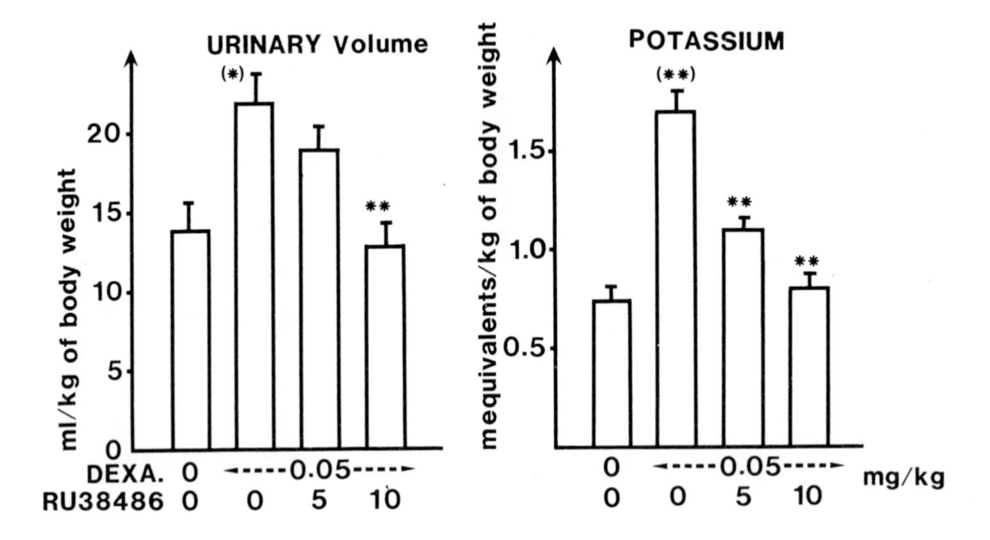

Fig 7 Antiglucocorticoid activity of RU 38486 on diuresis in rats

Adx male rats, fasted for 17 hours, received an oral administration of RU 38486 1 hour before a subcutaneous injection of dexa in a volume of 2 ml/kg (1% ethanol-normal saline) as well as a surcharge of 50 ml/kg of normal saline by i.p. route. The animals were immediately placed individually in diuresis cages. Urine was collected for 4 hours, the volume was measured and the urinary sodium and potassium were determined by flame photometry. The results were expressed in ml/kg and mequivalents/kg for groups of 6 rats.

The total antiglucocorticoid effect exerted by 10 mg/kg of RU 38486 was rapidly reversed (fig 8) by increasing doses of dexamethasone given subcutaneously one hour later ; 1.25 mg/kg of dexamethasone produced the same effect on urinary volume and potassium excretion as 0.05 mg/kg of dexamethasone injected alone. Thus, 25 times greater amount of dexamethasone was needed to induce the same response if rats were pretreated with RU 38486 as compared to control rats given dexamethasone alone.

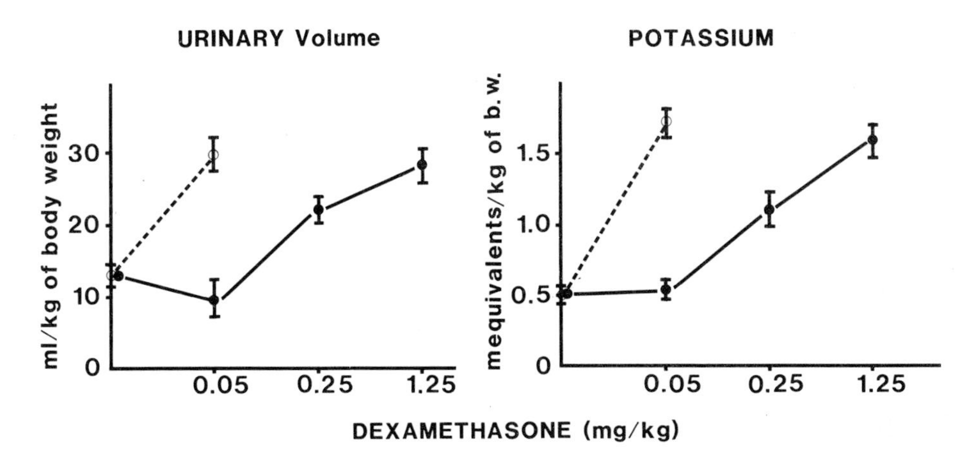

Fig 8 Reversibility of the antiglucocorticoid effect of RU 38486 on diuresis in Adx rats

Adx male rats received a single oral administration of 10mg/kg RU 38486 ; 1 hour later 4 groups of 6 animals were given by subcutaneous route increasing doses of dexamethasone (0-1.25 mg/ kg) ●——● . Control animals received first the vehicle (CMCP) and 1 hour later each received either the vehicle (1% ethanol-normal saline) or 0.05 mg/kg of dexamethasone o----o. For further details see legend to fig 7

Thymus

In chronic tests, RU 38486 also exhibited potent antidexamethasone activity on thymus weight in intact rats. The thymus weight decrease caused by 0.05 mg/kg of dexamethasone given orally for 4 days, was completely inhibited by 10 mg/kg of RU 38486 (fig 9). Cortexolone (S), $\Delta^{1,9(11)}$-11-deoxycortisol (S_1) or Medroxyprogesterone acetate (MPA), known to be effective in certain tests (14,17,18), were devoid of any antithymolytic effect at doses up to 100 mg/kg. RU 38486 administered alone, did not show any agonist effect up to 100 mg/kg. Furthermore, RU 38486 fully antagonized, at 10 mg/kg, the effect of 50 mg/kg of corticosterone administered orally.

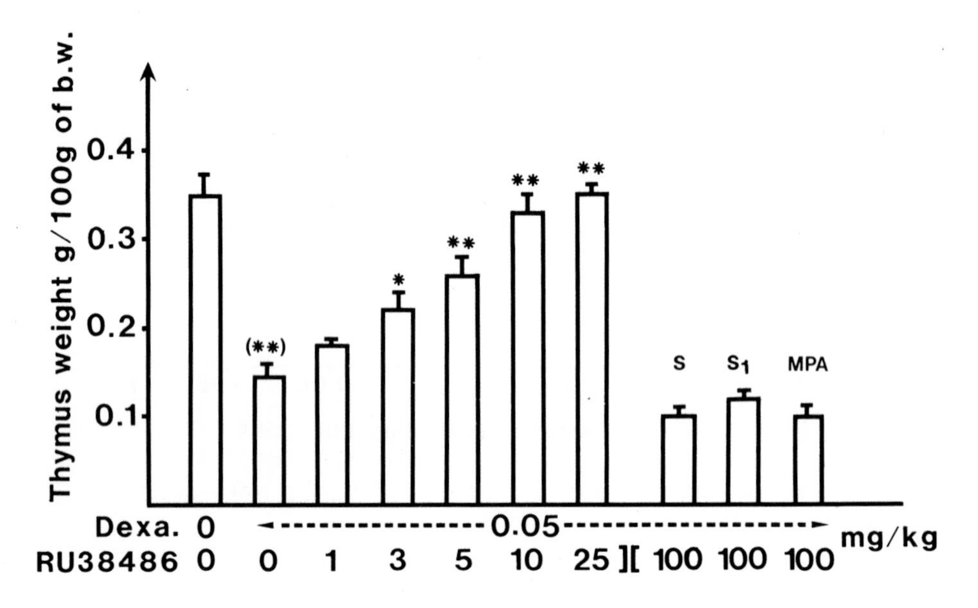

Fig 9 Antiglucocorticoid activity of RU 38486 on thymus weight in intact rats

Intact rats weighing about 100 g received daily for 4 days a subcutaneous injection of dexamethasone 15 min after either an oral dose of RU 38486, or S(cortexolone), S_1($\Delta^{1,9(11)}$-11-deoxycortisol), or MPA (Medroxyprogesterone acetate); 24 hours after the last administration, the animals were killed and the thymus was removed and weighed. Results are expressed as g per 100 g of body weight for 5 rats.

This dose of corticosterone displayed the same thymolytic effect as 0.05 mg/kg of dexamethasone (not shown). Similar results were obtained with the granuloma model and with adrenal and body weights.

II) Antiprogesterone activity

Endometrial proliferation

One of the bioassays commonly used for evaluating progestomimetic activity is the endometrial proliferation in immature rabbits pretreated with estradiol (57).In this test, 0.2 mg/kg

Table 2 Effect of RU 38486 on endometrial proliferation induced by progesterone in rabbits

ENDOMETRIAL PROLIFERATION (McPHAIL UNIT : MPU)					
P.O.	S.C.		ADMINISTRATION IN UTERO		
RU 38486	PROGESTERONE	MPU	RU 38486	PROGESTERONE	MPU
(MG/KG)			(μG/HORN)		
0	0	0	0	0	0.3
0.3			30	0	0.3
↓	0	0			
100			500	0	0.3
0	0.2	3.2	0	10	2.5
0.3	0.2	2.8	1	10	1.6
1	0.2	2.1	3	10	1.3
3	0.2	1.4	10	10	1.0
10	0.2	0.6	30	10	0.6
20	0.2	0	90	10	0.6

Groups of 3 immature rabbits weighing about 1 kg received from day 1 to day 5 a subcutaneous injection of 5 μg/kg of estradiol. Progesterone was then injected subcutaneously from day 7 to day 10 either alone or in combination with increasing doses of RU 38486 p.o. .On day 11, the rabbits were killed and the uterus were removed and fixed in a Bouin solution for histological analysis. Transformation of the endometrium was graded on the McPHAIL scale (58), the maximum of which is 4.
For studies <u>in utero</u>, groups of 3 immature female rabbits received on day 1 a deposit of 10μl of ethanol containing 25μg of estradiol on the previously shaved dorsal skin. On day 4, under ether anesthesia, 10 μl of progesterone with or without RU 38486, dissolved in sesame oil containing 5% benzylic alcohol, was introduced in the lumen between 2 ligatures of the uterus. The animals were killed on day 6 and the uterus removed and analyzed as described above.

of progesterone induced a strong transformation of the endometrium graded 3.2 MPU on the McPHAIL scale (table 2).
RU 38486 administered alone was totally devoid of any progestomimetic effect at doses from 0.3 to 100 mg/kg. Administered orally one hour before 0.2 mg/kg of progesterone, it inhibited in a dose dependent manner the action of progesterone ; its ED_{50} was about 3 mg/kg, whereas at 20 mg/kg its antihormonal effect was total.
When the 2 compounds were introduced in utero (59) the ED_{50} of RU 38486 was about 1/3 the dose of progesterone (10μg/horn). These last results are in good agreement with the RBA's of these two steroids for the rabbit uterine progestin receptor (table 1).

Mitochondria of uterine gland cells

More recently it was shown that progesterone induced in 3 weeks Ovx rat an increase in the volume density (Vv) of the mitochondria, and the appearance of giant mitochondria in the uterine glandular cells (60-62). This response, unlike many other biological effects of progesterone, does not necessitate a priming by estrogens.
As illustrated in fig 10, progesterone at 20 mg/kg increased 3 times the volume density of mitochondria whereas RU 38486 alone up to 30 mg/kg had no effect. When RU 38486 was administered in combination with progesterone, it completely blocked the action of the natural hormone at a dose of 10 mg/kg ($ED_{50} \simeq 3$ mg/kg).
Furthermore, RU 38486 has been shown to possess antinidatory and abortive activity in rats at all times of pregnancy (27). In normal or artificially cycled monkeys it induced menstruation when administered during the mid-luteal phase (27,29). Thus, RU 38486 was effective in antagonizing both exogenous and endogenous progesterone activity.

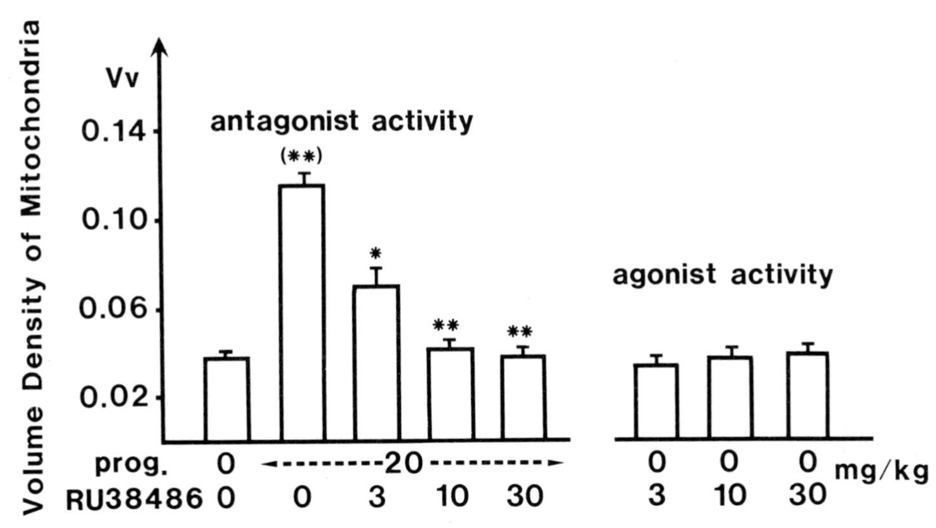

Fig 10 Antiprogesterone activity of RU 38486 on volume density of mitochondria of uterine gland cells in Ovx rats

3 week ovariectomized rats weighing about 250 g in groups of 5 were treated for 3 days with progesterone (20 mg/kg) by s.c. route, or with RU 38486 by oral route, alone or with added progesterone. The animals were killed on day 4, and the uterus removed and ultrathin sections from each uterus were examined at a magnification of 30000. The volume density of mitonchondria was quantified by stereology in the basal part of the glandular cells (62). Published with the permission of Drs. Secchi J.and Lecaque D.

III) Antiandrogenic activity

As expected by its relative binding affinity for the rat prostate androgen receptor, RU 38486 displayed an antiandrogenic activity. In fact, in male castrated rats, testosterone propionate induced at a dose of 0.5 mg/kg a sharp increase in both seminal vesicles and prostate weights (fig 11). These effects were inhibited in a dose dependent fashion by RU 38486. Androgen antagonism was more important on seminal vesicles than on prostate, 85% and 55% inhibition respectively at the higher dose of 100 mg/kg. At this dose, RU 38486 was totally devoid of any agonist activity.

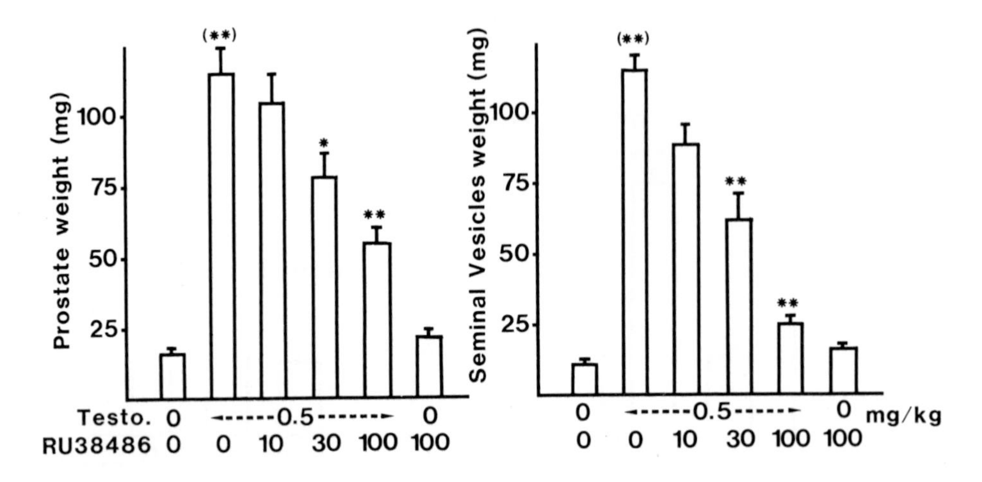

Fig 11 Antiandrogenic activity of RU 38486 in castrated male rats

Groups of 5 castrated male rats of about 100 g received simultaneously during 8 days RU 38486 orally and 0.5 mg/kg of testosterone propionate subcutaneously. The animals were killed 24 hours after the last treatment. The seminal vesicles and the prostate were removed, fixed for 24 hours in a normal saline solution containing 10% of formol, then dissected and weighed.

IV) Percentage of free receptor binding sites after oral administration of RU 38486 in rats

RU 38486 appears to be totally effective, as an antiglucocorticoid and as an antiprogesterone, at an oral dose of 10 mg/kg This dose is relatively high compared to the antagonist effect of RU 38486 on cellular models *in vitro*. In order to explain this difference of activity we determined the percentage of free binding sites of thymus glucocorticoid, uterine progestin, and kidney mineralocorticoid receptors in Adx and Ovx rats after oral administration of increasing doses of RU 38486. As shown in fig 12, this compound provoked a similar dose dependent decrease in percent free binding sites of glucocorticoid and progestin receptors, and at 10 mg/kg it totally saturated these two receptors. Furthermore, it did not modify as expected, considering its RBA (table 1), the level of free binding sites of mineralocorticoid receptor. This is in good agreement with its absence of mineralocorticoid or antialdosterone effect at an oral dose level up to 30 mg/kg in the urinary sodium and potassium excretion test of KAGAWA (63). It appears therefore that the full antagonist activity of RU 38486 is elicited at a dose level which saturates completely the corresponding receptor.

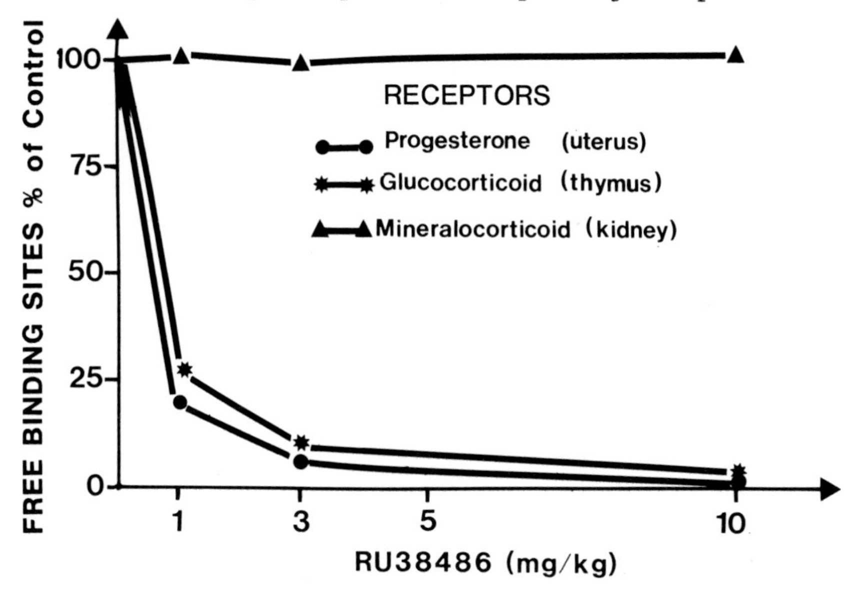

Fig 12 Percentage of free binding sites after oral administration of RU 38486 in rats

5 days Ovx and Adx female rats weighing about 200 g, received one s.c. injection of 10 µg estradiol. 40 hours later, increasing doses of RU 38486 were administered orally and the animals were killed 2 hours later. Target organs were removed, the cytosol prepared and the number of the binding sites of thymus glucocorticoid, uterine progestin and kidney mineralocorticoid receptors were determined using short incubation time with adequate radioligands and the DCC adsorption technique (see table 1) : results were expressed as percent relative to the number of binding sites of control animals.

Conclusion

RU 38486 appears to be the most potent antiglucocorticoid and antiprogesterone, known to this day, both in vitro and in vivo. Its novelty lies in the fact that it acts directly at the level of specific steroid receptors, without exhibiting any agonist effect. CHROUSOS et al have recently listed all known antiglucocorticoids (4,5) such as cortexolone (8,14), steroidal 17β carboxamides (11), 6β - bromoprogesterone (20), $\Delta^{1,9(11)}$-11-deoxycortisol (17), medroxyprogesterone acetate (18), 21-mesylate derivatives of cortisol and dexamethasone (12,64) and Δ^{1}-11 oxa-11-deoxycortisol (21). All these steroids display a clearly weaker antagonist effect in vitro than RU 38486. In vivo, they are active at very high doses, only in certain tests with a weak agonist effect in most cases.This could be explained by the fact that they show an association constant for the glucocorticoid receptor which is 10-100 fold lower than that of RU 38486 (4,5). Furthermore,in spite of the very slow dissociation rate of RU 38486 from this receptor, preliminary results indicate that the antagonist action of this compound is rapidly reversible.
As for the progesterone antagonists, KENDLE, who had recently reviewed the effective compounds in this field (24,25), showed that only one, R 2323 (22,23), clearly acts by interacting withthe receptor but this compound also displays only a partial antagonism. Other products such as RMI 12936 seem to be active by interfering on the biosynthesis of progesterone.
RU 38486, is therefore the first example of a true antagonist in progestin and glucocorticoid fields. It also possesses a moderate antiandrogenic activity which is 20-30 times weaker than its two other antagonist properties evaluated in rats. In addition, this steroid is devoid of androgen, estrogen,

mineralocorticoid and antimineralocorticoid actions. At very high doses, it induces a slight uterotrophic effect which is histologically different from the response observed with an estrogen.
Due to its antiprogesterone component, RU 38486 can therefore be used in fertility control (termination of pregnancy, induction of menstruation). It may also be useful in the treatment of certain hormono-dependent tumors (breast cancer, cerebral tumor....).
Its antiglucocorticoid potency may conceivably be applied in the treatment of diseases due to hypercorticism. Much wider clinical applications may be envisaged in cases where glucocorticoids play a role such as : hypertension, glaucoma, obesity, diabetes, immune depression, sleep and behavorial disorders.

Aknowledgments

I am grateful to Dr Moguilewsky, M., Tournemine, C., Drs Secchi, J. and Lecaque, D. for helpful discussion. Bouchoux, F., Branche, C., Bremaud, J., Cerede,E., Coutable, D., Humbert,J., Guery, C., and Viet, S. for technical assistance.
I would like to thank Magne, M. and Vercambre, A. for typing the manuscript.

References

1. Addison, G.M., Wirenfeldt Asmussen, N.N., Corvol, P., Kloppenborg, P.W.C., Norman, N., Schroder, R., Robertson, J.I.S., eds.: in aldosterone antagonists in clinical medicine, Excerpta Medica, Amsterdam, (1978)

2. Legha, S.S., Davis, H.D., Muggia, F.M.: Ann. intern. Med., 88, 69-77 (1978)

3. Martini, L., Motta, M. eds.: In androgen and antiandrogens. Raven Press, New-York, (1977)

4. Chrousos, G.P., Cutler, G.B.,Jr., Simons, S.S.,Jr., Pons, M., John, L.S. Moriarty, R.M., Loriaux, D.L.: In progress in research and clinical application of corticosteroids (Lee, H.J., Fitzgerald, T.J. eds) pp 152-176, Heyden and Son Co. Philadelphia (1982)

5. Chrousos, G.P., Cutler, G.B.,Jr., Sauer, M., Simons, S.S., Jr., Loriaux, D.L. : Pharmc. Ther. 20, 263-281 (1983)

6. Munck, A., Brinck-Johnsen, T.: J. Biol. Chem. 243, 5556-5565 (1968)

7. Samuels, H.H., Tomkins, G.M.:J.Mol.Biol. 52, 57-74 (1970)

8. Kaiser, N., Milholland, R.J., Turnell, R.W., Rosen, F.: Biochem. Biophys. Res. Commun. 49, 516-521 (1972)

9. Rousseau, G.G., Schmit, J.P.: J. Steroid Biochem. 8, 911-919 (1977)

10. Dausse, J.P., Duval, D., Meyer, P., Gaignault, J.C. Marchandeau, C., Raynaud, J.P.: Mol. Pharmacol. 13, 948-955 (1977)

11. Rousseau, G.G., Kirchhoff, J., Formstecher, P., Lustenberger, P.: Nature 279, 158-160 (1979)

12. Simons, S.S.,Jr., Thompson, E.B., Johnson, D.F.: Proc. Natl. Acad. Sci. U.S.A. 77, 5167-5171 (1980)

13. Duncan, M.R., Duncan, G.R.: J. Steroid Biochem. 10, 245-259 (1979)

14. Cutler, G.B.,Jr., Barnes, K.M., Sauer, M.A., Loriaux, D.L. Endocrinology 104, 1839-1844 (1979)

15. Agarwal, M.K., Sekiga, S., Lazar, G. : In Antihormones (Agarwal, M.K., ed.), pp. 51-73. Elsevier/North Holland Biomedical Press, New-York (1979)

16. Naylor, P.H., Gilani, S.S., Milholland, R.J., Rosen, F.: Endocrinology 107, 117-121 (1980)

17. Chrousos, G.P., Barnes, K.M., Sauer, M.A., Loriaux, D.L., Cutler, G.B., Jr.: Endocrinology 107, 472-477 (1980)

18. Guthrie, G.P., Jr, John, W.J.: Endocrinology 107, 1393-1396 (1980)

19. Naylor, P.H., Gilani, S.S., Milholland, R.J., Ip, M., Rosen, F. : J. Steroid Biochem., 14, 1303-1309, (1981)

20. Naylor, P.H., Rosen, F.: in Hormone Antagonists (Agarwal, M.K., ed.), pp. 365-380, De Gruyter, Berlin, New-York, (1982)

21. Chrousos, G.P., Sauer, M.A., Cutler, G.B., Jr. Loriaux, D.L. : Steroids 40, 425-431 (1982)

22. Sakiz, E., Azadian-Boulanger, G. : In Proc.III Int. Congr. on Horm. Steroids Hamburg 1970 (James V.H.T., Martini, L. eds),pp.865-871,ICS 219 Excerpta Medica,Amsterdam,(1971)

23. Azadian-Boulanger, G., Secchi, J., Sakiz, E. : In Proc.VII World Congr. on Fertil.Steril. Tokyo 1971 (Hasegawa, T., Hayashi, M., eds), pp.129-131, ICS 278 Excerpta Medica, Amsterdam, (1973)

24. Kendle, K.E.: In antihormones (Agarwal, M.K.ed.), pp.239-305, Elsevier/North Holland Biomedical Press (1979)

25. Kendle, K.E.: In hormone antagonists (Agarwal, M.K.ed), pp. 233-246,De Gruyter, Berlin, New-York, (1982)

26. Philibert, D., Deraedt, R., Teutsch, G.: VIII Int. Congr. of pharmacology. Tokyo, abstr. n°1463 (1981)

27. Philibert, D., Deraedt, R., Teutsch, G., Tournemine, C., Sakiz, E.: Endocrine Society, 64th annual meeting, San Francisco, abstr. n°668, (1982)

28. Chobert, M.N., Barouki, R., Finidori, J., Aggerbeck, M., Hanoune, J., Philibert, D., Deraedt, R. : Biochem. Pharmacol. 3481-3483, (1983)

29. Healy, D.L., Baulieu, E.E., Hodgen, G.D.: Fertil. Steril. 40, 253-257 (1983)

30. Jung-Testas, I., Baulieu, E.E.: J. Steroid Biochem. 20, 301-306 (1984)

31. Sakly, M., Philibert, D., Lutz-Bucher, B., Koch, B.: J. Steroid Biochem. 20, 1101-1104 (1984)

32. Moguilewsky, M., Philibert, D.: J. Steroid Biochem. 20, 271-276 (1984)

33. Duval, D., Durant, S., Homo-Delarche, F. : J. Steroid Biochem. 20, 283-287 (1984)

34. Bourgeois, S., Pfahl, M., Baulieu, E.E., : the Embo journal 3 , 751-755 (1984)

35. Chasserot, S., Beck, G. : J. Steroid Biochem. (In press.)

36. Herrmann, W., Wyss, R., Riondel, A., Philibert, D., Teutsch, G., Sakiz, E., Baulieu, E.E.: C.R. Acad. Sc. Paris Série III 294, 933-938 (1982)

37. Gaillard, R.C., Riondel, A., Herrmann, W., Müller, A.F., Baulieu, E.E. : Endocrine Society, 65th annual meeting, San Antonio, abstr. n° 219, (1983)

38. Bertagna, X., Bertagna, C., Girard, F.: Endocrine Society, 65th annual meeting, San Antonio, abstr. n°40, (1983)

39. Bertagna, X., Bertagna, C., Luton, J.P., Husson, J.M., Girard, F. : J. Clin. Endocr. Met. In press, (1984)

40. Philibert, D., Moguilewsky, M. : Endocrine Society, 65th annual meeting, San Antonio, abstr. 1018 (1983)

41. Raynaud, J.P., Bonne, C., Bouton, M.M., Moguilewsky, M., Philibert,D., Azadian-Boulanger, G.: J. Steroid Biochem., 6, 615-622, (1975)

42. Ojasoo, T., Raynaud, J.P.:Cancer Res. 38, 4186-4198 (1978)

43. Raynaud, J.P., Ojasoo, T., Bouton, M.M., Philibert, D.:in Drug design, vol VIII (Ariens, E.J., ed.) pp.169-214. Academic Press, New-York, (1979)

44. Raynaud, J.P. : In Progesterone receptors in normal and neoplastic tissues (McGuire, W.L., Raynaud, J.P., Baulieu, E.E., Eds) pp.9-21, Raven Press, New-York.(1977)

45. Philibert, D., Raynaud, J.P.: Steroids 22, 89-98 (1973)

46. Philibert, D., Ojasoo, T., Raynaud, J.P.: Endocrinology 101, 1850-1861 (1977)

47. Philibert, D., Raynaud, J.P.: in Progesterone receptors in normal and neoplastic tissues (McGuire, W.L., Raynaud, J.P., Baulieu, E.E., eds.), pp. 227-243, Raven Press, New-York, (1977)

48. Walters, M.R., Clark J.H., : in progesterone receptors in normal and neoplastic tissues. (McGuire, W.L., Raynaud, J.P., Baulieu, E.E., eds) pp 271-285, Raven Press, New-York.(1977)

49. Dunnett, C.W.: Am. Stat. Assoc. J., 60, 1086, (1955)

50. Giesen, E.M., Bollack, C., Beck, G. : Mol. Cell Endocrinol. 22, 153-168 (1981)

51. Wyllie, A.H. : Nature 284, 555-557 (1980)

52. Bia, J.M., Tyler, K., Defronzo, R.A. : Endocrinology 111, 882-888 (1982)

53. Meier, R., Schuler, W., Desaulles, P., : Experientia 6, 469-471 (1950)

54. Schmidt, F.H., : Klin Wschr. 39, 1244-1247 (1961)

55. Knox, W.E., Auerbach, V.H. : J. biol. chem. 214, 307-313

56. Diamondstone, I.: Anal. Biochem. 16, 395-401 (1966)

57. Clauberg, G.: Zentralbl. Gynaekol. 54, 2557-2770 (1930)

58. McPhail, M.K. : J. Physiol. London 83, 145-156 (1934)

59. McGinty, D.A., Anderson, L.P., McCullough, N.B.: Endocrinology 24, 829-832 (1939)

60. Ljunkvist, I.: Acta Soc. Med. Uppsalien. 76, 110-126 (1971)

61. Nilsson, O.: Acta Endocr. 78, 349-352 (1975)

62. Secchi, J., Lecaque, D.: Cell and tissue research (in press)

63. Kagawa, C.M. : Endocrinology 67, 125-132 (1960)

64. Simons, S.S., Jr., Thompson E.B. : Proc. Natl. Acad. Sci. U.S.A. 78, 3541-3545 (1981)

THE MOLECULAR BASIS OF THE ANTAGONISM BETWEEN BACTERIAL ENDOTOXIN AND GLUCOCORTICOIDS

L. Joe Berry and Gregory M. Shackleford

Department of Microbiology
University of Texas at Austin
Austin, Texas 78712-1095

Introduction

The ability of glucocorticoids to antagonize some of the biological effects of endotoxin was first reported in the literature thirty years or more ago. Duffy and Morgan (1) noticed that the febrile response to endotoxin in rabbits treated with cortisone was less than in untreated controls while Geller et al (2) were able to protect mice against endotoxin lethality by concurrent injection of cortisone. Berry and Smythe (3) in 1959 determined that the conversion of the protein broken down in response to cortisone administration was quantitatively converted into carbohydrate in both fed and fasted mice. However. in mice given endotoxin, protein breakdown was unchanged from that in unpoisoned controls but the amount of increase in stored carbohydrate was less than half of that in normal mice. It was concluded that glyconeogenesis was blocked in mice given endotoxin. Obviously, this implied that some disruption in the complex sequence of enzymatic reactions required in the conversion of nonsugars into carbohydrates was responsible.

Prior to the work just described (3), Knox and Auerback (4) reported that adrenocortical hormones, specifically cortisone, caused the induction of the hepatic enzyme tryptophan oxygenase. Within two to three hours of injection, the activity of the enzyme in rat liver was approximately twice the control level. This enzyme converts tryptophan into formyl kynurenine, a compound that, in turn, is the first in a pathway that leads to nicotinamide synthesis. The question arose as to whether endotoxin would be able to interfere with this glucocorticoid activity and

Adrenal Steroid Antagonism

the answer was in the affirmative (5). The full induction of tryptophan oxygenase by cortisone was blocked by a concurrent injection of endotoxin. In the same year, Nordlie and Lardy (6) reported that another liver enzyme, phosphoenolpyruvate carboxykinase (PEPCK) was also induced by cortisone in a manner much like that of tryptophan oxygenase. While these studies were in progress, other hormonally inducible enzymes were identified. These included tyrosine aminotransferase (7), glucose-6-phosphatase (8), fructose 1,6 diphosphatase (8) and glycogen synthase (9). All of these except the two phosphatases have half-lives of about 2 hr and the glucocorticoid induction of all of them is known to be inhibited by endotoxin (10,11,12). Endotoxin also inhibited the induction of PEPCK by the stress of fasting, presumably because the action of endogenously released glucocorticoid is antagonized by the bacterial poison (12a). In fact, the diurnal rhythm this enzyme normally undergoes is eliminated by endotoxin. The one enzyme we have studied that is not inhibited by endotoxin is tyrosine aminotransferase (10). For reasons not yet understood, this glucocorticoid inducible enzyme appears to be uninfluenced by endotoxin. This is not only an observation that challenges our understanding but also establishes the fact that under experimental conditions where synthesis of other enzymes fails to occur, this one is fully induced. Failure of enzymes to induce is not attributable, therefore, to a liver so badly damaged by endotoxin that it is no longer capable of forming new protein.

In fact, Shtasel and Berry (13) used radiolabeled precursors to show that livers of mice 8 hr after an injection of endotoxin synthesized more protein and messenger ribonucleic acid than livers of control mice. Actinomycin D, on the other hand, inhibited both processes. Actinomycin D also blocked the synthesis of tryptophan oxygenase and PEPCK in a manner indistinguishable from that of endotoxin (10,14). On the one hand, actinomycin D and endotoxin block the hormonal induction of certain hepatic enzymes to the same degree but the antibiotic's action is far more generalized than that of endotoxin as far as all hepatic proteins are concerned. It becomes a fascinating puzzle to try to unravel this mystery. Rippe and Berry (14,15) used the radial immunodiffusion technique to prove that less tryptophan oxygenase and PEPCK were synthesized in livers of cortisone injected mice poisoned with either endotoxin or actinomycin D than in livers of mice given hormone alone. The enzyme activity measurements in these livers reflect the amount of enzyme (antigen) present and not some change that results in diminished activity. These findings force the conclusion that the action of endotoxin on inducible enzymes is highly selective and is not to be attributed to some generalized damage to the liver.

When one ponders how endotoxin might possibly intervene in the synthesis of enzymes inside hepatocytes, it seems evident that a substance with its structure could hardly act directly. In the first place, its molecular (particle) weight is in the millions (16) and besides it is not very soluble (at least the lipid A moiety which is believed to exert most of the biological effects of endotoxin). In addition, radiolabeled endotoxin is probably not sequestered in liver parenchymal cells (17), even though Zlydaszyk and Moon (18) have published data to the contrary. Due to these considerations, Berry in 1971 (19) postulated that the effect of endotoxin on enzyme regulation is due to the action of a mediator rather than to a direct intervention by the bacterial product. When this suggestion was made, there was no experimental evidence to support it. It merely seemed logical and reasonable.

Evidence for the mediation of reduced enzyme induction by endotoxin in mice.

The first results in support of a mediator were obtained with tolerant mice (20). An injection of endotoxin into these animals failed to inhibit PEPCK induction by hydrocortisone. When serum from mice bled after an injection of endotoxin was given with the hormone, enzyme induction was significantly depressed. This effect of the serum had to be due to the presence of a mediator since the tolerant mice would not have responded to any residual endotoxin that might have been present. The source of the mediator was indicated when conventional mice that gave the expected result when endotoxin and hormone were given concurrently failed to show inhibition of PEPCK when they were treated one hour before with anti-mouse macrophage serum raised in rabbits. The apparent destruction of the macrophages by the antiserum either eliminated or greatly diminished the release of mediator. The antiserum against macrophages was shown to be cytotoxic by permitting entry of trypan blue into the cells. Moreover, when the antiserum was given directly to mice, the time required for the removal of colloidal carbon from the blood more than doubled.

Since some of the biological effects elicited in laboratory animals by endotoxin are also seen following the administration of synthetic polynucleotides, especially polyriboinosinic-polyribocytidylic acid (poly I:C) (21-23), its effect on PEPCK induction by hydrocortisone was tested in mice (24). When 50 ug poly I:C was given 4 hr prior to the injection of 1 mg hydrocortisone, PEPCK induction was completely inhibited. Induction was not significantly reduced when poly I:C was given concurrently with or 2 hr prior to the hormone. The mode of action of the polynucleotide may be different from that of endotoxin but there are sufficient parallelisms to make it likely that a mediator is responsible for the effects of both.

Results obtained with nude mice provide further evidence that a mediator is responsible for blocking the hormonal induction of PEPCK by endotoxin (25). The data are presented in Table 1. Notice that endotoxin has no inhibitory effect on the induction of PEPCK by hydrocortisone but serum from Zymosan-primed mice gives complete blockage. The Zymosan injections, given as shown in the footnote to Table 1, result in proliferation of the reticuloendothelial system (RES) and, as will be established below, increases the yield of mediator. As further evidence that the mediator comes from cells of the RES, peritoneal exudate cells given along with a small (20 ug) dose of endotoxin block enzyme induction.

Nude mice are known to lack thymus-derived lymphocytes and, on this basis, one might be tempted to infer that the mediator with which we are concerned is a T-cell product. However, our findings with anti-macrophage serum (20), make this doubtful. There is no reason to believe that macrophages in nude mice are normal. They too could be faulty and this was suggested in the paper (25). Results not included in Table 1 established that heterozygous (nu/+) litter mates of nude mice respond in the

Table 1

Inducibility of PEPCK in Nude Mice Given Endotoxin or Serum Rich in a Mediator

Treatment	No. of Mice	PEPCK Activity ± S.E.M.[a]
No injections	7	187 ± 20
1 mg hydrocortisone	7	267 ± 20
1 mg hydrocortisone + 50 μg endotoxin	7	271 ± 11
1 mg hydrocortisone + 0.3 ml Zymosan-endotoxin serum[b]	6	157 ± 11
1 mg hydrocortisone + 20 μg endotoxin + $6x10^7$ peritoneal exudate cells[c]	4	151 ± 16

[a] μmoles PEP/g dry wt liver/6 min ± standard error of the mean

[b] serum collected from mice injected with 0.5 mg Zymosan on day 0, 1 mg Zymosan on day 2 and they were bled 2 hr after 25 μg endotoxin given on day 5.
Modified from (25).

[c] obtained from heterozygous (nu/+) mice

same manner as outbred mice to endotoxin; i.e. the hormonal induction of PEPCK was blocked by endotoxin.

Notice that the control level of activity of PEPCK in nude mice is much higher than the values obtained in conventional animals. This suggests that the endogenous level of the mediator in nude mice is lower than in conventional animals. As a consequence, a normal regulatory mechanism for enzyme control is lacking.

Additional studies were conducted with the nonresponder inbred C3H/HeJ mouse (26). This animal was shown by Sultzer (27) to be remarkably refractory to the biological actions of endotoxin. As the results in Table 2 make clear, endotoxin injected at the same time as hydrocortisone

Table 2

Response of C3H/HeJ Mice to Endotoxin, to Serum Rich in Mediator and to Peritoneal Exudate Cells as Judged by PEPCK Induction

Treatment	PEPCK Activity ± S.E.M.[a]
Controls	84 ± 10
4 hr after 1 mg hydrocortisone	208 ± 10
4 hr after 1 mg hydrocortisone + 50 μg endotoxin	202 ± 9
4 hr after 1 mg hydrocortisone + 0.5 ml of serum from Zymosan-primed mice bled after endotoxin	130 ± 8
4 hr after 1 mg hydrocortisone + 1×10^7 peritoneal exudate cells + 20 μg endotoxin	171 ± 20

[a] μmoles PEP/g dry wt liver/6 min ± standard error of the mean. All values from groups of at least 7 mice.
Modified from (26).

does not inhibit the induction of PEPCK. Serum from Zymosan-primed mice bled 2 hr after endotoxin (Zymosan endotoxin serum or ZES) results in a highly significant reduction in induction. Administration of peritoneal exudate cells from conventional mice plus endotoxin and hydrocortisone results in a decrease in the level of induction even though it is less complete than the result obtained with ZES. With these inbred mice, conclusions similar to those reached with nude mice seem justified. The C3H/HeJ Mice have faulty macrophages and do not form the mediator in normal manner. These animals can respond, however, to the mediator when it is obtained from an exogenous source or is formed in vivo by cells taken from mice that respond normally. The mediator was named glucocorticoid antagonizing factor (GAF) (26).

Indomethacin, a nonsteroidal anti-inflammatory agent, was used in outbred mice to see if it altered the release or synthesis of GAF in mice injected with endotoxin. The results in Table 3 make it apparent that treatment with the drug alone resulted after 5 hr in an activity of PEPCK

Table 3

The Effect of Indomethacin on PEPCK Induction in Mice Given Endotoxin and Hydrocortisone

Treatment 1 Hr after 0.5 mg indomethacin	PEPCK Activity ± S.E.M.[a] (No. of Mice in Group)
None	285 ± 30 (7)
1 mg hydrocortisone + 10 μg endotoxin	385 ± 22 (8)
1 mg hydrocortisone + 0.2 ml normal mouse serum	402 ± 33 (15)
1 mg hydrocortisone + 0.2 ml serum collected from mice 2 hr after 25 μg endotoxin	263 ± 11 (7)

[a]μmoles PEP/g dry wt liver/6 min ± standard error of the mean. Modified from (28).

that was larger than that seen in conventional mice injected with cortisol (controls about 135 and after cortisol about 225, as shown in many publications). Apparently, indomethacin blocks endogenous GAF production and thereby eliminates one of the regulatory controls for the enzyme. Table 3 also shows that PEPCK is inducible to a remarkably high level in indomethacin pretreated mice and the induction is blocked, however, by the administration of serum from mice injected with endotoxin. GAF must have been present in sufficient amount in this serum to stop the inductive process.

Table 4 provides confirmatory evidence that indomethacin blocks GAF formation as judged by the absence of inhibition of PEPCK by serum

Table 4

The Effect of Indomethacin on the Ability of Mice to Release GAF into Their Serum after an Injection of Endotoxin

Treatment	PEPCK Activity ± S.E.M.[a] (No. of Mice in Group)
Saline in tolerant mice	180 ± 21 (5)
Tolerant mice given 0.2 ml serum collected from mice 2 hr after 25 μg endotoxin + 1 mg hydrocortisone	205 ± 21 (5)
Tolerant mice given 0.2 ml serum collected from mice given indomethacin 1 hr before 25 μg endotoxin & then bled 2 hr later + 1 mq hydrocortisone	287 ± 27 (5)

[a]μmoles PEP/g dry wt liver/6 min ± standard error of the mean. Modified from (28).

collected from mice that had been pretreated with the drug for 1 hr and were then bled 2 hr after an injection of endotoxin. A large PEPCK induction followed in these mice when cortisol was given 4 hr earlier. Serum similarly collected from mice that had been given no indomethacin was significantly inhibitory of the hormonal induction of the enzyme. The different values for PEPCK in this Table are smaller than those seen in Table 3. All the mice used for the results in Table 3 had been pretreated with indomethacin while those for Table 4 had not been pretreated. Only the mice that served as serum donors (and failed to produce much, if any, GAF) had been given indomethacin prior to bleeding.

Two final points should be made with regard to the experiments with indomethacin. This compound depends for its action, at least in part, on its ability to suppress the synthesis of prostaglandins (29). When 100 μg of prostaglandin E was injected intravenously, the cortisol induction of PEPCK was not inhibited (data not shown) (28). Moreover, 1 mg of arachidonic acid, a precursor that is rapidly converted into prostaglandins in mice (30), was also non-inhibitory at a dosage of 1 mg iv. The action of indomethacin in the experiments that yielded the data of Tables 3 and 4 does not appear to be related to prostaglandins unless the dosages were inadequate.

Indomethacin does not have the ability to protect mice against the lethal effect of endotoxin. It prolonged the time of death of mice sensitized to endotoxin with lead acetate (28) but their survivorship was unaltered. Other workers have been unable to protect cats against endotoxin lethality using indomethacin (31). This suggests that GAF does not contribute directly to the death of laboratory animals.

GAF not only interferes with the hormonal regulation of PEPCK and other hepatic enzymes mentioned previously in the paper, it also, after a delay, blocks the ability of cortisol to protect mice against the lethal effect of endotoxin. When mice are given an LD_{100} of endotoxin, concurrent injection of 1 mg cortisol reduces the percentage of deaths to only about 20%. If the cortisol injection is delayed for 1,2,3, or 4 hr, the percentage mortality, respectively, is 80%, 90%, 80%, and 100% (32). The time when cortisol loses its protective ability varies with dose of endotoxin and probably with the strain of mice being used.

The effect of endotoxin and GAF on PEPCK induction in hepatoma cells.

Mice have served as the model for nearly all work done by this laboratory on the interaction between glucocorticoids and endotoxin. There are difficulties, however, when intact animals are being used, especially since adrenocortical hormones are known to be released following an injection of endotoxin (33). The release occurs promptly while the peak amount of GAF in serum does not appear until 2 hr after endotoxin is administered (26). The early presence of endogenous hormones could very well lead to changes in the host that are not readily discerned. The fact that Wicks et al (34) were able to use Reuber H35 hepatoma cells in culture to study the mechanism of induction of PEPCK by glucocorticoid prompted Goodrum and Berry (35) to use these cells in experiments now to be described. Table 5 summarizes the relevant findings. It is important to note that endotoxin alone has no effect on the hormonal induction of PEPCK and the numerical agreement between groups indicates a consistent behavior of the hepatoma cells in culture. It is also apparent that GAF in the form of serum from Zymosan-primed mice bled 2 hr after an injection of endotoxin completely inhibited induction of PEPCK. This total suppression of the inductive process was not seen in mice, presumably because the release of endogenous glucocorticoid initiated induction before GAF was able to block its action.

Several additional observations were made in experiments with hepatoma cells (35). It was possible to establish that macrophages treated with endotoxin release into the medium the factor (GAF) that blocks the hormonal induction of PEPCK. This confirms results obtained with mice as described above. It also became possible to use the hepatoma cells to assay serum from mice for its GAF content (36). This shows that serum from Zymosan-primed mice bled 2 hr after an injection of endotoxin had more than double the GAF content that serum from normal mice had 2 hr after endotoxin. Moreover, GAF is present in detectable amounts in normal mouse serum. Serum from Zymosan-primed mice given no endotoxin prior to bleeding had the same amount of GAF as normal mouse serum. Thus, endotoxin triggers the appearance of GAF in serum but details of the

Table 5

Effect of Endotoxin and GAF on PEPCK Induction by Hydrocortisone in Hepatoma Cells in Culture

Treatment	PEPCK Activity ± S.E.M.[a]
None	40 ± 2
Endotoxin - 10 μg/ml	41 ± 2
Cortisol - 1 μM	96 ± 2
Cortisol - 1 μM + Endotoxin - 10 μg/ml	94 ± 2
Serum from Zymosan-primed mice bled 2 hr after endotoxin 2% (v/v)	41 ± 1
Same as above + Cortisol - 1 μM	39 ± 1

[a]Activity measured as nmoles $NaH^{14}CO_3$ fixed/min per mg protein ± standard error of the mean. The period of induction was 8 hr and all values were derived from 6 duplicate experiments.

Modified from (35).

process are lacking. Indomethacin and cortisone treatment each lowers the amount of GAF in serum of control and Zymosan-primed mice following an injection of endotoxin. This helps to explain some of their effects. Tolerant mice had the least titer of GAF in their serum compared to that of any other's and it was not elevated following injection of endotoxin. The action of GAF on hepatoma cells was partially reversible if the cells were washed 2 hr after exposure (36).

The hepatoma cells permitted some characterization of GAF (36). It is a large molecular weight protein as judged by its characteristics when filtered through Sephadex G200 and it is inactivated by heating at 56°C for 1 hr or 70°C for 30 min. Finally, there are suggestions in the literature (37,38) that the changes in carbohydrate metabolism following endotoxin injection may be due to insulin release or to a factor from macrophages that has insulin-like activity. The hepatoma cells made it possible to distinguish between the action of GAF and insulin. GAF does not block the induction of PEPCK by dibutyryl cyclic AMP (in the presence of theophylline) while insulin does. Moreover, insulin was not affected by the heat that inactivated GAF. In addition, insulin does not inhibit the glucocorticoid induction of renal PEPCK and GAF does (unpublished observation). These differences between GAF and insulin make it seem highly improbable that they are the same.

How does GAF inhibit the glucocorticoid induction of PEPCK?

It is conceivable that glucocorticoids do not increase the synthesis of regulatory enzymes but rather enhance their catalytic action. Experiments in several laboratories (39-41), in addition to those of Rippe and Berry (14,15) have shown, however, that de novo synthesis of the enzymes does occur after steroid administration. Pulse-labeling and immunoprecipitation experiments have been used to reach these conclusions. Hydrocortisone does, therefore, enter the cell and result in an increase in *in vitro* translatable enzyme mRNA (42,43). Glucocorticoid hormones must stimulate the transcriptional process (44) and GAF must interfere in either some pretranscriptional event or else alter a posttranscriptional change that is necessary for induction.

The events in the inductive process have been relatively well-characterized (45). Fig. 1 shows diagramatically what they are. The first step is the entry of the hormone into the target cell, presumably by means of diffusion, where it binds with high affinity to a cytoplasmic protein. A conformational change in this complex, referred to as activation, step 2, permits entry into the nucleus where it binds to putative specific acceptor sites on the chromatin. The next step is an increase in the transcription rate of a small subset of genes encoding the enzymes induced. The subset is believed to be less than 1% of active genes. Finally, translation of the mRNA occurs in the cytoplasm and the new enzyme appears.

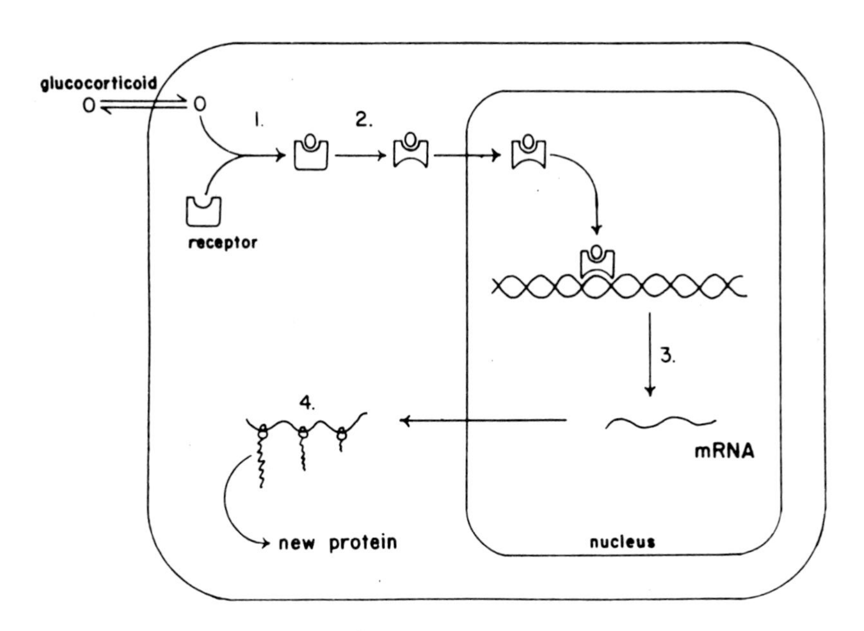

Figure 1. Events in the glucocorticoid induction of proteins. (1) Glucocorticoid-receptor complex formation; (2) activation of the complex; (3) induced transcription of specific genes after binding of the complex to chromatin; (4) translation of the induced quantity of specific mRNA.

Glucocoritcoid receptors were quantitated in the livers of control and test animals by measuring the binding of [^{3}H] dexamethasone to cytoplasmic receptors in vitro. For this purpose, the mice were killed, their livers were perfused in situ with cold 0.9% NaCl solution, then weighed and homogenized. Supernates after high speed centrifugation (100,000 xg) were used to measure total binding by adding [^{3}H] dexamethasone to a final concentration of 100 nM. Nonspecific binding was determined by adding unlabeled dexamethasone to the same concentration in addition to the labeled compound. Specific binding was calculated by subtracting nonspecific binding from total binding.

Bound [^{3}H] dexamethasone can be determined in several ways. The one employed primarily in these studies (46) is a modification of the technique developed by Pratt et al. (47). A G-50 minicolumn was used to separate the protein-bound labeled hormone from the unbound steroid. The receptor complexes were then eluted and counted.

An assay described by Beato and Feigelson (48) was also used. In this procedure the unbound [^{3}H] dexamethasone is adsorbed to charcoal. Dextran is included with the charcoal to prevent the protein from also adsorbing. After an appropriate period of incubation, the charcoal is removed by centrifugation and the radioactive content of the supernatant is determined.

A third method is a modification of a procedure described by Maniatis et al. (49). This entails applying the sample to a minicolumn of Sephadex G-50 and then accelerating the chromatography by centrifugation at 1600 xg. The sample thus eluted is counted.

Another technique much like the first substituted ultrogel AcA 34 for the Sephadex G-50 column and made use of a different column buffer. Fractions were collected and counted.

The results obtained with different doses of endotoxin using the centrifugation chromatography technique are summarized in the second column of Table 6. Notice that livers came from adrenalectomized mice sacrificed

Table 6

Effect of Dose of Endotoxin on Binding of $[^3H]$ Dexamethasone in Hepatic Cytosol

Treatment	Specifically Bound $[^3H]$ Dexamethasone (cpm/mg protein)[a] as Determined by		
	Centrifugation Chromatography	Charcoal-Dextran Method	Sephadex G-50 Mini-Column
Untreated	3536 ± 816 (6)[b]	2610 ± 374 (6)	4097 ± 274 (7)
150 µg endotoxin	4085 ± 668 (5)	2681 ± 430 (4)	4069 ± 201 (9)
50 µg endotoxin	3539 ± 458		
16 µg endotoxin	3120 ± 541 (5)		
5 µg endotoxin	3749 ± 425 (7)		

[a]Adrenalectomized mice were sacrificed 4 hr after treatment and their livers were processed for the assay.

[b]Number of duplicate tests run.

4 hr after treatment. With the amounts of endotoxin administered, no change in the binding of dexamethasone appeared. Experimental values were not significantly different from controls nor were they different from each other. The charcoal-dextran method (third column in Table 6) gave values much lower than those found by the other two methods while the Sephadex G-50 minicolumn procedure, last column in Table 6, yielded results in agreement with those observed with the method shown in the second column.

These data are not in agreement with those recently reported by Stith and McCallum (50) who found that endotoxin resulted in a decrease in dexamethasone binding but at a time when the mice were beginning to die. This was at 6 hr after an injection of 100 ug of Boivin endotoxin. For adrenalectomized mice, this is a large dose. So is our largest dose, 150 μg, but phenol-water derived endotoxin is less toxic than Boivin endotoxin and our mice were sacrificed after 4 hr. Stith and McCallum measured binding by the charcoal dextran method and though they failed to give data for PEPCK activity under the conditions when no decrease in binding was measurable, they stated that induction of the enzyme was blocked under all relevant conditions. We too found that all doses shown in Table 6, blocked PEPCK induction to about the same extent. These data will be available in reference (46). In summary, the results of Stith and McCallum differ from our data since their mice were killed 6 hr after endotoxin and ours were sacrificed at 4 hr. The time difference may have resulted in their mice being in poorer condition than ours. Additional experiments are needed to clarify the apparent contradictions in results.

The time course of binding of dexamethasone to the cytosol receptor was determined in liver supernatants from control and endotoxin poisoned mice as was the rate of dissociation of the radioactive complex when cold dexamethasone was added. The results, not shown here, further support our conclusion that binding of the hormone to its receptor protein is uninfluenced by endotoxin.

The entry of glucocorticoids into liver cells represents the earliest step where endotoxin might exert an inhibitory effect on the hormonal induction of PEPCK. To evaluate this possibility, the binding of [^{3}H] dexamethasone to cytosols prepared from control and endotoxin-poisoned adrenalectomized mice was compared to that found for intact animals. Hydrocortisone was also given to intact mice prior to the time when cytosols were prepared. The results are shown in Fig. 2. As reported in Table 6, endotoxin fails to alter the binding of the hormone in adrenalectomized mice while in intact mice it does. Not only is there less binding of dexamethasone in intact mice given endotoxin, the binding in control mice is lower than in adrenalectomized animals. Moreover, mice injected with 1 mg of hydrocortisone found less labeled dexamethasone. It is believed that the lower values in intact mice is due to the presence of

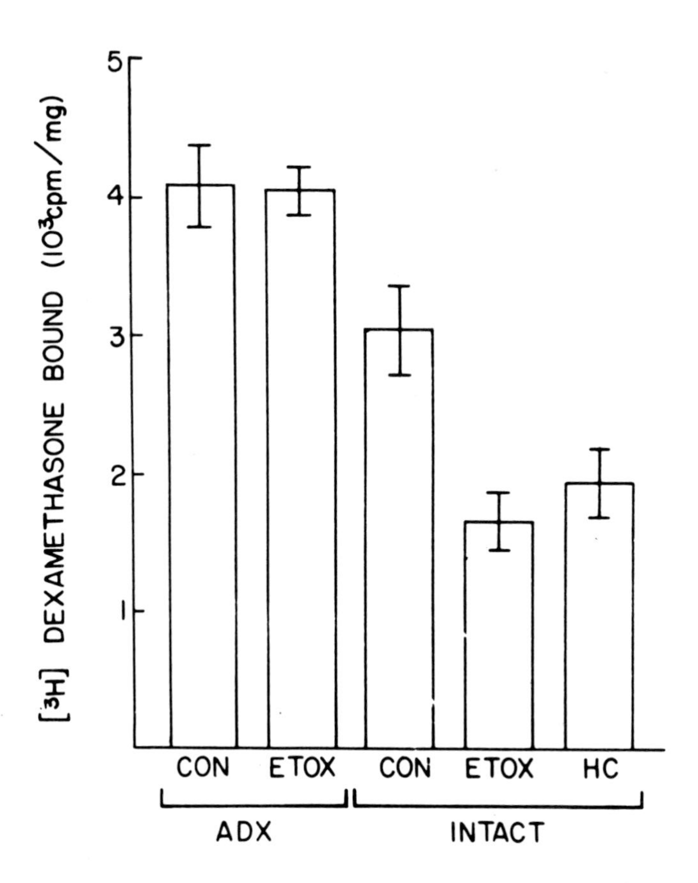

Figure 2. Specific binding of [^{3}H] dexamethasone to hepatic cytosols from untreated and treated intact and adrenalectomized mice. Cytosols from untreated mice, or mice injected with endotoxin (50 μg, ip, 4 hr) or hydrocortisone (1 mg, sc, 4 hr) were incubated with 100 nM [^{3}H] dexamethasone for 4 hr and specific binding was determined on the cytosol of each animal by the G-50 minicolumn assay. Values are expressed as averages of binding per mg protein ± SEM. Groups consisted of at least 5 animals.

---

(1) endogenous glucocorticoid release from the adrenals of control mice, (2) the even larger release from the adrenals of endotoxin injected mice and (3) the effect of exogenous glucocorticoid. Under all of these conditions, the receptor proteins must have been occupied in part by either the endogenous or exogenous hormone leaving less labeled hormone to be bound. These data very strongly suggest that endotoxin has no effect on

the free entry of the steroids into hepatic cells.

As mentioned previously, activation is the process by means of which the glucocorticoid receptor complex gains the ability to translocate from the cytoplasm into the nucleus and bind to chromatin acceptor sites. This process has now been demonstrated to occur in vivo (51) and is believed to involve a conformational change. Chromatographic resolution of activated and unactivated receptors on DEAE-cellulose has been well-characterized by Schmidt et al. (52) and was used for the results shown in Fig. 3. Panel A contains the data for cytosols from untreated adrenalectomized mice and

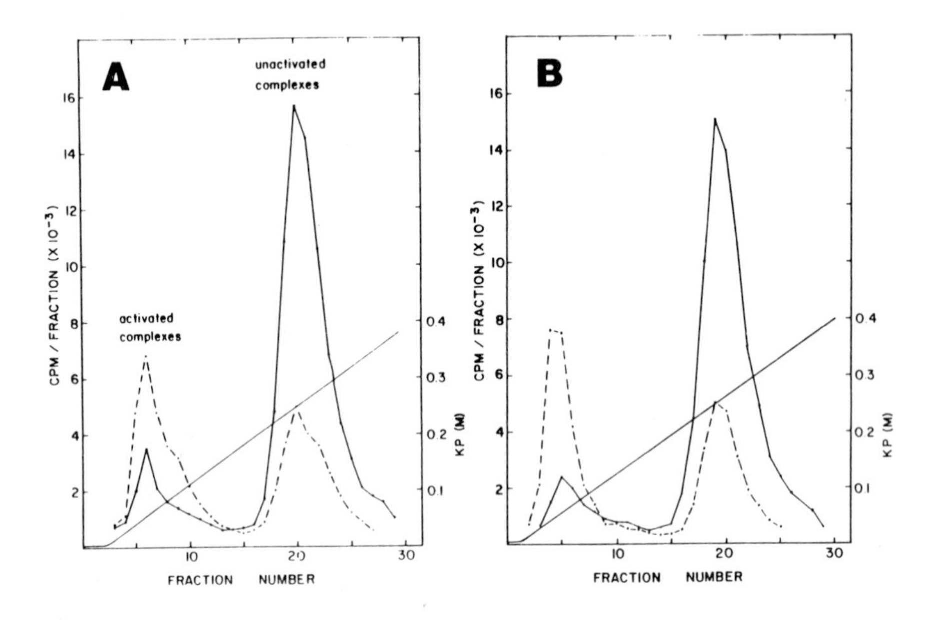

Figure 3. Activation of hepatic [^{3}H] dexamethasone-receptor complexes from untreated and endotoxin-injected adrenalectomized mice. The 33% (w/v) cytosols from pooled livers of untreated (A) or endotoxin-injected (50 μg, ip) (B) mice were allowed to bind 100 nm [^{3}H] dexamethasone and then split into two equal aliquots, one of which was activated (23°C, 30 min) and the other not activated (0°C, 30 min). These were desalted to remove unbound hormone and applied to parallel DEAE-Sephadex columns. After washing, activated and unactivated complexes eluted sequentially with a 5-400 mM potassium phosphate gradient. Activated cytosol (·----·); unactivated cytosol (•——•).

panel B for cytosols from endotoxin injected (50 ug) adrenalectomized mice. The cytosols from each group of mice were divided into two equal parts. One was activated by heating at 23°C for 30 min and the other was maintained at 0°C for 30 min. All were desalted to remove unbound hormone and then chromatographed. The solid line in each panel shows the results obtained with the unactivated complexes and the dashed line the activated complexes. Elution was accomplished with a 5-400 mM potassium phosphate gradient. The concentration of the salt solution at which activated and unactivated receptors elute (0.05 and 0.25 M, respectively) agree well with reported values (0.05 and 0.21 M) (52). The data show that each receptor preparation undergoes activation and shifts its profile to the first peak. Endotoxin, therefore, does not appear to influence the activation process.

In the next set of experiments, the active PEPCK mRNA was measured. Rippe and Berry (14,15) demonstrated earlier that endotoxin inhibits the glucocorticoid induction of PEPCK as determined by radial immunodiffusion. This implies that a transcriptional or pretranslational step is the site of intervention. To make this determination, the rabbit reticulocyte cell-free translation system (53) was utilized to translate mRNA, isolated from livers of treated and untreated mice. The reticulocyte lysate was treated with micrococcal nuclease to degrade endogenous mRNA (54). Total liver RNA was isolated by a procedure based on methods by Deeley et al. (55) and Chirgwin et al. (56). The translation reaction mixture contained [^{35}S] methionine to label the products. The PEPCK produced was immunoprecipitated. This procedure required the purification of PEPCK from mouse liver (57,58), raising antibody against it in rabbits and the IgG purified from it. The protein A containing *Staphylococcus aureus*, Cowan strain, was prepared (59) and used to co-precipitate the labeled PEPCK from the *in vitro* reaction mixture. Both total and immunoprecipitated proteins were subjected to SDS-polyacrylamide gel electrophoresis and the results with total proteins are shown in Fig. 4. The autoradiograms show no significant differences in the amounts of protein or the range of molecular weights synthesized. All mice were sacrificed 4 hr after treatment.

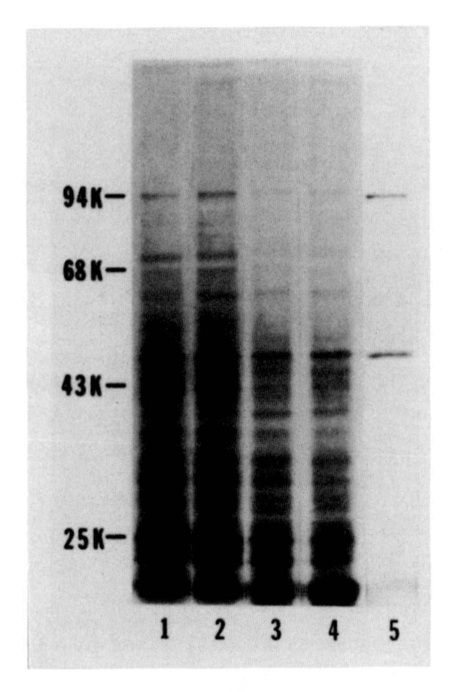

Figure 4. Fluorogram of in vitro translation products of RNA isolated from hydrocortisone and/or endotoxin treated adrenalectomized mice. Isolated total cellular RNA (750 μg/ml of reaction mixture) was used as translation template in the rabbit reticulocyte lysate in vitro translation system. Aliquots (3 μl) of post-ribosomal supernatants were subjected to SDS-PAGE and the gel fluorographed with EnHance for 3 days. Treatment groups: untreated control (lane 1); hydrocortisone, 1 mg, sc, 4 hr (lane 2); hydrocortisone (same dose) plus endotoxin, 50 μg, ip, 4 hr (lane 3); endotoxin (lane 4); no RNA added (lane 5).

Fig. 5 is a typical fluorogram in the designated treatment groups of the immunoprecipitated PEPCK synthesized in vitro under the direction of RNA, isolated from livers of mice killed 4 hr after treatment. The results show that hydrocortisone induces PEPCK while concurrent injection of hormone and endotoxin almost totally blocked synthesis of PEPCK. The same shows for mice given endotoxin alone. Although the data are qualitative, they are highly significant and provide evidence that endotoxin exerts its effects on the synthesis of PEPCK mRNA. The absence of any apparent effect of endotoxin on total protein synthesis (Fig. 4) testifies to the specificity involved in this phenomenon. The radioactive band

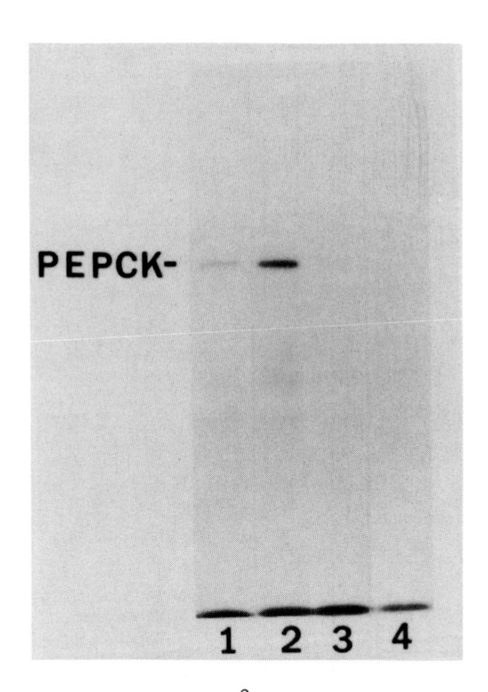

Figure 5. Typical fluorogram of [^{3}H] dexamethasone-labeled PEPCK synthesized *in vitro*, immunoprecipitated and separated on SDS-PAGE. Total cellular RNA (hepatic) was isolated from the following treatment groups of adrenalectomized mice, translated in the rabbit reticulocyte *in vitro* translation system immunoprecipitated with anti-PEPCK and separated by SDS-PAGE: untreated controls (lane 1); hydrocortisone (lane 2); hydrocortisone plus endotoxin (lane 3); endotoxin (lane 4). Doses are the same as Fig. 4. RNA for each reaction was pooled from 2 animals.

---

labeled PEPCK was identified by (i) its ability to be precipitated by specific antibody (Fig. 5), (ii) its inducibility by glucocorticoid (Fig. 5), (iii) its exact comigration with highly purified PEPCK from mouse liver (data not shown), and (iv) the ability of purified PEPCK to compete with the *in vitro* synthesized PEPCK in the immunoprecipitation reaction (data not shown).

Figure 6 provides additional evidence for the indicated conclusions reached on the basis of experiments just described. Here the gels

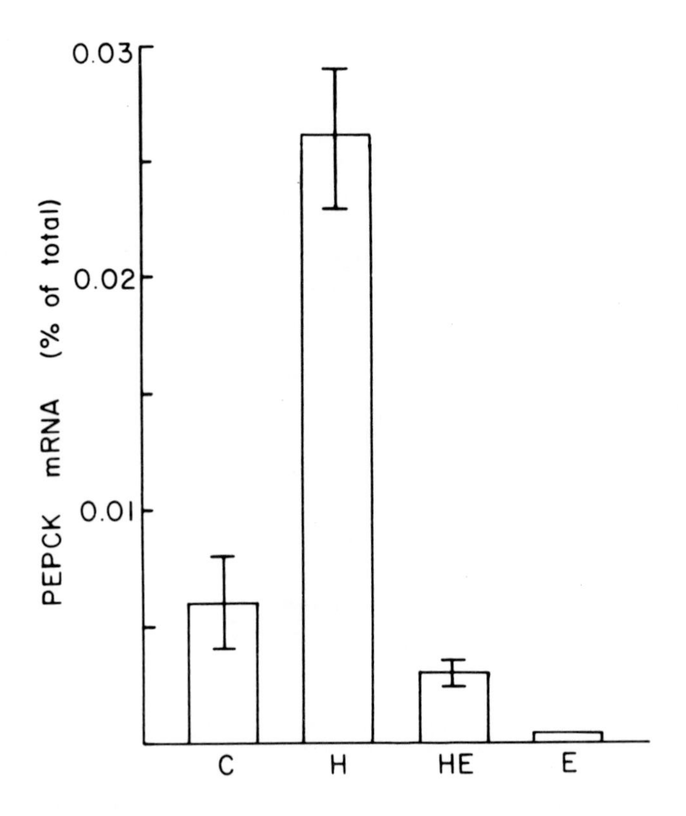

Figure 6. Effect of endotoxin on hydrocortisone induction of PEPCK mRNA. PEPCK mRNA is quantitated according to its translational activity _in vitro_. Data was accumulated from two experiments using 6 RNA preparations per group and is expressed as the average ± standard error of the mean.

---

obtained with the RNA isolated from the livers of different treatment groups were cut at the location of PEPCK and their radioactive counts determined. The expected induction of PEPCK mRNA occurred in the hydrocortisone treated group while endotoxin given with the hormone or alone blocked the synthesis of PEPCK mRNA.

Discussion

This overview of the antagonism between endotoxin and glucocorticoid hormones adduces evidence for the fact that at least one of the sites of conflict is at the level of transcription or, possibly, some posttranscriptional event necessary for the appearance of functional mRNA. The antagonism is not a direct one but is rather between an endotoxin mediator and the hormone. The mediator appears to be of macrophage origin but other cells of the RES may play a role (25).

How GAF interferes with mRNA synthesis remains to be answered as does the nature of its specificity. The fact that GAF has no measurable effect on PEPCK induction by cAMP (35) nor on the induction of tyrosine aminotransferase by glucocorticoid (10) also remains enigmatic. There is still uncertainty about how glucocorticoid and cAMP act in increasing enzyme synthesis. The hormone, according to the current literature, stimulates mRNA synthesis (60) while cAMP is thought to stimulate translation (34,39). In actuality both systems may be in operation (61,62) even though this has not been firmly established. Were this true, it would help provide a mechanism for the "permissive" action of glucocorticoids (63). The permissive effect is when two inducers used simultaneously yield a larger induction than the sum of the two used separately (64).

Macrophages are responsible for producing a number of mediators. Filkins (65) has identified a factor he calls macrophage insulin-like activity which is produced in response to endotoxin. It may be identical to GAF since it blocks certain glucocorticoid effects. Mishell et al. (66) have detected a macrophage factor which protects _in vitro_ immune responses against the immunosuppressive effects of glucocorticoids. The factor has been named glucocorticoid response-modifying factor (GRMF) and is said to be induced by cell wall constituents of both Gram positive and Gram negative bacteria. Since Shackleford in unpublished observations has found that the hydrocortisone induction of PEPCK is not inhibited in mice by Gram positive organisms one has to conclude that GRMF and GAF are distinct. Lowitt et al. (67) found that dexamethasone induction of tryptophan oxygenase in hepatocytes is inhibited only when nonparaenchymal cells are also present in the culture. This suggests that a mediator originating in cells other than hepatocytes was responsible. Lipoprotein lipase activity and immunoprecipitable enzyme from mouse preadipocytes is decreased more than 95% by conditioned medium from mouse peritoneal exudate cells incubated with endotoxin (68,69). The relationship between GAF and these agents will have to await their purification. The ability of factors derived from macrophages to interfere with protein (enzyme) synthesis is becoming to be recognized as a general phenomenon. It is likely that continued study of the mode of action of these substances will help to broaden our understanding of specific biological responses to such important substances as endotoxin but it may well advance our insight into cell regulatory processes.

Acknowledgment

This work was supported in part by Public Health Service Grant AI 10087 from the National Institutes of Health.

References

1. Duffy, B. J., Morgan, H. R.: Proc. Soc. Exp. Biol. & Med. 78, 687-689 (1951).
2. Geller, P., Merrill, E. R., Jawetz, E.: Proc. Soc. Exp. Biol. & Med. 86, 716-719 (1954).
3. Berry, L. J., Smythe, D. S.: J. Exp. Med. 110, 407-418 (1959).
4. Knox, W. E., Auerbach, V. H.: J. Biol. Chem. 214, 307-313 (1955).
5. Berry, L. J., Smythe, D. S.: J. Exp. Med. 118, 587-603 (1963).
6. Nordlie, R. C., Lardy, H. A.: J. Biol. Chem. 238, 2259-2263 (1963).
7. Kenney, F. T., Flora, R. W.: J. Biol. Chem. 236, 2699-2702 (1961).
8. Kvam, D. C., Parks, R. E., Jr.: Am. J. Physiol. 198, 21-24 (1960).
9. Hers, H. G., DeWulf, H., Stalmans, W., Van Den Berghe, G.: Adv. Enzyme Regul. 8, 171-190 (1970).
10. Berry, L. J., Smythe, D. S. Colwell, L. S.: J. Bacteriol. 96, 1191-1199 (1968).
11. McCallum, R. E., Berry, L. J.: Infect. Immun. 6, 883-885 (1972).
12. McCallum, R. E., Berry, L. J.: Infect. Immun. 7, 642-654 (1973).
12a. Phillips, L. J., Berry, L. J.: Amer. J. Physiol. 218, 1440-1444 (1970).
13. Shtasel, T., Berry, L. J.: J. Bacteriol. 97, 1018-1025 (1969).
14. Rippe, D. F., Berry, L. J.: Infect. Immun. 6, 766-772 (1972).
15. Rippe, D. F., Berry, L. J.: Infect. Immun. 8, 534-539 (1973).
16. Lüderitz, O., Westphal, O., Staub, A. M., Nikaido, H.: Microbial Toxins, Vol. 4, ed. Weinbaum, G., Kadis, S., Ajl, S. J., Academic Press, New York, p. 145 (1971).
17. Braude, A. I., Carey, F. J., Sutherland, D. and Zalesky, M.: J. Clin. Invest. 34, 850-857 (1955).
18. Zlydaszyk, J. C., Moon, R. J.: Infect. Immun. 14, 100-105 (1976).
19. Berry, L. J. in Bacterial Toxins, Vol. 5, ed. Kadis, S., Weinbaum, G., Ajl, S. J., Academic Press, New York, p. 165 (1971).
20. Moore, R. N., Goodrum, K. J., Berry, L. J.: J. Reticuloendothelial Soc. 19, 187-197 (1976).
21. Stinebring, W. R., Youngner, J. S.: Nature, London 204, 712-713 (1964).
22. Field, A. K., Tytell, A. A., Lampson, G. P., Hilleman, M. R.: Proc. Natl. Acad. Sci., U.S.A. 58, 100-41010 (1967).
23. Lindsay, H. L., Trown, P. W., Brandt, J., Forbes, M.: Nature, London 223, 717-718 (1969).
24. Moore, R. N., Berry, L. J.: Experientia 32, 1566-1567 (1976).

25. Moore, R. N., Goodrum, K. J., Berry, L. J., McGhee, J. R.: J. Reticuloendothelial Soc. 21, 271-278 (1977).

26. Moore, R. N., Goodrum, K. J., Couch, R. E., Jr., Berry, L. J.: Infect. Immun. 19, 79-86 (1978).

27. Sultzer, B. M.: Nature, London, 219, 1253-1254 (1968).

28. Goodrum, K. J., Moore, R. N., Berry, L. J.: J. Reticuloendothelial Soc. 23, 213-212 (1978).

29. Horton, E. W.: Monograph on Endocrinology, Vol. 7, p. 63, Springer-Verlag, New York (1972).

30. Collier, H.O.J., Saeed, S. A., Schneider, C., Warren, B. T.: Advances in the Biosciences, Vol. 9, (S. Bergström, ed.) p. 413, Pergamon Press, New York (1973).

31. Parratt, J. R., Sturgess, R. M.: Br. J. Pharmacol. 533, 485-492 (1975).

32. Moore, R. N., Goodrum, K. J., Couch, R. E., Jr., Berry, L. J.: J. Reticuloendothelial Soc. 23, 321-332 (1978).

33. Berry, L. J., Smythe, D. S.: J. Exp. Med. 114, 761-778 (1961).

34. Wicks, W. D., Barnett, C. A., McKibbin: Fed. Proceed. 33, 1105-1111 (1974).

35. Goodrum, K. J., Berry, L. J.: Proc. Soc. Exp. Biol. Med. 159, 359-363 (1978).

36. Goodrum, K. J., Berry, L. J.: Lab. Invest. 41, 174-181 (1979).

37. Filkins, J. P.: Circulatory Shock 9, 269-280 (1982).

38. Yelich, M. R., Filkins, J. P.: Amer. J. Physiol. 239, E156-E161 (1980).

39. Gunn, J. M., Tilghman, S. M., Hanson, R. W., Reshef, L., Ballard, F. J.: Biochem. 14, 2350-2357 (1975).

40. Iynedjian, P. B., Ballard, F. J., Hanson, R. W.: J. Biol. Chem. 250, 5596-5603 (1975).

41. Tilghman, S. M., Hanson, R. W., Reshef, L., Hopgood, M. F., Ballard, F. J.: Proc. Natl. Acad. Sci. U.S.A. 71, 1304-1308 (1974).

42. Iynedjian, P. B., Hanson, R. W.: J. Biol. Chem. 252, 8398-8403 (1977).

43. Mencher, D., Reshef, L.: Eur. J. Biochem. 94, 581-589 (1979).

44. Yamamoto, K. R., Alberts, B. M.: Ann. REv. Biochem. 45, 721-746 (1976).

45. Higgins, S. J., Gehring, U.: Adv. Cancer Res. 28, 313-397 (1978).

46. Shackleford, G. M., Berry, L. J. Submitted for publication.

47. Pratt, W. B., Kaine, J. L., Pratt, D. V.: J. Biol. Chem. 250, 4584-4591 (1975).

48. Beato, M., Feigelson, R.: J. Biol. Chem. 247, 7890-7896 (1972).

49. Maniatis, T., Fritsch, E. F., Sambrook, J.: Molecular Cloning, Cold Spring Harbor Laboratory, Cold Spring Harbor, New York 346-347 (1982).

50. Stith, R. D., McCallum, R. E.: Infect. Immun. 40, 613-621 (1983).

51. Munck, A., Foley, R.: Nature 278, 752-754 (1979).

52. Schmidt, T. J., Harmon, J. M., Thompson, E. B.: Nature 286, 507-510 (1980).

53. Crystal, R. G., Elson, N. A., Anderson, W. F. In Nucleic Acids and Protein Synthesis, Moldave, K., Grossman, L. eds., Academic Press, New York 101-105 (1974).

54. Pelham, R. B., Jackson, R. J.: Eur. J. Biochem. 67, 247-256 (1976).

55. Deeley, R. G., Gordon, J. I., Burns, A.T.H., Mullinix, K. P., Binastein, M., Goldberger, R. B.: J. Biol. Chem. 252, 8310-8319 (1977).

56. Chirgwin, J. M., Przybyla, A. E., MacDonald, R. J., Rutter, W. J.: Biochem. 418, 5294-5299 (1979).

57. Colombo, G., Carlson, G. M., Lardy, H. A.: Biochem. 17, 5321-5329 (1978).

58. Iynedjian, P. B: Enzyme 24, 366-373 (1979).

59. Kessler, W. S.: Methl. Enz. 73, 442-459 (1981).

60. Baxter, J. D., MacLead, K. M. In Metabolic Control and Disease, Bandy, P. K., Rosenberg, L. E. eds. Saunders, Philadelphia, 103-160 (1980).

61. Noguchi, T., Diesterhaft, M., Granner, D.: J. Biol. Chem. 253, 1332-1335 (1978).

62. Schudt, C.: Biochem. Biophys. Acta. 628, 277-285 (1980).

63. Salavert, A., Iynedjian, P. B.: J. Biol. Chem. 257, 13404-13411 (1982).

64. Lewis, E. J., Colie, P., Wicks, W. D.: Proc. Natl. Acad. Sci. U.S.A. 79, 5778-5783 (1982).

65. Filkins, J. P.: J. Reticuloendothelial Soc. 27, 507-510 (1980).

66. Mishell, R. I., Bradley, L. M., Chen, Y. U., Grabstein, K. H., Shiigi, S. M. in Microbiology, Schlessinger, D. ed. Amer. Soc. Microbiol., Washington, D.C. 82-86 (1980).

67. Lowitt, S., Szentivanyi, A., Williams, J. F.: Biochem. Pharmacol. 30, 1999-2006 (1981).

68. Kawakami, M., Pekala, P. H., Lane, M. D., Cerami, A.: Proc. Natl. Acad. Sci. U.S.A. 79, 912-916 (1982).

69. Pekala, P. H., Kawakami, M., Angus, C. W., Lane, M. D., Cerami, A.: Proc. Natl. Acad. Sci. U.S.A. 80, 2743-2747 (1983).

MODULATION OF RAT LIVER GLUCOCORTICOID RECEPTOR BY INHIBITORS

Virinder K. Moudgil, Noriko Murakami, Thomas E. Eessalu, Virginia M. Caradonna, Virendra B. Singh, Shaun P. Healy, Therese M. Quattrociocchi

Molecular Endocrinology Laboratory, Department of Biological Sciences, Oakland University, Rochester, Michigan 48063

Introduction

Steroid hormones are known to bring their effects by entering the target cells to interact with intracellular cytoplasmic proteins termed "receptors." The binding of hormone to its receptor is thought to cause conformational change(s) in the complex collectively known as "activation" or "transformation." Activation of the steroid-receptor complex leads to its relocation in the nucleus via a process termed "translocation." In the nucleus the steroid-receptor complex is known to interact with certain chromatin sites which subsequently result in the alteration of gene expression (1,2). The exact molecular mechanism of steroid hormone action has remained speculative.

After their initial discovery in the Thymus (3), receptors for glucocorticoids have been target of numerous investigations. Majority of published data has accumulated from observations made in cell-free systems. When receptors for steroid hormones are extracted from target cells at low temperature, they are primarily obtained in the high-speed supernatant ("cytosol") of tissue homogenates. These cytosolic receptors can be complexed with glucocorticoids to form complexes which are said to be in nonactivated form with little affinity for nuclei isolated from target cells. The steroid-receptor complexes can be activated in vitro by

Adrenal Steroid Antagonism

incubation at elevated temperature (4) or with ATP at 0-4°C (5-8) and with various other treatments (9-12). Upon these treatments, the activated steroid receptor complexes acquire an increased affinity for isolated nuclei (13,14), DNA-cellulose (15,16) and ATP-Sepharose (17,18), and an altered mobility on ion-exchange resins (19,20). Although the knowledge about the process of activation of steroid receptors at physiologic conditions has remained limited, activation of glucocorticoid receptors has been reported in intact cells at physiologic conditions (21-24).

Some recently published reports have generated much interest about the cellular distribution of steroid receptors. Two laboratories have independently demonstrated the presence of unoccupied estrogen receptors in the target cell nuclei (25, 26). These observations are in disagreement with the generally held belief that in the absence of steroid hormones, their receptors reside primarily in the cytoplasm from where they migrate into the nucleus in the presence of hormone. A two-step model had been thus proposed to explain the action of steroid hormones (27). The differences in cellular distribution of unoccupied receptor may be thus attributed to differences in the techniques used for tissue homogenization. The receptor generally recovered in the cytosol fraction of a homogenate may, therefore, represent receptor that is weakly bound with the nucleus; and its binding with hormone may lead to a stronger association of the hormone-receptor complex with the nuclear sites. Considering that the above findings (25,26) might have more general applications, the process of activation of steroid hormone receptors may then simply represent intranuclear events.

In order to gain a better understanding of the process of receptor activation, a variety of approaches have been applied. One of these approaches has been to study the interaction of steroid receptors with isolated nuclei, chromatin, or with other cellular constituents. Consequently, various compounds have been identified which block the interaction of glucocorticoid-receptor complexes with isolated nuclei, DNA and ATP-Sepharose (28-35). In this chapter we shall describe effects of certain chemical agents whose inhibitory actions on the properties of rat liver glucocorticoid receptor have been studied in this laboratory over the past several years.

Inhibitors of Receptor Activation

In order to gain insight into the structure and function of steroid hormone receptors, their interaction with various cellular constituents has been studied widely. Steroid receptors are relatively large proteins whose function can be regulated or altered by cellular or synthetic chemical agents. Because of limited availability of appropriate biological systems, interaction or influence of modifying agents has been studied in cell-free systems. Extensive analysis of avian progesterone receptor with inhibitors of activation or ATP, DNA and nuclear binding has been reported (1,36,37). We present here a detailed account of our investigations on the effects of various inhibitors which block glucocorticoid receptor function.

A) Transition metal ions of groups Vb and VIb (molybdate, tungstate and vanadate).

a) Effects of sodium molybdate. Initial studies had reported use of sodium molybdate as a stabilizing agent for glucocorticoid receptors (38,39). It was subsequently shown that addition of molybdate blocks the activation process of steroid receptors (5,36,40-42). However, no inhibitory effect of molybdate was evident in these studies when molybdate was added to preparations that contained an already activated receptor. Our studies indicate that molybdate blocks the activation in vitro of glucocorticoid receptor, but it also inhibits the binding of activated glucocorticoid-receptor complexes to acceptors such as isolated rat liver nuclei, DNA-cellulose and ATP-Sepharose (31). Therefore, both nonactivated and activated receptor forms are sensitive to molybdate action. In addition, activated glucocorticoid-receptor complexes bound to acceptors could be extracted by molybdate, thus suggesting an interaction between glucocorticoid receptor and molybdate.

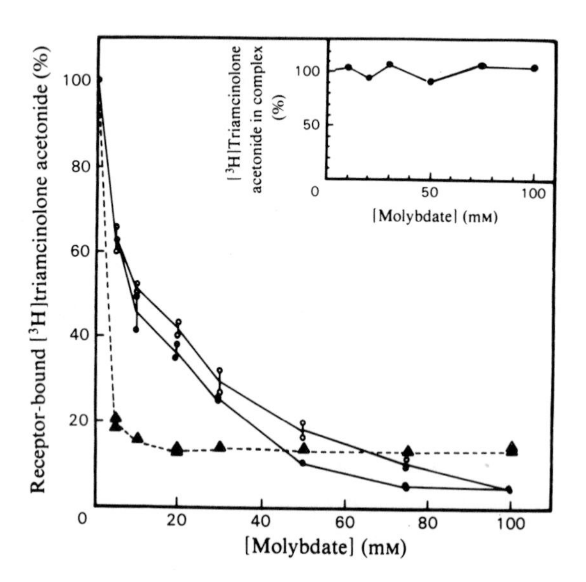

FIG. 1. Effect of molybdate on the activation of glucocorticoid-receptor complex. Cytosol aliquots containing [^{3}H]-triamcinolone acetonide-receptor complexes were incubated at 23°C for 40 min with Tris-HCl buffer (control) or different concentrations of sodium molybdate. The extent of receptor activation was determined by measuring nuclear (O), DNA-cellulose (●) and ATP-Sepharose binding (▲) as described (31). Inset: Effect of molybdate on the dissociation of [^{3}H]-triamcinolone acetonide from the activated complexes. A 100% value represents the binding of complexes to the acceptors in the absence of inhibitor, or the number of complexes present in the control sample before the addition of molybdate. Taken from Ref. 31.

b) Effects on receptor activation. Figure 1 illustrates a typical profile of heat-activation of rat liver glucocorticoid-receptor complexes in the presence of different concentrations of sodium molybdate. Three different acceptors were used to measure receptor activation. The acceptors appeared to differ somewhat in their affinity for molybdate treated preparations. A 50% inhibition in the binding of receptor to nuclei and DNA-cellulose was seen at 10mM sodium molybdate, whereas receptor binding to ATP-Sepharose was completely eliminated by pre-treatment with 5mM Na_2MoO_4. Above treatments did not alter the number of complexes that were present before and after molybdate treatment.

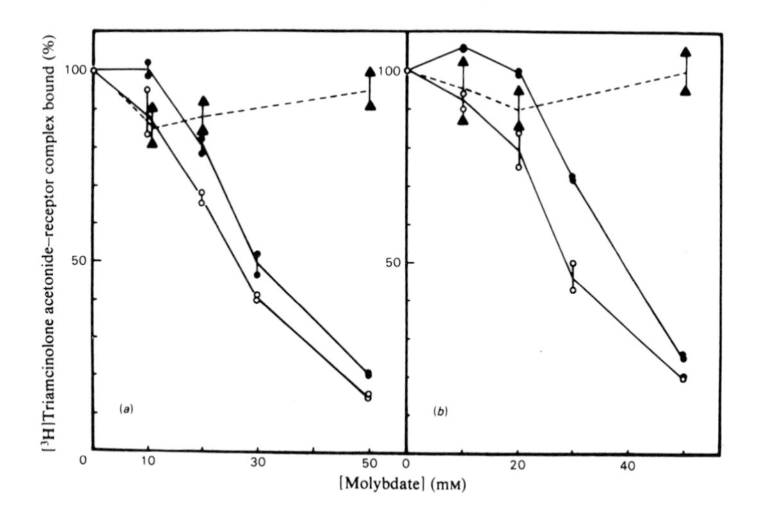

FIG. 2. Effect of molybdate on the binding of heat-activated glucocorticoid-receptor complexes to acceptors. The $[^3H]$-triamcinolone-receptor complexes from rat liver cytosol were activated by warming at 23°C for 40 min (a) or by fractionation with $(NH_4)_2SO_4$ (b). Aliquots were incubated with buffer (control, 100% binding) or with different concentrations of molybdate. The binding of receptor to nuclei (O), DNA-cellulose (●), and ATP-Sepharose (▲) was measured as described (31). Taken from Ref. 31.

Since most previous studies had indicated that effects of molybdate were only limited to preparations that contained nonactivated steroid-receptor complexes (36,41,42), we examined effects of molybdate on the acceptor binding ability of heat-activated glucocorticoid-receptor complexes. Results of Fig. 2 demonstrate that activated receptor is sensitive to molybdate action, and that a treatment of activated receptor complexes with 20-50 mM Na_2MoO_4 resulted in their reduced binding to rat liver nuclei and DNA-cellulose. Interestingly, only a minor inhibitory effect of molybdate was seen on the receptor binding to ATP-Sepharose suggesting that the activated receptor either possesses a stronger affinity for ATP-Sepharose or the sites of ATP and molybdate action are presumably different; ATP-binding site unaltered by treatment of activated receptor with molybdate.

The suggestion, that the effects of molybdate illustrated in Fig. 2 were actually due to presence of certain population of nonactivated or partially activated receptor in these preparations, was considered. To eliminate this possibility, cytosol preparations were either heat-activated (Fig. 2a) or fractionated with 35% saturation of $(NH_4)_2SO_4$ (Fig. 2b). The receptor precipitated by $(NH_4)_2SO_4$ has been reported to be in an activated state (14) and cannot be activated further by heat (13). Similar pattern of inhibition in the DNA and nuclear binding capacity of molybdate-treated preparations was observed whether molybdate was added to heat activated (Fig. 2a) or $(NH_4)_2SO_4$-fractionated cytosol preparations. Since the conductivity of sodium molybdate was determined and found to be comparable to that of KCl on an equimolar basis, KCl was included in parallel series of samples. The inhibitory effects of molybdate seen in results of Fig. 2 are, therefore, not due to its contribution to the ionic strength of buffer solutions. The ion may be interacting with the activated glucocorticoid-receptor complex at a site(s) which may partially survive the process of receptor activation and thus require a higher concentration (20-50mM Na_2MoO_4) to block the acceptor binding of activated receptor.

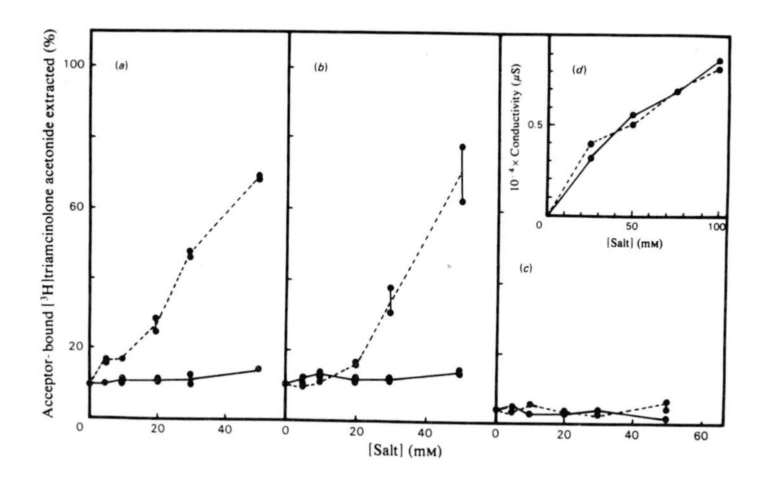

FIG. 3. Extraction of glucocorticoid-receptor complexes from nuclei, DNA-cellulose and ATP-Sepharose by equimolar concentrations of molybdate and KCl. Portions of rat liver nuclear preparation (a), DNA-cellulose (b), and ATP-Sepharose (c) were incubated with aliquots containing preformed [^{3}H]-triamcinolone acetonide-receptor complexes. The receptor-charged acceptors were subsequently incubated at 0°C with varying concentrations of KCl (●—●) or molybdate (●---●). Inset: Conductivity determination of KCl and molybdate solutions. Taken from Ref. 31.

c) Extraction of receptor from acceptors. Further support for the hypothesis, that effects of molybdate involve a direct interaction between the receptor and the ion, comes from the results in Fig. 3. When the DNA-cellulose and the nuclear preparations, charged with activated glucocorticoid-receptor complexes, were incubated with different concentrations of sodium molybdate: the complexes could be extracted from these acceptors in a concentration-dependent manner.

Once again, ATP-Sepharose-bound receptor showed insensitivity toward molybdate presence and could not be extracted. When the complexes were formed in vivo by an administration of [^{3}H] triamcinolone acetonide, molybdate treatment (20-50mM) of isolated nuclei resulted in the extraction of majority of the nuclear glucocorticoid-receptor complex (not shown) (31).

d) Reversibility of molybdate effects. The effects of molybdate on the binding of activated receptor to nuclei were found to be reversible. A thorough dialysis of molybdate-treated preparations allowed recovery of nuclear-binding capacity of glucocorticoid-receptor. Similarly glucocorticoid receptor extracted from acceptors (DNA, nuclei) by molybdate retained the ability to rebind the acceptors upon removal of molybdate by dialysis (31). These observations indicate that treatment of receptors with molybdate or their extraction from acceptors, does not destroy the capacity of glucocorticoid-receptor to bind to nuclei or DNA.

e) Mechanism of molybdate action. Molybdate is a known inhibitor of phosphatase activities, and the observations that phosphatase inhibitors like molybdate, tungstate and vanadate stabilize steroid receptors and block transformation of cytosol receptor suggested that these compounds may act indirectly by blocking the action of phosphatases (29). It was also reported that the rate of activation of glucocorticoid-receptor complex is stimulated by incubation with calf intestine alkaline phosphatase (30). Based on these findings it was proposed that activation of steroid receptors involves a dephosphorylation of receptor itself or of some related regulatory component(s) of cytosol (29,30). The studies in this laboratory had suggested that the effects of molybdate and related compounds are due to their interaction with the glucocorticoid receptor (31). This argument is based on the following lines of evidence. (a) Although Na_2MoO_4 and Na_2WO_4 block the process of receptor activation, other phosphatase inhibitors, such as levamisole, fluoride and arsenate, which are specific inhibitors of alkaline phosphatases (43), do not prevent activation (30,31,42). (b) Molybdate is able to inhibit binding of activated receptor to nuclei, DNA-cellulose and can extract the acceptor bound receptor in a concentration-dependent manner. (c) Relatively high concentrations of exogenous calf-intestine alkaline phosphatase are required to stimulate activation (30). (d) Several phosphorylated compounds, including ATP, increase the rate and extent of glucocorticoid receptor activation (1,5-8). Finally, we have recently obtained evidence that chick oviduct progesterone receptor can be retained on columns of molybdate-Sepharose (44). More detailed analysis of the direct interaction between molybdate and steroid receptors should be forthcoming.

Based on the results described in Figs. 1-3, we propose a model to explain the effects of molybdate on different aspects of receptor function related to activation. Figure 4 illustrates a construction of speculative domains of glucocorticoid receptor that are influenced by molybdate. The receptor contains a molybdate binding site which is exposed both before and after activation. Addition of the ion prior to activation causes it to bind in a manner that the sites for binding to ATP, DNA and nuclei are no longer available. When molybdate is added to a preparation that contains an activated receptor, it binds to it in a manner which reduces its DNA and nuclear binding capacity without altering the ATP-binding site. Alternatively, when receptor is first activated and then adsorbed on these acceptors, DNA and nuclear binding still allows molybdate to extract the receptor, whereas, once bound to ATP-Sepharose, the receptor becomes insensitive to elevated molybdate concentrations (20-50mM). The results (Figs. 1-3) and the model (Fig. 4) depicting them suggest that ATP, DNA and nuclear binding sites that are involved in the uptake of activated glucocorticoid receptor are qualitatively different but may appear similar during quantitative analysis.

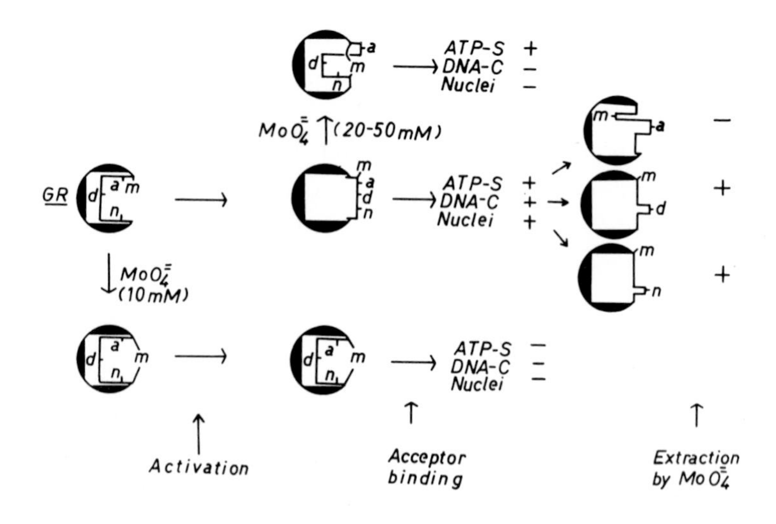

FIG. 4. A speculative model of glucocorticoid receptor interaction with molybdate, ATP-Sepharose, DNA-cellulose and nuclei. GR, glucocorticoid receptor; d, DNA-cellulose binding site; a, ATP-Sepharose binding site; n, nuclear site; m, molybdate binding site; ATP-S, ATP-Sepharose; DNA-C, DNA-cellulose; +, binding or extraction; -, no binding or extraction.

f) Effects of sodium tungstate. The effects of the transition element ions were noted to be pH-dependent. Nishigori and Toft (42) pointed out that molybdate is 10 times more effective against avian progesterone receptor at pH 7 than at pH 8. Barnett et al. (30) observed that at pH 8 tungstate presence up to 10mM had no effect on the DNA-cellulose binding of glucocorticoid receptor. In this laboratory, we have been interested in exploring the potential of these ions in affecting glucocorticoid receptor properties under different pH conditions. The studies to be discussed in the next section were designed mainly to examine: (a) whether or not the effects of molybdate and tungstate are mediated via similar mechanisms, (b) if these ions were effective equally at comparable concentrations and pH, and finally (c) whether both nonactivated and activated receptor forms were equally sensitive to these ions.

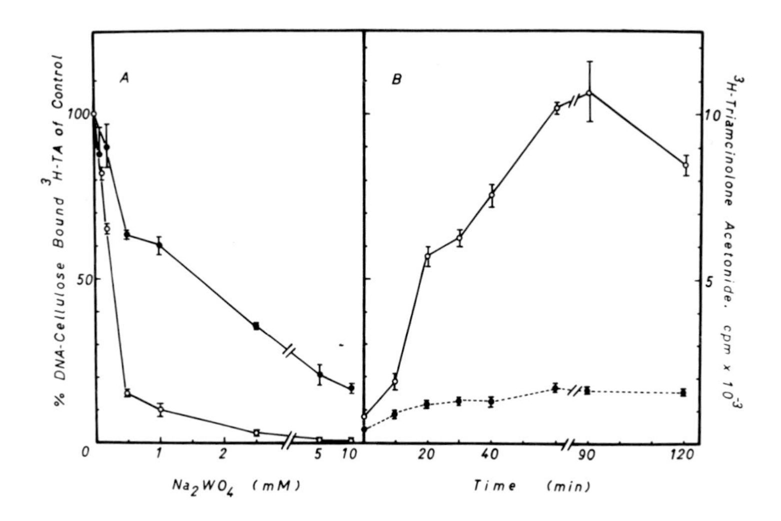

FIG. 5. Effect of sodium tungstate on the rate and extent of heat activation of glucocorticoid-receptor complex. (A): Cytosol and all other solutions were prepared at pH 7 (o—o) or pH 8 (●—●). Aliquots of cytosol were incubated with different concentrations of sodium tungstate at pH 7 or pH 8 for 40 min at 23°C. The extent of activation was measured by determining the DNA-cellulose uptake by receptor. (B): Different aliquots of cytosol containing glucocorticoid-receptor complexes were incubated at 23°C without and with sodium tungstate for time period shown.

g) Effects of tungstate on receptor activation. Heat activation of glucocorticoid-receptor complex can be blocked by treatment of cytosol with sodium tungstate, an observation related to effects of molybdate described earlier in this chapter. However, tungstate appears to be a more effective agent for blocking receptor activation. Its presence inhibited the heat-activation of [^{3}H]-triamcinolone acetonide-receptor complex with $I.D._{50}$ values of 0.2mM and 1.5mM at pH 7 and pH 8, respectively (Fig. 5A). Furthermore, when the effects of tungstate were examined on the rate of activation by incubating the inhibitor with receptor aliquots at 23°C for different time periods, tungstate completely inhibited the receptor activation throughout the incubation period suggesting that it inhibits the extent of receptor activation rather than the rate at which activation occurs (Fig. 5B).

h) DNA-binding and tungstate effects. Figure 6 illustrates the effects of tungstate on the DNA-cellulose binding of activated glucocorticoid-receptor complex. At pH 8 tungstate had only a minor effect on the DNA binding of receptor but at lower pHs tungstate blocked the DNA binding of glucocorticoid-receptor complexes in a concentration-dependent manner. The inhibitory effects of tungstate were not caused due to its complexing with DNA-cellulose, since a pre-treatment of the resin with up to 50mM inhibitor had no effect on the capacity of DNA-cellulose to retain glucocorticoid-receptor (not shown).

Since tungstate, like molybdate, is a very effective inhibitor of phosphatases, other compounds either structurally similar to tungstate (metals of element group Vb and VIb) or with demonstrable ability to inhibit phosphatase activities were tested to determine the specificity of their effects on DNA-cellulose binding. The compounds were dissolved in buffer (50mM Tris-HCl, 12mM thioglycerol, pH 7) and were incubated at 0°C for 30 min at 5mM with the cytosol preparations containing activated [^{3}H]-triamcinolone acetonide-receptor complexes. Figure 7 illustrates that while monochromate and dichromate inhibited (35-55%) DNA binding, levamisole, arsenate, fluoride, phosphate, molybdate and vanadate had no effects on this process at 5mM concentration. Tungstate (5mM), therefore, was effective at pH 7 in selectively blocking the DNA binding of activated receptor.

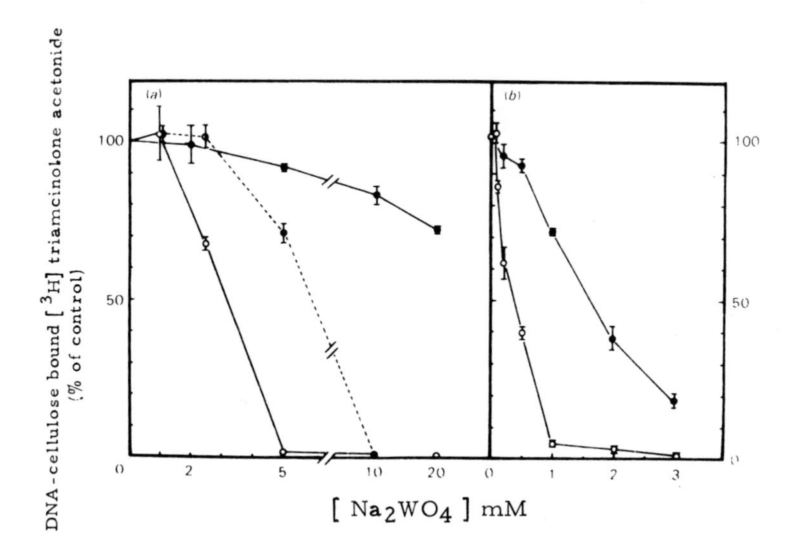

FIG. 6. Inhibition in the DNA-cellulose binding of activated glucocorticoid-receptor complex as a function of pH. (a) Cytosol was complexed at pH 7 (o—o), pH 7.5 (●--●) & pH 8 (●—●) with 10nM [^{3}H]-triamcinolone acetonide for 4 h at 0°C. The complexes were heat-activated for 40 min at 23°C. Portions of cytosol were then incubated at 0°C for 40 min with DNA-cellulose suspensions in the presence of various concentrations of tungstate. The control samples did not contain tungstate and their binding to DNA-cellulose represented a 100%. The DNA-bound complexes were subsequently extracted with buffers at respective pHs. (b) Cytosol preparations containing glucocorticoid-receptor complexes were fractionated with 35% saturation of $(NH_4)_2SO_4$ at pH 7 (O) and pH 8 (●). The precipitated receptor was dissolved in 10mM Tris-HCl buffer, pH 7 or pH 8. Following an incubation with different concentrations of sodium tungstate, DNA-cellulose binding assays were performed as described (35). Taken from Ref. 35.

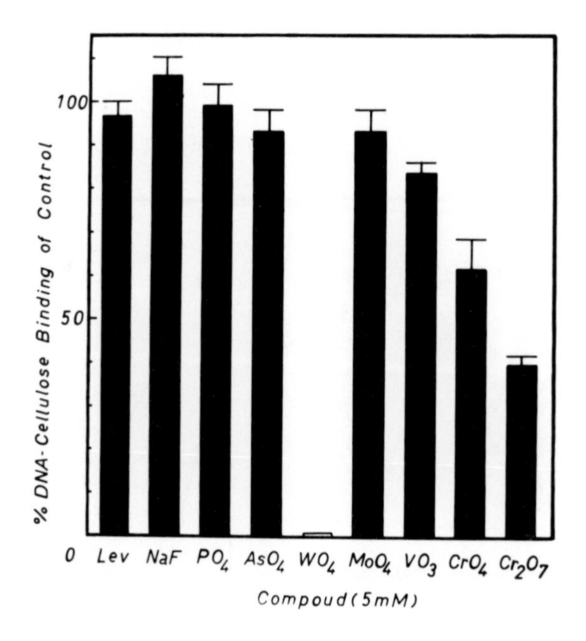

FIG. 7. Effects of different compounds on the DNA-cellulose binding of activated glucocorticoid-receptor complex. The compounds were dissolved in 10mM Tris-HCl, pH 7 and added to cytosol preparations containing glucocorticoid-receptor complexes. After 40 min at 0°C, the mixtures were incubated with pellets of DNA-cellulose to determine the uptake of glucocorticoid receptor. Lev, levamisole; NaF, sodium fluoride; Po_4, phosphate; AsO_4, arsenate; WO_4, tungstate; MoO_4, molybdate; VO_3, vanadate; CrO_4, chromate; Cr_2O_7, dichromate.

i) Extraction of glucocorticoid receptor from DNA-cellulose by tungstate. Since activation and DNA binding of glucocorticoid receptor was found to be blocked by relatively low tungstate concentrations, its potential on the extraction of DNA-bound receptor was examined. The heat-activated cytosol complexes were adsorbed on DNA-cellulose and the resin was washed to rid of free steroid and the unadsorbed complexes. The above DNA-cellulose suspensions were then mixed with buffer (50mM Tris-HCl, 12mM thioglycerol, pH 7) containing different concentrations of Na_2WO_4. Figure 8 demonstrates that about 60% of the adsorbed complexes can be extracted from the resin by 50uM tungstate; the release of the entire receptor could be accomplished with 1-2mM Na_2WO_4. The pH dependency of tungstate effects was evident as 100 times higher tungstate concentrations were required for receptor extraction from DNA-cellulose at pH 8 (Fig. 8, inset). This strong action of tungstate in releasing DNA-bound receptor was very selective. Other related compounds including some phosphatase inhibitors were ineffective in extracting any appreciable quantities of glucocorticoid-receptor complexes from DNA-cellulose (Fig. 9). The results of Figs. 5-9 strongly suggest that at pH 7 tungstate interacts with the glucocorticoid-receptor complex in a specific manner.

Previous studies had shown that effects of molybdate and vanadate on the stabilization and activation of steroid receptors are reversible (29,42,45). In our studies, addition of tungstate to the activated complexes, and the extraction of complexes from DNA-cellulose by tungstate resulted in irreversible change in the properties of glucocorticoid-receptor complex leading to a loss of its DNA-binding capacity (35).

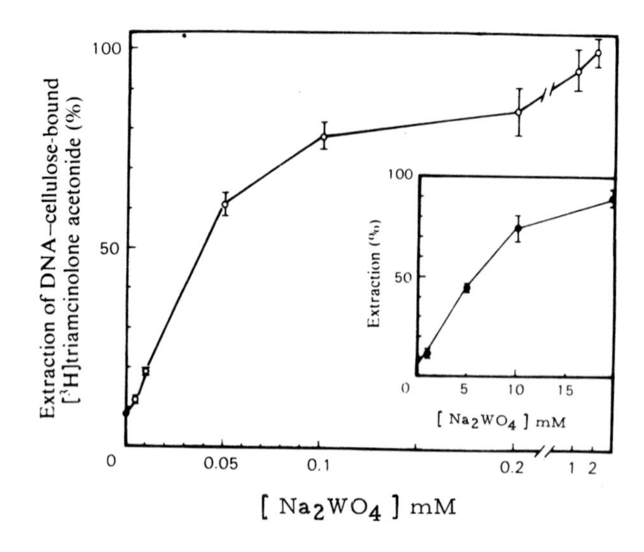

FIG. 8. Extraction of [^{3}H]-triamcinolone acetonide-receptor complex from DNA-cellulose by sodium tungstate. DNA-cellulose suspensions charged with glucocorticoid-receptor complex were incubated with Tris-HCl buffer containing 1M KCl or different concentrations of sodium tungstate. All solutions and preparations were either pH 7 (o—o) or pH 8 (●—●). The amount of receptor extracted by KCl represented a 100%. Taken from Ref. 35.

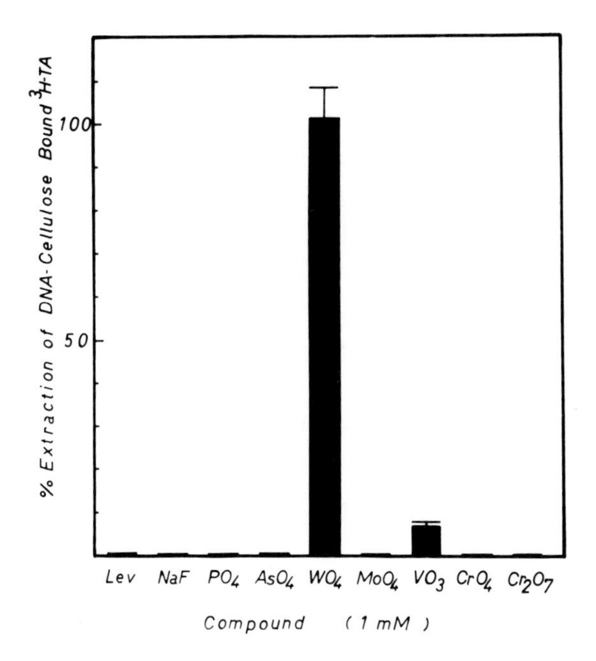

FIG. 9. Extraction of DNA-cellulose-bound glucocorticoid-receptor complex by different phosphatase inhibitors. Cytosol from rat liver was prepared at pH 7 and complexed with [^{3}H]-triamcinolone acetonide for 4 h at 0°C. Aliquots of above complexes were heat-activated (23°C, 1 h) and mixed with DNA-cellulose pellets. The resin-bound complexes were then extracted by incubation with different phosphatase inhibitors. Lev, levamisole; NaF, sodium fluoride; Po_4, phosphate; AsO_4, arsenate; WO_4, tungstate; MoO_4, molybdate; VO_3, vanadate; CrO_4, chromate; Cr_2O_7, dichromate.

j) Behavior of tungstate-treated preparations on ion-exchange resins. The observations on the irreversibility of tungstate action and the resultant loss of DNA binding capacity of glucocorticoid receptor led us to examine whether the ion was converting an activated receptor to a nonactivated non-DNA-binding form. Since the experimental designs normally used for studying effects of inhibitors on steroid receptors provide limited information on the alterations in receptor forms, we employed ion-exchange chromatography (19,20) to determine the resolution of activated, non-activated and tungstate-treated preparations. Results illustrated in Fig. 10 demonstrate that cytosolic nonactivated glucocorticoid-receptor complexes bind more strongly to DEAE-Sephacel (a positively charged resin), and a majority of them elute with buffers of higher ionic strength (Fig. 10, top panel) whereas heat-activated, more positively charged steroid-receptor complexes elute mainly with low ionic strength buffer (bottom panel). When tungstate was present during the warming of cytosol preparations, it blocked activation of receptor allowing a major portion of receptor to elute in the region of nonactivated receptor. However, when tungstate was added to activated receptor preparations, the receptor remained in its activated form eluting with low-salt buffer solutions. The results of Fig. 10 confirm our earlier postulations that the inhibitory effects of tungstate on the nonactivated and activated receptor forms are direct and due to its interaction with the receptor protein. Although activated receptor treated with tungstate remains in its activated form (Fig. 10), its DNA binding ability is completely lost (Figs. 6,7).

k) Sedimentation analysis of tungstate and molybdate-treated glucocorticoid receptor. Action of chemical modifiers of receptor activation can be studied by sedimentation analysis to monitor changes in molecular size of receptor. Cytosolic nonactivated steroid-receptor complexes generally sediment in 7-9S region (1). Following treatments with activation inducing agents (heat, salt, ATP), the receptor complexes sediment at a slower rate (3.5-5S). The profiles of rates of sedimentation of glucocorticoid-receptor complexes from cytysol prepared at 0°C (nonactivated) and incubated at 23°C (activated) are shown in Fig. 11. Nonactivated glucocorticoid-receptor complex migrated farther than the 7.9S internal marker (glucose oxidase) in 5-20% linear sucrose gradients sedimenting approximately as 9S. Upon heat-activation, the complex migrates as a 4S moiety, characteristic of activated/transformed steroid-receptor complexes. When the cytosol containing glucocorticoid-receptor complexes was warmed at 23°C in the presence of 10mM sodium molybdate and sodium tungstate, the receptor remained in its nontransformed, larger form sedimenting in ∿9S region. It is clear, therefore, that inclusion of molybdate and tungstate in the cytosol prior to or during activation of glucocorticoid-receptor complex blocks the activation process preventing a reduction in the size of receptor which may be required for its binding to various acceptors (DNA-cellulose, ATP-Sepharose, nuclei). Such a receptor form (nonactivated) also exhibits charge distribution on the surface of the protein that is characteristic of less positively charged nontransformed receptors.

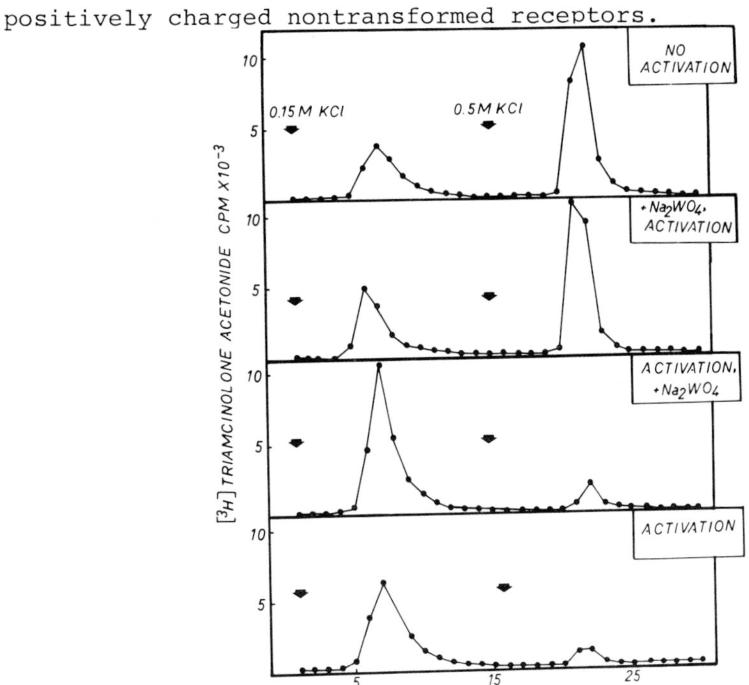

FIG. 10. Ion-exchange chromatography of tungstate-treated glucorticoid-receptor complex. Aliquots of rat liver cytosol were complexed with 20nM [^{3}H]-triamcinolone acetonide for 4h at 0°C. Portions were heat-activated (23°C, 1 h) with and without 10mM tungstate. In some cases Na_2WO_4 was added following heat-activation of the complexes. Samples were then chromatographed over 5ml columns of DEAE-Sephacel equilibrated with 20mM Tris-HCl, 12mM thioglycerol, 20% glycerol, pH 7.5. No activation, cytosol receptor prepared and kept at 0°C; activation, cytosol + [^{3}H]-triamcinolone acetonide, 23°C, 1h.

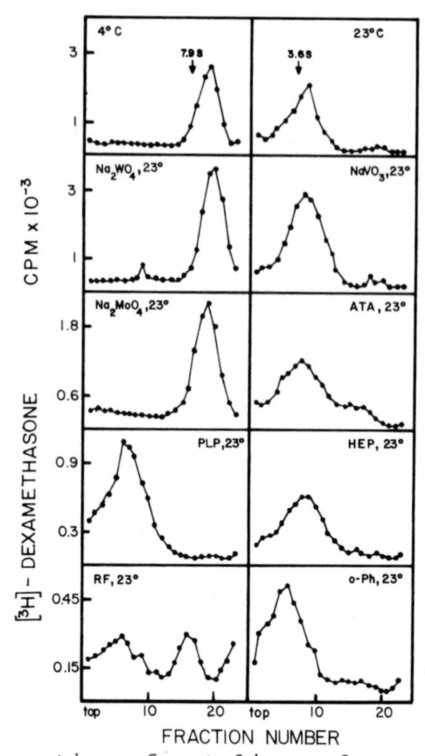

FIG. 11. Transformation of rat liver glucocorticoid-receptor complex: Determination of the rate of sedimentation of receptor in presence of various chemical agents. Freshly excised liver from bilaterally adrenalectomized adult male rats was homogenized in 2 volumes (w/v) of Tris buffer (20mM Tris-HCl, 12mM thioglycerol, 10% glycerol and 0.3mM phenylmethylsulfonyl fluoride (PMSF)]. The homogenate was centrifuged at 150,000 x g for 1h. The resulting cytosol was complexed with 20nM [^{3}H]-dexamethasone for 2 h at 0°C and the unbound excess steroid was removed by treatment with charcoal. Aliquots of cytosol were either incubated at 0-4°C (control for nonactivated receptor) or heat-activated by exposure to 23°C for 60 min with buffer (10mM Tris-HCl, pH 7.5) (control for activated receptor) or in presence of different inhibitors. Portions (200ul) were layered over 5-20% sucrose gradients containing 20mM Tris-HCl, 12mM thioglycerol, 1mM EDTA, 0.15M KCl and 20mM Na_2MoO_4, pH 7.4. The gradients were centrifuged at 55,000 rpm in a TV-865 Sorvall vertical rotor for 2 h at 4°C ($W^2t=2.4 \times 10^{11}$). The tubes were pierced at the bottom and 0.2ml fractions were collected. Aliquots (10ug) of peroxidase (3.6S) and glucose oxidase (7.9S) were included in the gradients as internal markers. Na_2WO_4, Na_2MoO_4 and $NaVO_3$ were at 10mM; ATA, 0.01mM aurintricarboxylic acid; PLP, 5mM pyridoxal 5'-phosphate; HEP, 300ug/ml heparin; RF, 175ug/ml rifamycin AF/013; O-pH, 3mM o-phenanthroline.

Presence of vanadate failed to block the 9S→4S conversion and the receptor apparently remained in its activated form (Fig. 11). It is possible that vanadate only affects the rate of activation of glucocorticoid receptor and that under the conditions of the experiment the receptor activating factors overcame the inhibitory effect(s) of vanadate. Alternatively, at the concentrations employed, vanadate may not be effective in blocking receptor transformation.

l) Use of tungstate in the purification of glucocorticoid receptor. There was sufficient indirect evidence (Fig. 9) to suggest use of sodium tungstate in the purification of glucocorticoid receptor from rat liver. Since binding of receptor to DNA-cellulose itself allows removal of many cytosolic non-receptor proteins, extraction of receptor from DNA-cellulose by incubation with sodium tungstate offered premise of sufficient purification. This approach has two apparent advantages. First, extraction of DNA-cellulose-adsorbed receptor occurs in a selective manner (Fig. 9); other compounds at comparable con- - centrations are unable to release the complexes from the resin. Secondly, use of tungstate should be preferred over the conventional mode of receptor extraction by high ionic strength buffers; the latter may alter the size of receptor or may cause other subtle changes such as dissociation of receptor subunits (if applicable) or may influence cellular factors required for receptor function.

Table 1 shows that cytosolic glucocorticoid-receptor complex can be purified to a sufficient degree with a good yield. The receptor was activated and subsequently adsorbed on DNA-cellulose. The resin was washed thoroughly with salt-free buffer to remove non-DNA-binding proteins. An extraction of receptor with tungstate yielded above 500-fold purification with a 30% recovery. We are now exploring avenues to increase the fold-purification of receptor by combining this step with other conventional techniques of protein fractionation. Work is also in progress to characterize the receptor obtained from DNA-cellulose by extraction with tungstate.

TABLE 1. Purification of Rat Liver Glucocorticoid Receptor

Step	Vol. (ml)	Activity $1x10^{-5}$cpm)	Protein (mg/ml)	Spec. Activity ($1x10^{-3}$)	Purification (X-fold)	Yield (%)
A. Extraction of Receptor from DNA-Cellulose by 10mM Sodium Tungstate						
Cytosol	50	12.8	18	1.42	-	100
DNA-Cellulose Extract (n=3)	6	4	0.09	749.1	520	30
B. Extraction of DNA-Cellulose-Bound Receptor by Pyridoxal 5'-Phosphate						
Cytosol	40	11.2	19	1.47	-	100
DNA-Cellulose Extract	5	3.2	0.05	1281	910	28

Bilaterally adrenalectomized rats were killed by cervical dislocation and perfused with 20ml of 10mM HEPES, 10% glycerol, 10mM MoO_4, 12mM monothioglycerol, 1mM EDTA, pH 7.2. The tissue was homogenized with 2 vol of this buffer and the 150,000xg supernatant (cytosol) was passed through a phosphocellulose column (2.5x7cm) equilibrated in homogenization buffer. The flow thru was incubated with 20nM [^{3}H]-dexamethasone (40-50Ci/mmole) for 3 h at 0°C and the preparation was brought to 45% $(NH_4)_2SO_4$. The resulting pellets were dissolved and dialyzed (A) in 10mM Tris-HCl, 10% glycerol, 1mM EDTA, 12mM thioglycerol, 10% glycerol, 1mM EDTA, pH 7.8 (buffer C). Dialyzed samples were heat activated for 30-40 min at 23°C and brought to 0°C. Two columns (4ml) of DNA-cellulose (0.8-0.9mg DNA/ml cellulose) were equilibrated with buffer B and buffer C and loaded with dialyzed receptor samples. Column I was washed with buffer B and receptor was eluted with 10mM sodium tungstate. Column II was washed with a buffer containing 50mM sodium borate, 2mM $MgCl_2$, 10% glycerol, pH 8 and receptor was eluted with buffer containing 10mM pyridoxal 5'-phosphate. Fractions (0.67ml) were collected and 20µl used for measurement of radioactivity. Peak fractions were pooled and protein was assayed using micro BioRad assays.

B) Inactivation of glucocorticoid receptor by pyridoxal 5'-phosphate.

Pyridoxal 5'-phosphate is the active form of pyridoxine (vitamin B_6) and has been documented as an important cofactor that regulates coenzymatically a number of reactions involved in amino acid metabolism. The general mechanism of the pyridoxal 5'-phosphate action involves formation of a "Schiff base" with an amino acid group localized at the active site of an enzyme. It has been shown that addition of pyridoxal 5'-phosphate to receptor preparations inhibit their DNA, nuclear and ATP-Sepharose binding (28,46,47). The inhibition in the binding of receptor to the acceptors listed above is apparently reversible and could be removed by the addition of a primary amine, e.g. Tris.

Activation of receptor in the presence of pyridoxal 5'-phosphate reduces the affinity of receptor toward DNA-cellulose or ATP-Sepharose (Table II). Pyridoxal 5'-phosphate was equally effective when added to preparations containing activated glucocorticoid receptor. Its ability to inhibit DNA-binding of glucocorticoid receptor suggests that the inhibitor either binds to DNA-binding site(s) of receptor or to a site whose occupancy by pyridoxal 5'-phosphate alters the DNA binding domain. Interestingly, at 5mM concentrations this inhibitor did not prevent the transformation of 9S receptor to the 4S form (Fig. 11). These results suggest that pyridoxal 5'-phosphate is an inhibitor of the DNA binding properties of the glucocorticoid receptor and that it may not interfere with the process of receptor activation when used at 5mM concentration.

TABLE 2. Effects of Various Inhibitors on the Binding of $[^3H]$-Dexamethasone-Receptor Complex to DNA-Cellulose and ATP-Sepharose

Compound	Concentration Used	% specific $[^3H]$-dexamethasone binding to DNA-Cellulose	ATP-Sepharose
Control (no inhibitor)	-	100	100
$NaVO_3$	10mM	75	74
Na_2WO_4	10mM	3	0
Na_2MoO_4	10mM	0	0
Pyridoxal 5'-Phosphate	5mM	17	60
Aurintricarboxylic acid	0.01mM	14	34
Heparin	300μg/ml	0	0
o-Phenanthroline	3mM	0	7
Rifamycin AF/013	175μg/ml	0	0

Freshly excised livers from bilaterally adrenalectomized adult male rats were homogenized at 0°C in 2 volumes (w/v) of Tris-buffer (20mM Tris-HCl, 12mM thioglycerol, 10% glycerol and 0,3mM phenylmethylsulfonyl fluoride, pH 7.5). The homogenate was cleared by centrifugation at 150,000xg for 60 min. The resulting cytosol was complexed with 20nM $[^3H]$-dexamethasone for 2 h at 0°C. The excess steroid was removed by treatment with a charcoal suspension and aliquots of cytosol were incubated with the compounds listed in the table and were heat-activated (23°C, 1 h). Portions (0.5ml) were used to measure DNA-cellulose and ATP-Sepharose binding as described previously (8,18,31-35).

Assuming that DNA binding site of receptor is modulated or competitively occupied by pyridoxal 5'-phosphate, it should be possible to extract DNA-bound glucocorticoid receptor by incubation with buffers containing pyridoxal 5'-phosphate. Results of Table II show that the inhibitor can be used to purify glucocorticoid receptor adsorbed on columns of DNA-cellulose. In a single step, a 900-fold purification can be accomplished by adsorbing the activated receptors on DNA-cellulose followed by their elution with pyridoxal 5'-phosphate.

There are three reasons for considering pyridoxal 5'-phosphate as an ideal tool for receptor characterization. (a) Effects of pyridoxal 5'-phosphate are relatively specific, as other similar compounds are ineffective in blocking binding of receptors to acceptors (47); (b) the mode of its action is better understood than that of other inhibitors described in this chapter; (c) it is an active form of vitamin B_6, thus making it possible that it may be a biologic modifier of steroid receptors. Indeed, DiSorbo et al. (48) have reported that a deficiency of pyridoxine increases the proportion of glucocorticoid-receptor complexes capable of heat-activation. Furthermore, incubation of rat hepatoma cells (FAZA) in pyridoxine-free medium was reported (49) to result in a decreased intracellular level of pyridoxal 5'-phosphate with a concomitant enhancement in the induction of tyrosine aminotransferase, a process which is known to be influenced by glucocorticoids.

C) Inhibition of glucocorticoid receptor by o-phenanthroline.

RNA and DNA polymerases are known to be zinc-metalloenzymes based on their sensitivity to 1,10-phenanthroline (o-phenanthroline). The non-chelating analog of o-phenanthroline, m-phenanthroline is unable to bring similar effects. Chelation of enzyme-associated zinc has been suggested as a possible mechanism of inhibition of the polymerases by o-phenanthroline in vitro (50). Lohmar and Toft (13) originally demonstrated that a 23°C preincubation of chick oviduct cytosol containing progesterone-receptor complex with o-phenanthroline inhibited the binding of the complex to oviduct nuclei. Subsequently, a variety of receptor functions were described to be inhibited by treatment with o-phenanthroline (1,36,37,51-53). Based on these reports it has been proposed that steroid receptors are metalloproteins, and require a metal ion for adequate activation or nuclear binding.

Results of Table II indicate that o-phenanthroline is a potent inhibitor of both DNA-cellulose and ATP-Sepharose binding of glucocorticoid receptor. Addition of inhibitor to cytosol preparation containing preformed complexes prevents the binding of receptor to DNA-cellulose and ATP-Sepharose. We examined the effect of this compound on the rate of sedimentation of cytosol receptor subjected to heat activation. Figure 11 shows that although the DNA-cellulose and ATP-Sepharose binding or glucocorticoid receptor was completely blocked, the receptor remained in its activated form. The o-phenanthroline-treated receptor in the cytosol could be transformed to a slower sedimenting form. These results indicate that o-phenanthroline is not an inhibitor of receptor activation but it prevents binding of receptor to acceptors by complexing to DNA or ATP binding site or to both.

D) Receptor inactivation by rifamycin AF/013.

Rifamycin AF/013 is an antibiotic derivative which is capable of blocking the RNA and DNA polymerase activities in eukaryotic systems. The compound was initially shown to inhibit binding of progesterone receptors to nuclei and ATP-Sepharose (1,8,13,36,53). The DNA binding by glucocorticoid receptor was also reported to be reduced by pretreatment of activated receptor with rifamycin AF/013. Results of our study show that rifamycin AF/013 addition to rat liver cytosol prior to heat activation, prevents its ATP-Sepharose and DNA-cellulose binding (Table II). Analysis of glucocorticoid-receptor complex activated in presence of rifamycin AF/013 indicates that the compound may be partially effective in blocking 9S → 4S transformation of glucocorticoid receptor. A more detailed analysis is needed to categorically state whether or not this compound blocks receptor activation, or whether its effects are limited exclusively to binding to the acceptor site(s) without interfering with the process of glucocorticoid receptor activation. Although effects of this compound are apparently applicable to steroid receptor in general, its mode of action remains obscure.

E) Inhibitory effects of heparin.

Heparin is a highly sulfated dextrorotatory mucopolysaccharide. It possesses specific anticoagulant properties and is used pharmacologically for different clinical conditions. Effects of heparin on the properties of a steroid receptor were initially recognized by Moudgil et al. (54) who demonstrated that ATP-binding of progesterone-receptor complex fractionated by $(NH_4)_2SO_4$ can be blocked by treatment of receptor by heparin. Yang et al. (55) subsequently noted that chick oviduct progesterone receptor could be activated at 0°C by heparin. The transformed receptor exhibited properties of an activated receptor with a slow dissociation and sedimentation rates. Recently, inactivation of hepatic glucocorticoid receptors by heparin was reported (56). Heparin addition was shown to enhance the rate of inactivation of unbound receptor but had no apparent inactivating effect on prebound glucocorticoid-receptor complexes at 4°C.

Since some of the above studies apparently yielded results which were inconsistent, we examined the effects of heparin on the activation and subsequent binding of receptor to

ATP-Sepharose and DNA-cellulose. Table II shows that when cytosol was activated by warming at 23°C in presence of heparin, there was a total loss of DNA and ATP binding of glucocorticoid receptor. The analysis on the rate of sedimentation showed that glucocorticoid-receptor complexes heat-activated in presence of heparin were able to transform into a 4S form from a nonactivated 9S receptor form (Fig. 11). Results summarized above suggest that heparin does not interfere with the process of receptor activation but inhibits DNA-cellulose and ATP-Sepharose binding of activated glucocorticoid receptors. The actions of heparin, therefore, are mediated via its binding to or near the acceptor site(s). These observations are consistent with earlier report (54) on the heparin effects on oviduct progesterone receptor.

F) Receptor inactivation by aurintricarboxylic acid.

Aurintricarboxylic acid (ATA) is a synthetic triphenylmethane dye that was first employed for the quantitation of aluminum ion (57) and was introduced by Grollman and Steward (58) as an inhibitor of protein synthesis in cell-free systems. ATA has been employed to block a variety of cellular processes which depend on the formation of a protein-nuclei acid complex (59). An interference by ATA with the activities of nucleic acid binding proteins was reported that was presumably due to be complexing of ATA with the template binding sites of proteins (60). ATA has also been recently used to examine chromatin structure (61). Because of the generally known interaction between steroid receptors and target-cell nuclear sites, use of ATA offered an attractive approach to investigate site(s) of receptor involved in the processes such as nuclear and DNA binding. With that premise, a detailed characterization of estrogen and progesterone receptor-ATA interaction has been undertaken in this laboratory over the last several years (8,34,36,62-65).

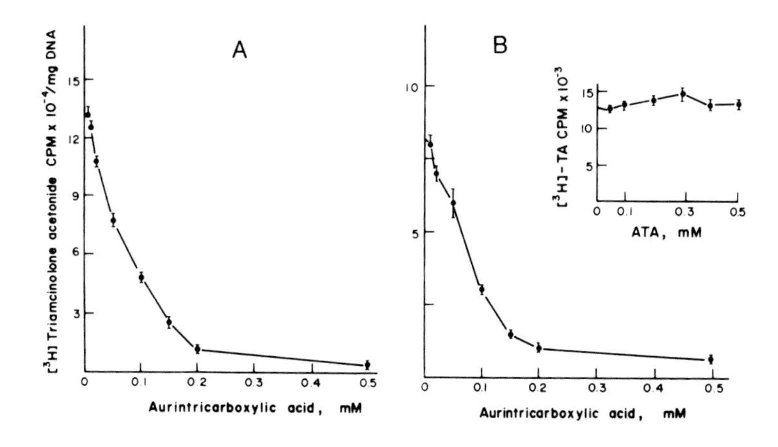

FIG. 12. Effect of aurintricarboxylic acid on the activation and nuclear binding of glucocorticoid-receptor complex from rat liver cytosol. Freshly prepared cytosol from rat liver was incubated with 10nM [^{3}H]-triamcinolone acetonide for 12 h at 0°C. Portions of above cytosol were incubated at 23°C for 60 min with buffer (control) or with varying concentrations of aurintricarboxylic acid (A). In a separate experiment, cytosol containing glucocorticoid-receptor complex was first incubated at 23°C for 1 h and then at 0°C with different concentrations of aurintricarboxylic acid (B). In both the cases (A,B), the nuclear binding assays were performed by incubating the samples with isolated nuclei (200μg/DNA/0.2ml). The details of procedures are described in (34). Taken from Ref. 34.

Glucocorticoid receptor appears equally sensitive to the presence of ATA. When rat liver glucocorticoid-receptor complexes were incubated with ATA and subsequently heat-activated, the presence of ATA inhibited the nuclear binding in a concentration-dependent manner showing an $I.D._{50}$ at 70uM ATA. Treatment of receptor with higher concentrations (>0.2mM) of

ATA eliminated nuclear binding completely (Fig. 12A). Similar results were obtained when glucocorticoid receptor activation was measured with DNA-cellulose or ATP-Sepharose as acceptors (Table II). The activated receptor retains its susceptibility to the inhibitory action of ATA; the nuclear uptake of heat-activated glucocorticoid-receptor complexes was abolished upon treatment with 0.1-0.5mM ATA (Fig. 12B) without any apparent effect on the dissociation of the complex under these conditions (Fig. 12B, inset). Presence of ATA during heat-activation of receptor does not seem to interfere with the conversion of 9S receptor to a 4S form suggesting that ATA does not interfere with the process of activation but rather brings its effects by affecting nuclear, DNA or ATP binding processes of activated glucocorticoid receptor.

If the prediction made in the preceding paragraph were to be true, it should be possible to extract glucocorticoid-receptor complexes from rat liver nuclei or other acceptors. The extraction of nuclear-bound [^{3}H]-triamcinolone acetonide-receptor complex by ATA is shown in Fig. 13. The extent of the release or extraction of receptor by ATA was seen to be comparable to that obtained by incubation of nuclear preparations with high-salt buffers. The mode of ATA-extraction of glucocorticoid receptor from nuclei is unclear at present. It may act competitively by binding to nuclear sites of the receptor or it may exert its influence allosterically by binding to a site which is in close proximity. An indirect action of ATA seems unlikely but can not be ruled out. Results of some of our preliminary studies showing a retention of steroid-receptor complexes over columns of ATA-Sepharose suggest the presence of an ATA binding site (44).

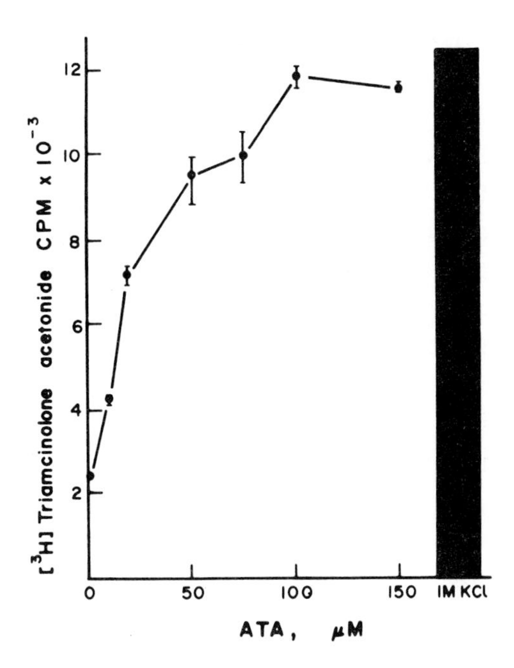

FIG. 13. Extraction of [^{3}H]-triamcinolone acetonide-receptor complex from rat liver nuclei by treatment with aurintricarboxylic acid. Small (0.5ml) aliquots of cytosol containing heat-activated glucocorticoid-receptor complexes were incubated for 40 min at 0°C with 0.3ml portions of rat liver nuclear preparations (150ug DNA/tube). After rinsing the suspension with buffer (10mM Tris-HCl, 1mM $MgCl_2$, 10% glycerol, pH 7.5), the nuclear-bound complexes were extracted by a 40 min incubation at 0°C with different concentrations of aurintricarboxylic acid (10-150uM). Duplicate aliquots of mixtures containing nuclear-bound complexes were also incubated with 1ml of buffer containing 1M KCl. Taken from Ref. 34.

Summary and Conclusions

We have described the effects of several chemical agents on the activation and acceptor binding properties of rat liver glucocorticoid-receptor complexes. The results, summarized in Fig. 14, indicate that molybdate and tungstate at concentrations <10mM are the only true inhibitors of process of receptor activation. Rifamycin appears to block receptor activation as well as inhibit acceptor binding of activated receptors. Molybdate and tungstate are also able to inhibit acceptor binding of activated receptors but at higher concentrations than needed to block activation. ATA, pyridoxal, 5'-phosphate, o-phenanthroline, heparin, preferentially interfere with the acceptor binding than process of receptor activation. Such distinction between the site of action was aided by the employment of density gradient centrifugation which allowed observation of transformation of one receptor form to the other without involving the acceptors in the picture. When this approach is not applied, it is difficult to speculate whether the inhibitors are blocking activation or merely interfering with the binding of receptor to acceptors since the compounds are still present during these measurements.

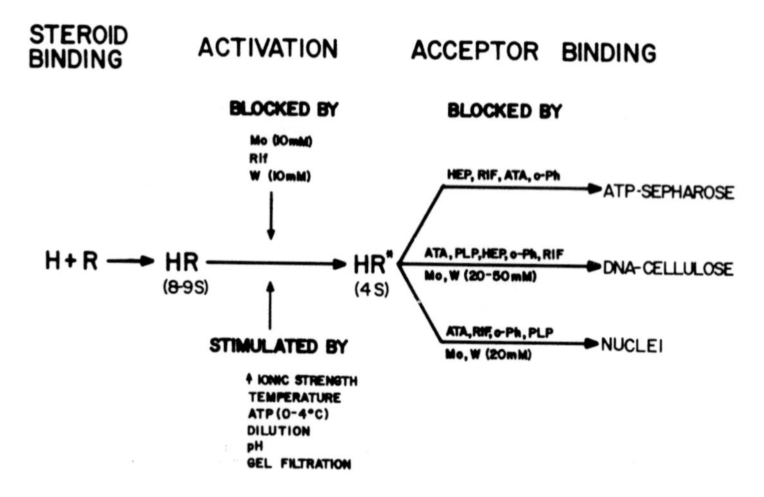

FIG. 14. A speculative model of steroid-receptor activation in vitro and the sites of action of different inhibitors.

Molybdate and tungstate are both potent inhibitors of activation and they are also known to stabilize unoccupied steroid receptors. In literature, their use has been mostly limited to studying activation and stabilization of receptors: effects which are apparent at concentration below or up to 10mM. However, there are subtle differences, mostly quantitative, between the effects of molybdate and tungstate. At lower pHs, tungstate is extremely effective in inhibiting receptor activation, DNA, nuclear or ATP binding even at 1mM concentrations. Relatively large molybdate concentrations would be required to block DNA or nuclear binding of activated receptors. Studies from this laboratory had initially postulated that these agents act directly via a binding site on the steroid receptors. This concept has subsequently received a wider acceptance. Inhibition by rifamycin AF/013, o-phenanthroline, pyridoxal 5'-phosphate, and ATA has one thing in common; all four compounds have been shown to inhibit activities of enzymes of nucleotide metabolism. Similarities between steroid receptors and polymerases have been suggested previously (1,13,36,51-53). Further analysis of steroid receptors employing use of these inhibitors should allow a better understanding of the nuclear, DNA and ATP-binding domains of receptors. Ability of some of these compounds to extract acceptor-bound receptor also has a great potential in the purification of steroid receptors.

Acknowledgments

The studies in this laboratory have been supported by NIH Grant AM-20893.

References

1. Moudgil, V.K.: In: Principles of Recepterology (M.K. Agarwal, ed.) p 273-379, Walter de Gruyter, Berlin-New York 1983.

2. Cake, M.H., Litwack, G.: In: Biochemical Actions of Hormones (G. Litwack, ed.) Vol. III, p 317-390, Academic Press, New York 1975.

3. Munck, A. Brinck-Johnsen, T.: J. Biol. Chem. 243, 5556-5565 (1968).

4. Milgrom, E., Atger, M., Baulieu, E.E.: Biochemistry 12, 519805205 (1973).

5. John, J.K., Moudgil, V.K.: Biochem. Biophys. Res. Commun. 90, 1242-1248 (1979).

6. Moudgil, V.K., John, J.K.: Biochem. J. 190, 799-809 (1980).

7. Moudgil, V.K., Eessalu, T.E.: FEBS Lett. 122, 189-192 (1980).

8. Moudgil, V.K., Kruczak, V.H., Eessalu, T.E., Paulose, C.S., Taylor, M.G., Hansen, J.C.: Eur. J. Biochem. 118, 547-555(198

9. Baxter, J.D., Rousseau, G.G., Bensen, M.C., Garcia, R.L., Ito, J., Tomkins, G.M.: Proc. Natl. Acad. Sci. U.S.A. 69, 1892-1896 (1972).

10. Higgins, S.J., Rousseau, G.G., Baxter, J.D., Tomkins, G.M.: J. Biol. Chem. 248, 5866-5872 (1973).

11. Goidl, J.A., Cake, M.H., Dolan, K.P., Parchman, L.G., Litwack, G.: Biochemistry 16, 2125-2130 (1977).

12. LeFevre, B., Bailly, A., Sallas, N., Milgrom, E.: Biochim. Biophys. Acta 585, 266-272 (1979).

13. Lohmar, P.H., Toft, D.O.: Biochem. Biophys. Res. Commun. 67, 8-15 (1975).

14. Buller, R.E., Toft, D.O., Schrader, W.T., O'Malley, B.W.: J. Biol. Chem. 250, 801-805 (1975).

15. Yamamoto, K.R., Alberts, B.M.: Proc. Natl. Acad. Sci., U.S.A. 69, 2105-2109 (1972).

16. Toft, D.O.: J. Steroid Biochem. 3, 515-522 (1972)

17. Miller, J.B., Toft, D.O.: Biochemistry 17, 173-177 (1978).

18. Moudgil, V.K., John, J.K." Biochem. J. 190, 809-818 (1980).

19. Sakaue, Y., Thompson, E.B.: Biochem. Biophys. Res. Commun. 77, 533-541 (1977).

20. Parchman, L.G., Litwack, G.: Arch. Biochem. Biophys. 183, 374-382 (1977).

21. Munck, A., Foley, R.: Nature 278, 752-754 (1979).

22. Markovic, R.D., Litwack, G.: Arch. Biochem. Biophys. 202, 374-379 (1980).

23. Schmidt, T.J., Harmon, J.M., Thompson, E.B.: Nature 286, 507-510 (1980).

24. Munck, A., Holbrook, N.J.: J. Biol. Chem. 259, 820-831 (1984).

25. Wilshons, W.V., Lieberman, M.E., Gorski, J.: Nature 307, 747-749 (1984).

26. King, W.J., Greene, G.L.: Nature 307, 745-747 (1984).

27. Jensen, E.V., Suzuki, T., Kawashima, T., Stumpf, W.E., Jungblut, P.W., DeSombre, E.R.: Proc. Natl. Acad. Sci. U.S.A. 59, 632-638 (1968).

28. Cake, M.H., DiSorbo, D.M., Litwack, G.: J. Biol. Chem. 253, 4886-4891 (1978).

29. Leach, K.L., Dahmer, M.K., Hammond, N.D., Sando, J.J., Pratt, W.B.: J. Biol. Chem. 254, 11884-11890 (1979).

30. Barnett, C.A., Schmidt, T.J., Litwack, G.: Biochemistry 19, 5446-5455 (1980).

31. Murakami, N., Moudgil, V.K.: Biochem. J. 198, 447-455 (1981).

32. Murakami, N., Moudgil, V.K.: Biochim. Biophys. Acta 676, 386-394 (1981).

33. Murakami, N., Quattrociocchi, T.M., Healy, S.P., Moudgil, V.K.: Arch. Biochem. Biophys. 214, 326-334 (1982).

34. Moudgil, V.K., Caradonna, V.M.: J. Steroid Biochem. 17, 585-589 (1982).

35. Murakami, N., Healy, S.P., Moudgil, V.K.: Biochem. J. 204, 777-786 (1982).

36. Moudgil, V.K., Nishigori, H., Eessalu, T.E., Toft, D.O.: In: Gene Regulation by Steroid Hormones (A. Roy and J.H. Clark, eds.) pp 103-119, Springer-Verlag, New York 1980.

37. Toft, D.O., Roberts, P.E., Nishigori, H., Moudgil, V.K.: Adv. Exptl. Med. Biol. 117, 329-342 (1979).

38. Nielsen, C.J., Sando, J.J., Pratt, W.B.: Proc. Natl. Acad. Sci. U.S.A. 74, 1398-1402 (1977).

39. Nielsen, C.J., Sando, J.J., Vogel, W.M., Pratt, W.B.: J. Biol. Chem. 252, 7658-7578 (1977).

40. Toft, D.O., Nishigori, H.: J. Steroid Biochem. 11, 414-416 (1979).

41. Leach, K.L., Dahmer, M.K., Hammond, N.J., Sando, J.J., Pratt, W.B.: J. Biol. Chem. 254, 1184-1190 (1979).

42. Nishigori, H., Toft, D.O.: Biochemistry 19, 77-83 (1980).

43. Fernley, H.N.: Enzymes 4, 417-477 (1977).

44. Samokyszyn, V., Moudgil, V.K.: unpublished results.

45. Nishigori, H., Alker, J., Toft, D.O.: Arch. Biochem. Biophys. 203, 600-604 (1980).

46. Nishigori, H., Toft, D.O.: J. Biol. Chem. 254, 9155-9161 (1979).

47. Nishigori, H., Moudgil, V.K., Toft, D.O.: Biochem. Biophys. Res. Commun. 80, 112-118 (1978).

48. DiSorbo, D.M., Phelps, D.S., Ohl, V.S., Litwack, G.: J. Biol. Chem. 255, 3866-3870 (1980).

49. DiSorbo, D.M., Litwack, G.: Biochem. Biophys. Res. Commun. 99, 1203-1208 (1981).

50. Chang, C.H., Yarbo, J.W.: Life Sci. 22, 1007-1010 (1978).

51. Shyamala, G.: Biochem. Biophys. Res. Commun. 64, 408-415 (1975).

52. Schmidt, T.J., Sekula, B.C., Litwack, G.: Endocrinology 109, 803-812 (1981).

53. Moudgil, V.K., Lohmar, P.H., Toft, D.O.: Proc. Endocrine Society meeting, San Francisco, Abstract (1976).

54. Yang, C.R., Mester, J., Wolfson, A., Renoir, J.M., Baulieu, E.E.: Biochem. J. 208, 399-406 (1982).

56. Hubbard, J.R., Kalimi, M.: Biochim. Biophys. Acta 755, 363-368 (1983).

57. Smith, W.H., Sager, E.E., Siewers, I.S.: Anal. Chem. 21, 1334-1338 (1949).

58. Grollman, A.P., Stewart, M.L.: Proc. Natl. Acad. Sci. U.S.A. 61, 719-725 (1968)

59. Gonzalez, R.G., Blackburn, B.J., Schleich, T.: Biochem. Biophys. Res. Commun. 55, 534-545 (1979)

60. Blumenthal, T., Landers, T.A.: Biochem. Biophys. Res. Commun. 55, 680-688 (1973).

61. Tsutsui, K., Yamaguchi, M., Oda, T.: Biochem. Biophys. Res. Commun. 55, 493-500 (1975).

62. Moudgil, V.K., Weekes, G.A.: FEBS Lett. 94, 324-326 (1978).

63. Moudgil, V.K., Eessalu, T.E.: Biochim. Biophys. Acta 627, 301-312 (1980).

64. Moudgil, V.K., Eessalu, T.E.: Life Sci. 27, 1159-1167 (1980).

65. Moudgil, V.K., Eessalu, T.E.: Arch. Biochem. Biophys. 213, 98-108 (1982).

SIMILARITY AND DISSIMILARITY IN REGULATORY MECHANISM BY CYTOPLASMIC MODULATORS BETWEEN GLUCOCORTICOID AND OTHER STEROID RECEPTORS

Keizo Noma, Bunzo Sato, Yasuko Nishizawa, Soji Kasayama, Susumu Kishimoto, Keishi Matsumoto

The Third Department of Internal Medicine and The Second Department of Pathology, Osaka University Hospital, Fukushima-ku, Osaka 553, Japan

Introduction

The study of mechanism of action of the direct gene regulatory proteins activated by a biological signal ligand is one of the central theme in the cell biology. These regulatory proteins have been identified to include repressor, catabolic gene activation protein in E. coli (1, 2), cyclic AMP binding protein in Walker 256 mammary carcinoma (3), chromatin-associated receptor for thyroid hormone (4) as well as 1,25 $(OH)_2D_3$ receptor (5) and steroid hormone receptors. The afferent pathways for these proteins to become biologically active are considered to have the common reaction process; that is, i) ligand-protein complex formation with high-affinity and limited capacity in the highly stereospecific manner correlated with biological responses, ii) activation process of the complexes which are defined as acquisition of the greater affinity to nucleus or DNA, and binding to the specific nuclear acceptor sites. Steroid receptor is one of the intracellular gene regulators in eukaryocytes extensively investigated for its structure and function. The distinct functional domains have been analyzed and can be separated by the limited proteolysis of of glucocorticoid (6), estrogen receptor (7), androgen receptor (8) and progesterone receptor (9). These structual analyses have revealed that all steroid receptors are composed of the similar and fundamental units; namely, ligand binding, DNA

Adrenal Steroid Antagonism

binding and the third portion of the unknown function. In addition to these proteolysis experiments, recent observations obtained with receptor-antibodies have also successfully demonstrated the presence of these domains (10, 11). The functional defect of these domains causing unresponsiveness to steroid hormone has been reported in transformed and nontransformed cells. Resistance to steroid associated with ligand binding abnormality has been demonstrated in glucocorticoid receptor (12, 13, 14, 15), estrogen receptor (16, 17, 18, 19) and androgen receptor (20, 21, 22, 23, 24). The evidence is also provided that steroid resistance is due to the abnormal DNA binding domain of the receptor molecules (25, 26, 27, 28). In some steroid insensitive cell lines of mouse lymphoma, it has also been confirmed by the hybrid cell study that the resistance to glucocorticoid was ascribed to the defect of the receptor molecule itself (29). Thus, the unresponsiveness due to defective receptor molecule seems to be well established in many steroid hormone target tissues.

On the other hand, the considerable evidence has been accumulated that the steroid receptor functions may be regulated by intracellular factors. The pioneer study by Munck et al. has clearly shown that the glucocorticoid binding capacity in intact rat thymic lymphocytes is markedly but reversibly reduced in anoxic and/or glucose-deprived state (30). Under the cell-free conditions, the glucocorticoid binding capacity in some target tissues has been shown to be drastically influenced by the addition of -SH reducing agents into assay tubes (31, 32, 33). A removal of the low molecular weight compounds by means of dialysis or gel filtration has been observed to induce instability of the steroid binding ability of glucocorticoid receptor (34, 35). All of these studies would critically suggest the important role of the enviromental factors for steroid receptor functions. We have recently reported that one of mouse Leydig cell tumor lines is resistant to hormonal manipulation probably due to the presence of the small molecule which renders estrogen receptor to be in the low-affinity

state for the ligand as well as nuclei (36). These results raised the interesting possibility that the presence of the abnormal modulator may be related to steroid resistance. In spite of the considerable efforts, however, the molecular mechanism of the cytoplasmic modulators to regulate the steroid receptor functions remains largely unknown.
In this review article, we would summarize recent observations from our and many other laboratories and try to obtain some clue leading to more clear understanding of this important relationship of the steroid receptor, especially glucocorticoid receptor, to intracellular modulators.

Evidence of the cytoplasmic modulators regulating steroid receptor function.

Before entering into the detailed discussion of glucocorticoid receptor (GR) function, our recent experimental results showing that subtle regulatory mechanism of estrogen receptor (ER) functions might exist at the target cell level would be stated. As summarized in Fig. 1, ER can be activated by the removal of the so-called low molecular weight inhibitor. This process is facilitated by complexing with estrogen, but its presence seems not to be obligatory to generate the activated form of ER (37, 38). A removal of the dialyzable compounds from the cytosol containing ER in the absence of estrogen can activate ER under the cell-free condition, resulting in translocation of unoccupied ER into nuclei. This activation of ER seems to be a direct consequence of dissociation of small molecules from ER, since unoccupied ER partially purified in the nonactivated state can interact with the low molecular weight inhibitor, preserving ER in the nonactivated state (39). These observations have raised the interesting possibility that some target cells possess unoccupied nuclear ER even in the absence of hormonal stimuli. The observation by Zava and McGuire (43) that cultured maligant

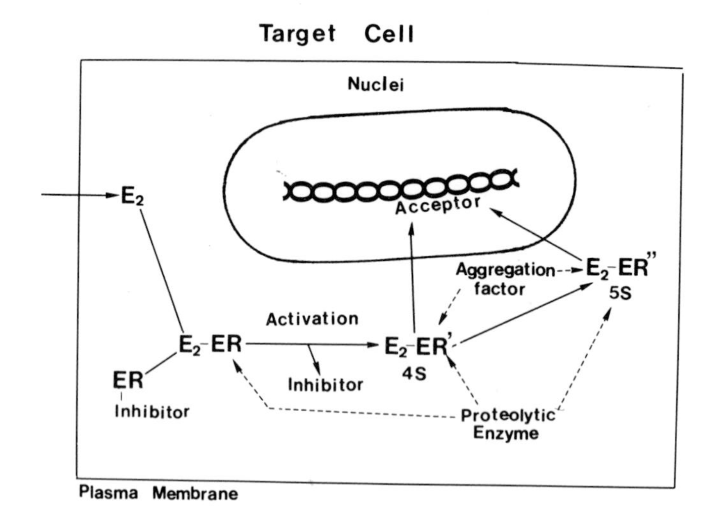

Fig. 1 Brief summary on of the pathway of ER translation nuclear acceptor site

This scheme is largely, based on our published results (37, 38, 39, 40, 41). The sedimentation constants of activated ERs in the high salt condition were also provided. The conversion from 4S to 5S was, however, not found to be essential as far as nuclear translocation of ER was concerned (37, 42). The abbreviations used here are: E_2, estradiol: ER, estrogen receptor.

human mammary cell (MCF-7) has unoccupied nuclear ER with some usual cytosolic receptor is the forerunner of a new concept on the subcellular localization of steroid receptor. In more recent articles (44, 45, 46), this subject has been extensively discussed without unequivocal answer. This might be related to the experimental results showing the simultaneous presence of the nuclear and the cytosolic ER in many target cells used in their studies. Thus, our laboratory tried to seek new target cells where the receptor is confined to the nuclear fraction. One of mouse Leydig cell tumor lines (Tumor 124958) has recently been found to contain nuclear ER without putative cytosolic ER which has been demonstrated under the cell-free (47) and the intact cell (48) conditions (Fig. 2). This unusual subcellular localization of ER is probably due to the decreased concentration of the low molecular weight inhibitor. The uterine cytosol containing [^{3}H] E_2-ER complexes was dialyzed against various tumor cytosols, followed by nuclear binding assays of these complexes. The results would indicate the diminished inhibitory activity against ER activation in this particular tumor (Tumor 124958) cytosol when compared with that of the other tumor (Tumor 134486), supporting the idea that the low molecular weight inhibitor identified and assayed in the cell-free condition is able to exert its effects in the intact cell (B. Sato and Y. Maeda: unpublished data). In addition, we have postulated that other modulators attack

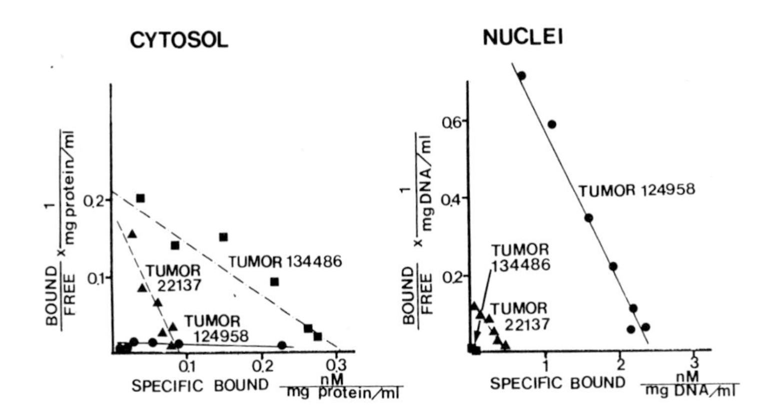

Fig. 2 The presence of unoccupied nuclear ER in mouse Leydig cell tumor line.

The cytosolic and nuclear estrogen binding components were examined. Two lines (Tumor 134486 and 22137) showed the presence of the high-affinity cytosolic binder, but Tumor 124958 did not contain the high-affinity E_2 binder. The incubation of the isolated nuclei, which were not exposed to estrogen in vivo, resulted in the demonstration of unoccupied nuclear ER in Tumor 124958.

only specialized forms of ER (40, 41). In view of these results, we have attempted to clarify the regulatory mechanism of GR functions. In the following sections of this review, the special attention will be made on the similarity and the dissimilarity in the regulatory mechanisms between GR and ER.

Endogenous modulators affecting hormone binding ability of GR.

During the course of the past few years, we were confronted with the difficulty concerning the instability of GR during purification procedures.
Recent evidence from several systems reveals the involvement of endogeneous modulators in the ability of GR to bind ligands in the specific manner. Much attention has been paid on the effect of sulfhydryl (-SH) reagents on GR function. Several groups have reported substantial stabilization and partial reacquisition of glucocorticoid binding capacity in rat thymocyte and lung cytosols by addition of dithiothreitol (DTT) (31, 32, 33). These results suggest that stabilization and reacquisition of glucocorticoid binding capacity of GR would at least partly require reduction of the receptor itself or of some other cytosolic components. The possibility has been raised from these data that the apparent glucocorticoid binding capacity in the target cells is under stringent control by endogeneous reducing equivalents since these have been reported to be markedly different among various tissues (49). In the first series of studies, we examined DTT effects on the glucocorticoid binding capacity. Table 1 summarizes data on DTT-dependent enhancement of glucocorticoid binding capacity in various cytosol preparations. The addition of DTT at the optimal concentration (10 mM) markedly enhanced the glucocorticoid binding ability in the thymus and spleen cytosols while that in the liver cytosol was only modestly affected by this treatment, suggesting that the glucocorticoid sensitivity can be modulated by -SH reducing capability in the target cell.

Table 1 DTT-enhanced glucocorticoid binding capacity in various cytosol preparations

source of cytosol	Specific Dex binding (fmol/mg protein)		ratio of DTT(+)/DTT(-)
	DDT(-)	DDT(+)	
Liver	461.6	420.4	0.91
Thymus	133.3	871.7	6.54
Spleen	21.7	539.1	24.84
Brain	135.0	192.9	1.43

The cytosol was prepared in 4 vol of 0.01 M Tris, 1.5 mM EDTA, pH 7.4 using various tissues from the adrenalectomized rats. The specific glucocorticoid binding capacity was determined by incubation with 40 nM $[^3H]$Dex ± 4 µM unlabeled Dex in the presence or the absence of 10 mM DTT.

Although these observations provide us with some clue for elucidating the molecular mechanism of GR function, several lines of evidence would indicate that the endogeneous factors other than reducing equivalents play important roles for the regulation of GR functions. For instance , the glucocorticoid binding ability in the liver cytosol treated with 10% dextran-coated charcoal, which has been known to remove the endogenous reducing equivalents, was reported to be fully reversed to the original binding level by the addition of 10 mM DTT (49). On the other hand, the binding ability of the liver cytosol predialyzed at 0°C was not fully recovered by the addition of 10 mM DTT (See below). These results would imply that the dialyzable compounds have some role for preserving GR in the steroid binding form. To further characterize this dialysis-induced loss of the ligand binding ability, we employed Na_2MoO_4. This anion was originally introduced into the steroid receptor research as a phosphatase inhibitor by Pratt and his colleage (50). However, its potential effect on purified steroid receptor even at the low temperature would suggest that Na_2MoO_4 directly interacted with the receptor molecule itself (51) . Using this anion, the role of the low molecular weight compounds was examined in relation to GR stability (Fig. 3). The small molecule-depleted liver cytosol was quite heat-labile. The addition of either Na_2MoO_4 or the dializable compounds into this cytosol could partly stabilize the binding ability. Marked stabilization was observed when the dialyzable compounds from the liver cytosol with Na_2MoO_4 were added to this cytosol. These results would suggest that the low molecular weight compound(s) act in conjunction

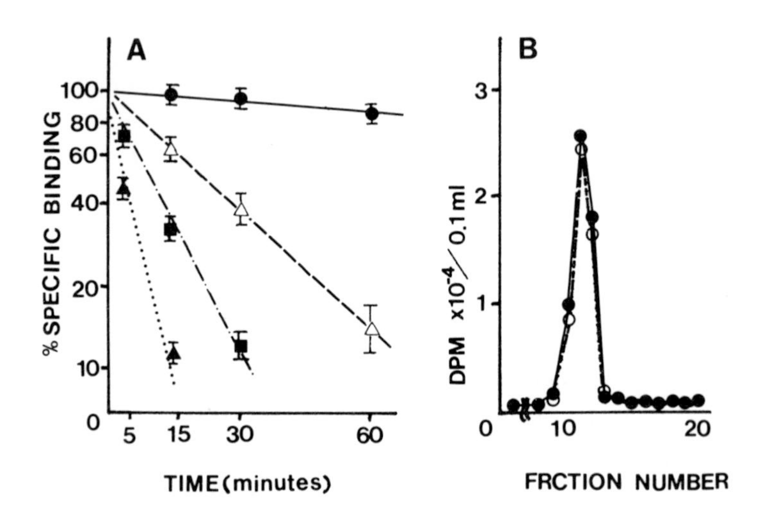

Fig. 3 Effects of dialyzable compounds or DTT on the glucocorticoid binding capacity of the gel-filtered cytosol.

(Panel A) The dialysate was prepared by dialysis of the liver cytosol against an equal volume of TM buffer (0.01 M Tris, 2 mM mercaptoethanol, pH 7.4) for 18 h. The liver cytosol, which had been gel-filtered through a small Sephadex G-25 column, was mixed with TM buffer (▲---▲), the dialysate (■–·–■), Na_2MoO_4 (Δ---Δ) or the dialysate + Na_2MoO_4 (●——●). These mixtures were heated at 25 C for the indicated periods of time. After heating, the aliquots were taken out to measure the amount of the specific binding of [^{3}H] Dex.
(Panel B) The liver cytosol was gel-filtered using a small Sephadex G-25 column. The specific [^{3}H] Dex binding capacity in each fraction was measured in the presence (●——●) or the absence (o---o) of 10 mM DTT. No significant difference between the two assay conditions was observed.

with Na_2MoO_4 to stabilize glucocorticoid binding ability. It should be specified that the stabilization activity in this fraction is not SH-dependent since the addition of DTT to the macromolecular fraction obtained by passing through a small Sephadex G-25 column does not enhance the glucocorticoid binding capacity (Fig. 3b). The additional evidence supporting the idea that Na_2MoO_4-dependent stabilization of the glucocorticoid binding capacity is potentiated by other cytoplasmic factor is shown in Fig. 4. This Na_2MoO4 stimulated stabilization was maximum at 2mM in the case of crude liver cytosol preparations. On the other hand, the partially purified GR by utilizing combination of DEAE cellulose and $(NH_4)_2SO_4$ fractionation required the higher (20-50 mM) concentrations of Na_2MoO_4 for the ligand binding ability to be stabilized against temperature-dependent degradation. When this partially purified GR was resuspended in the GR-free liver cytosol, the maximum effect of Na_2MoO_4 on the glucocorticoid binding ability was achieved at much lower concentration (1-5 mM) and and the stabilization effect was much more profound in the presence of the receptor-free liver cytosol compared with that in the absence of the cytosolic factors. This stabilization stimulated by the endogenous cytosolic factors might be different from phosphorylation-dependent reacquisition of the ligand binding capacity proposed by Pratt and his coworkers (52). In our experimental systems cited above, ATP was added at the final concentration of 0.1-10 mM without detectable effects. The detailed molecular mechanism of this stabilization process, however, remains to be elucidated. In addition, brief comments on the stability of the estrogen binding capacity of ER appear necessary. To our knowledge, marked destabilization induced by the removal of small molecular compounds has been not documented (37) and DTT addition is with mininum effects on the estrogen binding ability in many target cells. However, some reagents destroying -SH residue have been well known to diminish or nullify the estrogen binding capacity of ER (53). Therefore, it remains obscure whether the difference

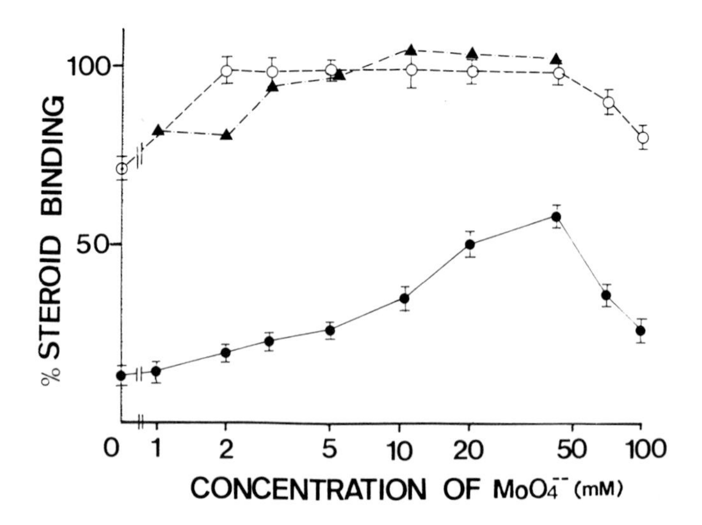

Fig. 4 The presence of the potentiation factor in the liver cytosol for Na_2MoO_4 effects on GR stabilization

The [^{3}H]Dex-GR complexes were partially purified using DEAE cellulose chromatography and $(NH_4)_2SO_4$ fractionation (40-60% saturation cut) in the presence of 10 mM Na_2MoO_4. The final pellet from $(NH_4)_2SO_4$ fractionation was resuspended in TM buffer containing various concentrations of Na_2MoO_4 with (▲---▲) or without (●——●) the receptor-free liver cytosol which had been prepared by heating at 37 C for 1 h. These resuspensions were relabeled with 40 nM [^{3}H] Dex, followed by heating at 25 C for 30 min. The specific binding was determined by a hydroxylapatite assay. As controls, the concentration dependency of Na_2MoO_4 on GR in the whole liver cytosol was also examined (o---o). The values presented here are percent, taking the values from nonheated samples as 100%.

between GR and ER systems is qualitative or quantitative.

Activation process of GR and ER

The receptor activation is a key step in steroid hormone action. Accordingly, many laboratories have focused on the study of the molecular mechanism of this process. As described in Fig. 1, the important role of the low molecular weight inhibitor involved in the activation step of ER has been observed in our laboratory (37, 38, 39). This information in turn has led to the identification of similar molecules regulating GR activation. As shown in Fig. 5, the dialysis of liver cytosol containing [^{3}H]Dexamehtasone(Dex)-GR complexes resulted in marked increase in their nuclear binding ability, suggesting the similarity in the activation process between ER and GR. The possibility of the activation of unoccupied GR was next assessed. When the uterine cytosol was dialyzed in the absence of estrogen for 3 h, followed by incubation with [^{3}H]E_2 at 4 C, marked increase in activation of ER was observed. We applied the same procedure to the liver glucocorticoid receptor. The significant but only modest increase in activation of GR was observed in response to predialysis before complexing with [^{3}H]Dex. Two possibilities should be taken into consideration: i) GR can be activated even in the absence of glucocorticoid; ii) removal of the dialyzable compounds is not enough for GR activation which can be initiated by the decreased concentration of the low molecular weight inhibitor only after a formation of complexes with Dex. To analyze these possibilities, the dialyzed liver cytosol was simultaneously mixed with [^{3}H]Dex and Na_2MoO_4 to block further activation after a formation of the ligand-GR complexes (54, 55). The data depicted in Fig. 5 clearly showed marked decrease in the amount of [^{3}H]Dex-GR complexes translocated into nuclei, resulting in the lack of demonstrable difference between predialyzed and nondialyzed samples. These observations would support the idea that the ligand binding is obligatory for GR activation. Furthermore, experiments involving the incubation of the dialyzed cytosol with isolated nuclei, followed by reisolation of nuclei, failed to show specific [^{3}H]Dex binding in these treated nuclei (data not illustrated). Additional evidence for the important role of the ligand for GR activation binding has come from the experiment using rat brain

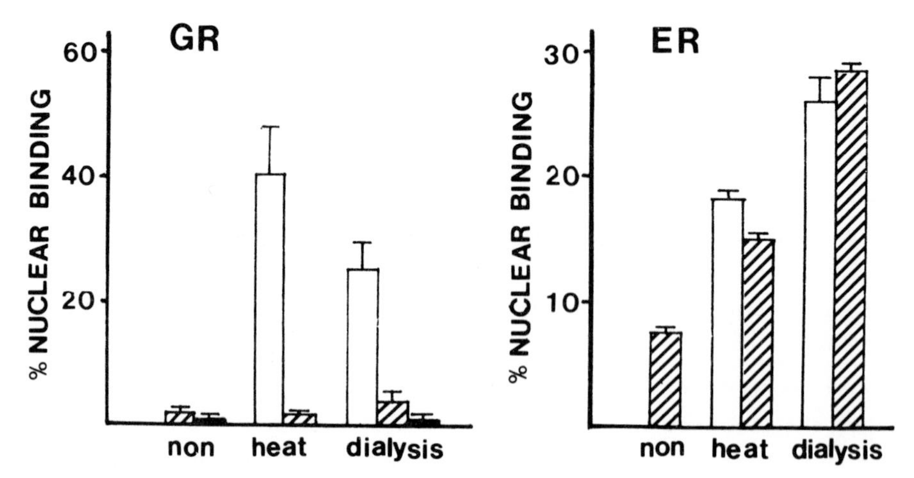

Fig. 5 Effect of the presence of steroid hormone on activation of GR and ER

The cytosol was prepared in 2 vol of TM buffer. The cytosol was heated at 25 C for 30 min, or dialyzed against TM buffer at 0 C for 3 h before labelling or after labelling with 40 nM [^{3}H] Dex ± 1,000 fold excess of unlabeled Dex. After these treatments, unoccupied GR was complexes with [^{3}H]Dex in the presence or absence of 10 mM molybdate. The aliquots of the treated cytosol were incubated with the purified liver nuclei at 0 C for 30 min for measurement of the amounts of the activated [^{3}H] Dex-GR complexes. Similar procedures were also applied to the rat uterine estrogen receptor systems. The data (mean ± S.E. were obtained in three separate experiments) were expressed as percent nuclear binding of the incubated complexes. [^{3}H]Dex prelabelling and then heating or dialysis; (□), heating or dialysis of unoccupied GR, followed by labelling with [^{3}H]Dex in the absence; (▨), or the presence; (▬) of molybdate.

GR. The whole brain cytosol prepared from the ovariectomized and adrenalectomized rats contained the high affinity progesterone binder (Kd ~5 x 10^{-9} M). This progesterone binding to the brain cytosol was found to be inhibited in a competitive manner by Dex. In addition, the number of the maximum binding sites obtained by [^{3}H]progesterone was quite similar to that with [^{3}H]Dex, suggesting that [^{3}H]progesterone is mostly associated with brain "GR". Under these conditions, the stability and the activation behavior of [^{3}H]progesterone-GR complexes were examined. As described in Table 2, an exposure of these complexes to the moderately high temperature (25 C) for 10 min resulted in a rapid loss of the ligand binding activity while Dex-GR complexes were found to be relatively stable. Based on their DNA binding ability, however, [^{3}H]progesterone-GR complexes were judged to be activated by heating. A similar type of experiment was quite difficult to carry out using liver cytosol because of its activity to metabolize progesterone into more polar steroids even at 0 C. The data on rat brain would stress the importance of the ligand binding for GR

Table 2 Temperature-dependent destabilization and activation of progesterone-GR complexes from the rat brain

[^{3}H]ligand	Receptor stability (DPM/0.1mlcytosol)		DNA binding (DPM/0.4mlcytosol/0.3mlDNA)	
	heat(-)	heat(+)	heat(-)	heat(+)
Dexamethasone	15078	10727	687	7147
Corticosterone	22182	15867	2042	6255
Progesterone	12062	5375	1576	4662

Whole brain cytosol was prepared in 4 vol of 0.01 M Tris, 1.5 mM EDTA, 2 mM mercaptoethanol, pH 7.4 using the adrenalectomized and ovariectomized rats. The cytosol was incubated with 20 nM [^{3}H]Dex, [^{3}H]corticosterone or [^{3}H]progesterone in the presence or the absence of a 100-fold molar excess of dexamethasone. These complexes were heated at 25 C for 30 min ([^{3}H]Dex) or for 10 min ([^{3}H]corticosterone and [^{3}H]progesterone), followed by assays of the specific ligand and DNA binding abilities.

activation even though partial antagonists were used as the ligand. These results are in quite contrast with those in ER systems as described previously (37). The cell-free analyses of rat uterine ER system have clearly demonstrated the capability of unoccupied ER to translocate into nuclei without any aid of estrogenic compounds. This striking difference in activation process between ER and GR system might imply the involvement of entirely different small molecules regulating GR activation. To test this possibility, both the

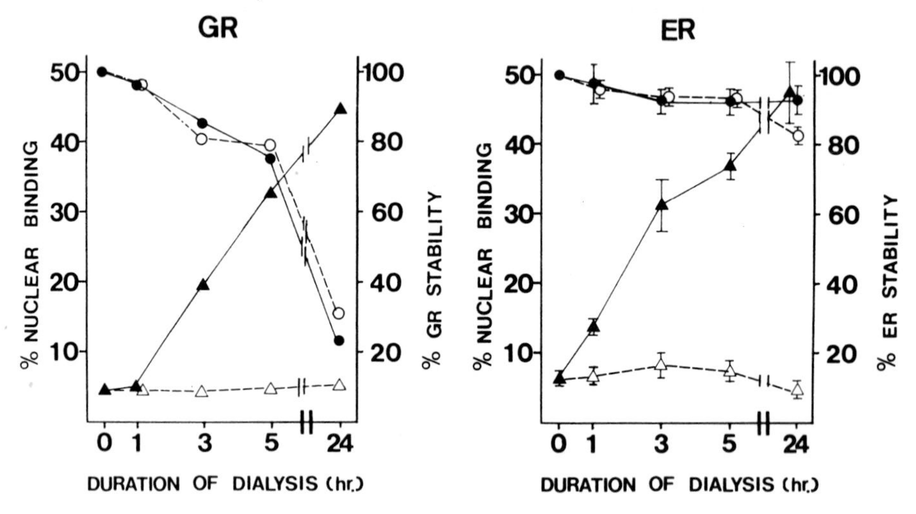

Fig. 6 The liver cytosol contains the small molecule inhibiting the activation of both GR and ER.

The liver cytosol was ultrafiltered through UM05 membrane (Amicon) to obtain the small molecular fraction with the molecular weight of less than 500. The liver cytosol containing nonactivated [^{3}H] Dex-GR complexes (right panel) or the uterine cytosol containing nonactivated [^{3}H] E_2-ER complexes (left panel) was dialyzed against 50 volumes of the ultrafiltrate or TM buffer at 0-4 C for the indicated periods of time. The percent ligand binding ability (ultrafiltrate; o---o, TM buffer; ●——●) was calculated, taking the value of nondialyzed sample as 100%. The percent nuclear binding (ultrafiltrate; Δ---Δ, TM buffer; ▲——▲) was obtained by dividing the amount of translocated receptor with that of total incubated receptor.

rat liver cytosol containing [^{3}H]Dex-GR complexes and the uterine cytosol containing [^{3}H]E_2-ER complexes were dialyzed against the small molecular fraction prepared by ultrafiltration of the liver cytosol through UM05. Fig. 6 shows that this ultrafiltrate contains the factors inhibiting the activation of [^{3}H]Dex-GR complexes as well as [^{3}H]E_2-ER complexes, suggesting that the similar molecule is involved in the activation process of both receptor systems. Next, an posibility was assessed that the liver cytosol contains the additional modulator which inhibits translocation of the

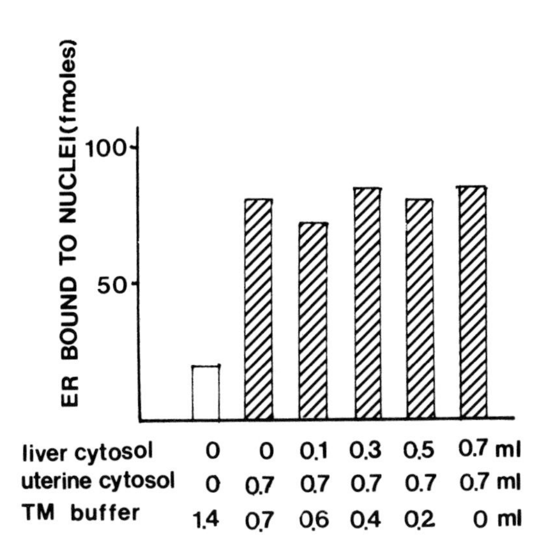

Fig. 7 Lack of effect of liver cytosol on translocation of unoccupied uterine ER into nuclei.

The uterine cytosol prepared in 4 volumes of TM buffer was mixed with various amounts of the liver cytosol which was also prepared in 4 volumes of TM buffer. These mixtures were dialyzed against 500 volumes of TM buffer at 0-4 C for 3 h. The aliquots (1.2 ml) of these treated cytosols were incubated with isolated nuclei at 0 C for 1 h. After incubations, nuclei were reisolated and incubated with 4 nM [^{3}H] E_2 ± 0.4 μM unlabeled E_2 at 0 C for 1 h. These incubated nuclei were reisolated to obtain the specific binding of [^{3}H] E_2. The procedures used here were published previously (37).

unoccupied activated receptor into nuclei. The mixture of the rat uterine cytosol and the liver cytosol was dialyzed and then incubated with the uterine nuclei. After incubation, the nuclei were reisolated, followed by incubation with $[^3H]E_2$ ± 100 fold excess of unlabeled E_2 at 0 C. As shown in Fig. 7, the addition of the liver cytosol did not decrease unoccupied nuclear ER content. Therefore, marked difference in hormone dependency in dialysis-induced receptor activation between ER and GR could not be explained by the environmental factors, although the possibility can not be excluded that the translocation inhibitor specific for unoccupied GR exists in the liver cytosol. This difference might reside in the receptor itself.

Does the identical low molecular weight compound regulate both stabilization and activation of steroid receptors?

The important role of the small molecules present in target cells has become evident from data cited in the previous sections of this review. As one can realize, however, the destabilization and the activation do not always progress in a parallel fashion. For instance, ER can be easily activated by removal of the small molecules without destabilization of the estrogen binding ability, while unoccupied GR is very sensitive to dialysis in terms of the glucocorticoid binding ability without activation. The experimental procedures achieving the lowered concentration of the dialyzable compounds in the cytosol allow clear distinction between destabilization and activation, although Na_2MoO_4 has been well known to block both destabilization and activation process (54, 55). These results favor the idea that different small molecule(s) separately regulate these two important processes. In order to obtain the final answer, however, the identification of these small molecules would be definitely required.

The other endogeneous modulators affecting steroid receptor functions.

Several other endogenous modulator, such as macromolecular translocation inhibiting factor (56) , ribonuclear protein (57) and proteolytic enzymes (6, 7, 8, 9) have been reported. Although their biological roles in the living cells have remained entirely obscure, these modulators provide us with a powerful tool to analyze the steroid receptor structure. In this section, we would focus on our recent studies on ER aggregation factor (40) and endoprotease in human mammary cancer (41). ER aggregation factor has been identified in the adult rat uterus but not in the liver. This factor could aggregate activated ER in a temperature- and

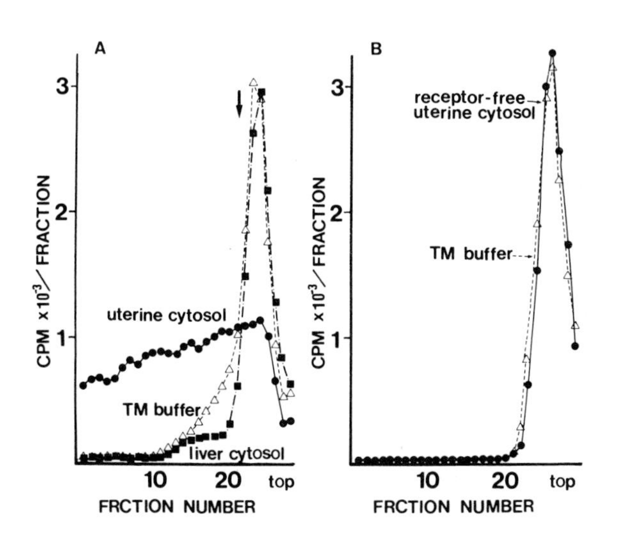

Fig. 8 ER specific aggregation factor in adult rat uteri.

The receptor-free uterine or liver cytosol prepared by heating at 37 C for 1 h was mixed with the uterine [^{3}H] E_2-ER complexes (panel A) or with the liver [^{3}H] Dex-GR complexes (panel B). As controls, these receptors were also mixed with TM buffer. These mixtures were heated at 25 C for 30 min in the presence of 0.15 M KCl. After being treated with dextran-coated charcoal, these complexes were analyzed by high salt (0.4 M KCl) sucrose density gradients. The arrow indicates the position of human gamma globulin.

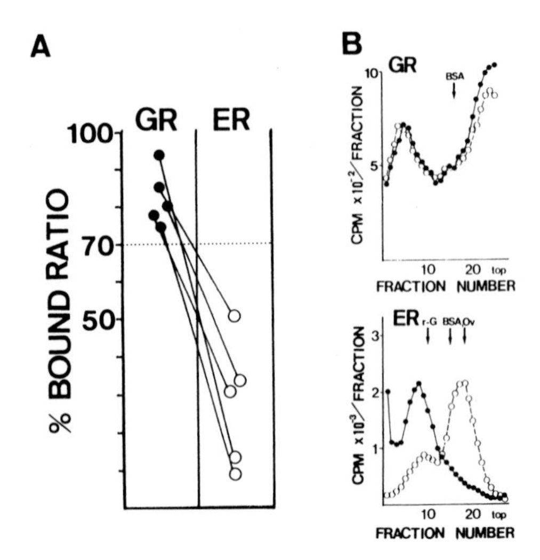

Fig. 9 The lack of the demonstrable effect of ER cleaving enzyme on GR.

Rat tissues (uteri and liver) and human breast cancer specimens were homogenized in 4 volumes of TM buffer. [^{3}H] E_2 and [^{3}H] Dex, respectively. These labeled cytosols were mixed with an equal volume of the tumor cytosol or TM buffer, followed by heating at 25 C for 30 min. Their DNA binding ability (panel A) and sedimentation profiles in low salt sucrose desity gradients (panel B) were monitered. DNA binding was expressed as % bound ratio, taking the DNA binding values of samples mixed with TM buffer as 100%.

KCl-dependent manner, concomitant with a loss of its nuclear binding ability. Interestingly, this factor was unable to attack nonactivated as well as activated GR, suggesting that some structural difference in DNA binding domain or the area near the DNA binding site would exist between ER and GR (Fig. 8). In addition to this aggregation factor, the use of endoprotease from human breast cancer resulted in the successful demonstration of the structural difference between ER and GR (Fig. 9). This enzyme is able to cleavage ER into a small fragment (MW ~38,000 in the high salt condition) which retains both steroid and DNA binding domain. But its DNA binding ability is strongly suppressed probably due to tight interaction with the low molecular weight inhibitor. This enzyme has failed to fragment GR, as judged by data on sucrose density gradient analysis and DNA binding assays. These two endogeneous modulators seem worthy of consideration in analysis of steroid receptor structure especially in relation to receptor specificity.

Summary

The aspects described in this review do not conflict with the recent reports concerning the importance of the environmental factors on GR functions. However, we would point out some unique results; i) the requirement of the small molecular components for molybdate effect on GR stabilization, ii) the dissimilarity in activation process between GR and ER. The mode of action of molybdate on steroid receptors has not been fully elucidated. The evidence showing the direct interaction of this anion with the receptor moelcule has been accumulating (51,58) although this anion had been originally considered to exert its stabilizing effect through inhibition of the endogeneous phosphatase (50). Currently, two different processes have been reported to regulate glucocorticoid binding capacity. First, the conversion of GR from steroid nonbinding form to the binding form can be achieved by dithiothreitol of other sulfhydryl reagents (30, 31, 32). Second, preservation of the binding form of the unoccupied GR and ligand-GR complex can be potentiated by molybdate and/or endogeneous stabilizer. Our dialyzable cytosol factors could be involved in GR stabilization. This factor itself is very weak in stabilization of Sephadex G-25 filtered unoccupied GR, but potentiates molybdate stabilizing effect on steroid binding. This evidence may suggest that the interaction between GR and molybdate is strengthened by endogeneous factors. This consideration would be supported by the observation that the addition of the cytosplasmic factor to the partially purified GR potentiates molybdate effect at very low concentration. The activation of GR, which can conceivably be induced by dissociation of the low molecular weight inhibitor from GR, seems to be dependent upon ligand binding. In our previous paper (38), the dissociation process was suggested to be promoted by dialysis even in the absence of the ligand. The present observation would indicate that GR activation occuring in the predialyzed cytosol can be initiated only after complexing with ligands. This conclusion was drawn by the experiment using molybdate on receptor labelling. This strict dependency of the ligand binding on GR activation might be related to the absence of unoccupied nuclear GR in the target cell. In relation to the significance of ligand binding for GR activation, progesterone-GR complexes seem to be interesting. Progesterone has been categorized as a partial agonist and antagonist of the glucocorticoid (59). The administration of the large dose of medroxyprogesterone acetate to treat patients with breast cancer, or other endocrine-related cancers, has resulted in the Cushingoid appearance and the lower level of plasma cortisol (60). The present results suggesting that progesterone can form the activated but unstable state of GR partly explain the antagonistic or agonistic effect in many glucocorticoid targets.

Finally, it would be important to note that some common regulatory mechanisms among all classes of steroid receptors exist, although some differences are evident: i) molybdate has the potential effect of stabilization and activation of all steroid receptors (55), ii) the small molecular fraction inhibits the activation of GR, ER and androgen receptor (38). These similarities would suggest that all steroid receptors may share a common regulatory domain which is different from a ligand or DNA binding domain. This possibility may be strengthened by development of a monoclonal antibody that recognizes a common non-hormone binding portion in four steroid hormone receptors (61). In conjunction with our aggregation factor and/or ER cleaving enzymes which recognize some special form of steroid receptor, these new tools such as monoclonal antibodies would be quite promising to further analyze the structure and function of steroid receptors.

Acknowledgements

We thank Drs. Maeda Y., Maeda K., Nakao M, and Ochi H. for their valuable contributions to this study. This work was partly supported by Grants-in aids from the Ministry of Education, Tokyo Japan; Japanese Cancer Research Foundation, Hisamitsu Cancer Research Fund, Naito for 1979 and Hirai Memorial Cancer Research Fund.

References

1. Barkley, M.D., Bourgeois, S.: The operon (ed Miller J.H., Reznikoff, W.S.) Cold Spring Harbor Laboratory pp 172-211 New York (1978)
2. Pastan, I., Adya, S.: Bacteriol. Rev. 40, 527-551 (1976)
3. Cho-Chung, Y.S., Clair, T., Porper, R.: J. Biol. Chem. 252, 6349-6355 (1977)
4. Oppenheimer, J.H., Schwartz, H.L., Surk, M.I., Koerner, D., Dillmann, W.H.: Recent Progress in Hormone Research 32, 529-565 (1976)
5. Norman, A.W., Roth, J., Orci, L.: Endocrine. Rev. 3, 331-366 (1982)
6. Wrange, O., Gustafsson, J-A.: J. Biol. Chem. 253, 856-865 (1978)
7. Sala-Trepat, J.M., Vallet-Strouve: Biochim. Biophys. Acta 371, 186-202 (1984)
8. Wilson, E.M., Trench, F.S.: J. Biol. Chem. 254, 6310-6319 (1979)
9. Sherman, M.R., Pickering L.A., Rollwagen, F.M., Miller, L.K.: Fed. Proc., 37, 167-173 (1978)
10. Gustafsson, J-Å, Carlstedt-Duke, J., Okret, S., Charlotte, W-Å: J. Steroid Biochem. 20, 1-4 (1984)
11. Borgan, J-L, Franque, S., Rochefert, H.: Biochemistry 23, 2163-2168 (1984)

12. Lippman,M.E., Halterman,R.H.,Leventhol,B.G., Perry,S., Thompson,E.B.: J. Clin. Inv. 52, 1715-1725 (1973)

13. Hollander,N., Chiw,Y-W.: Biochem.Biophys. Res. Commun. 25, 291-297 (1966)

14. Yamamoto,K.R., Gehring,U., Stampfer,M.R., Sibley,C.H.: Recent Prog. Horm. Res. 32, 3-32 (1976)

15. Chronsos,G.P., Vingerhoeds,A., Brandon,D.,Pugeot,M., DeVro ede,M., Loriaux,D.L., Lipsett,M.B.: J. Clin. Invest.69, 1261-1269 (1982)

16. Jensen,E.V., DeSombre,E.R., Jungblut,P.W.: Estrogeneous Fac tors influencing Host-Tumor Balance (ed. Wessler,P.W.,Dao, T.L., Wood Jr S.) University of Chicago Press, Chicago. pp 15-30 (1967)

17. McGuire,M .L., Carbone,P.P., Sears,M.E., Esher,G.C.: Estrogen Receptors in Human Breast Cancer (ed. McGuire,M.L., Carbone,P .P.) Raven Press New York pp 1-7 (1975)

18. DeSombre,E.R., Greene,G.L., Jensen,E.V.: Progress in Cancer Reseach and Therapy vol. 10 (ed. McGuire,M.L.) Raven Press, New York pp1-14 (1978)

19. Matsumoto,K., Ochi,H., Nomura, Y., Takatani, O., Izuo,M., Okamoto,R., Sugano,H.: Progress in Cancer Research and Therapy vol.10 (ed. McGuire,M.L.)Raven Press pp 43-58 (1978)

20. Keenan,B.S., Meyer,W.J.III., Hadjan,A.J., Jones,H.W.: J. Clin. Endocrinol. Metab. 38, 1143-1146 (1974)

21. Wieland, S.J., Fox, T.O.: Cell. 17, 781-787 (1979)

22. Amehein, T.A., Meyer, III.W.J., Jones, Jr H.W., Migeon: Proc. Natl. Acad. Sci. USA 73, 891-894 (1976)

23. Daufman, M., Straisfeld, C., Pinkey, L.: J. Clin. Invest. 58, 345-350 (1976)

24. Griffin, J.E., Duirant: J. Clin. Endociinol. Metab. 55, 465-474 (1982)

25. Stevens, J., Stenens, Y-W., Haubenztock: Biochemical actions of Hormones (ed. Litwack G.) Academic Press, New York p.383-446 (1983)

26. Schmidt, T.J., Harmon, J.M., Thompson, E.B.: Nature 286, 507-510 (1980)

27. Baskevitch, P.P., Vignon, F., Bousquet, C., Rochefort, H.: Cancer Res. 43, 2290-2297 (1983)

28. Collier, M.E., Griffin, J.E., Wilson, J.P.: Endocrinology 103, 1499-1505 (1978)

29. Gehring, U., Thompson, E.B.: Glucocorticoid Hormone Action (Baxter, J.D. and Rousseau, G.G.) Springer Verlag. Berlin Heiderberg New York, p.399-421 (1979)

30. Munk, A., Brink-Johnsen, T.: J.Biol Chem. 243, 5556-5565 (1968)

31. Granberg, J.P., Ballord, P.L.: Endocrinology 100, 1160-1168 (1977)

32. Rees, A.M., Bell, P.A.: Biochim. Biochys. Acta 411: 121-132 (1975)

33. McBlain, W.A., Shyamala, G.: J. Biol. Chem. 255, 3884-3891 (1980)

34. Bruchosky, N., Wilson, J.D.: J. Biol. Chem. 243, 5953-5960 (1968)

35. Cake, M.H., Goidl, J.A., Parchman, L.G., Litwack, G.: Biochem. Biophys. Res. Com. 71, 45-52 (1976)

36. Sato, B., Maeda, Y., Noma, K., Matsumoto, K., Yamamura, Y.: Endocrinol. 108, 612-619 (1981)

37. Sato, B., Nishizawa, Y., Noma, K., Matsumoto, K., Yamamura, Y.: Endocrinol. 104, 1474-1479 (1979)

38. Sato, B., Noma, K., Nishizawa, Y., Nakao, K., Matsumoto, K Yamamura, Y.: Endocrinol. 106, 1142-1148 (1980)

39. Sato, B., Nishizawa, Y., Maeda, Y., Noma, K., Honma, T., Matsumoto, K.: J. Steroid Biochem. 19, 315-321 (1983)

40. Nishizawa, Y., Maeda, Y., Noma, K., Sato, B., Matsumoto, K., Yamamura, Y.: Endocrinol. 109, 1463-1472 (1981)

41. Maeda, K., Tsuzimura, T., Nomura, Y., Sato, B., Matsumoto, K.: Cancer Res. 44, 996-1001 (1984)

42. Sato, B., Huseby, R.A., Samuels, L.: Cancer Res. 38, 2842-2847 (1978)

43. Zava, D.T., McGuire, W.L.: J. Biol. Chem. 252, 13703-13708 (1977)

44. Edwards, D.P., Martin, P.M., Horwitz, K.B., Chamness, G.C. McGuire: Exp. Cell. Res. 127, 197-213 (1980)

45. Geier, A., Haimsohn, M., Malik, Z., Lunenfeld: Biochem. J. 200, 515-520 (1981)'

46. King, W.J., Green, G.L.: Nature 307, 745-746 (1984)

47. Maeda, Y., Sato, B., Noma, K., Kishimoto, S., Koizumi, K., Aono, T., Matsumoto, K.: Cancer Res. 43, 4091-4097 (1983)

48. Sato, B., Maeda, Y., Nakao, M., Noma, K., Kishimoto, S., Matsumoto, K.: Eur. J. of Cancer & Clin. Oncol. in press

49. Grippo, J.F., Trienrungroj, W., Dahmer, M.K., Houseley, P.R. Pratt, W.B.: J. Biol. Chem. 258, 13658-13664 (1983)

50. Nielsen, C.J., Sando, J.J., Vogel, M.W., Pratt, W.B.: J. Biol. Chem. 252, 7568-7578 (1977)

51. Grandics, P., Miller, A., Schmitt, T.J., Mittman, D., Litwack, G.: J. Biol. Chem. 259, 3173-3180 (1984)

52. Sando, J.J., LaForest, A.C., Pratt, W.B.: J. Biol. Chem. 254, 4772-4778 (1979)

53. Jensen, E.V., Hurst, D.J., DeSombre, E.R., Jungblut, P.W., Science 158, 385-387 (1967)

54. Leach, K.L., Dahmer, M.K., Hammond, N.D., Sando, J.J., Pratt, W.B.: J. Biol. Chem. 254, 11884-11890 (1979)

55. Noma, K., Nakao, K., Sato, B., Nishizama, Y., Matsumoto, k. Yamamura, Y.: Endocrinol. 107, 1205-1211 (1980)

56. Chamness, G.C., Jennings, A.W., McGuire, W.L.: Biochemistry 13, 327-331 (1974)

57. Liao, S., Smythe, S., Tymoczko, J.L., Rossini, C.P., Chen, C., Hiipakka, R.A.: J. Biol. Chem. 255: 5545-5551 (1980)

58. Puri, R.K., Grandics, P., Dougherty, J.J., Toft, D.O.: J. Biol. Chem. 257, 10831-10837 (1982)

59. Rousseau, G.G., Baxter, J.D., Tomkins, G.M.: J. Mol. Biol. 67, 99-115 (1972)

60. Veelen, H., Willemse, P.H.W., Sleijfer, D.T., Pratt, J.J., Slaiter, W.J., Doorenbos, H.: Cancer Chemother. Pharmacol. 12, 83-86 (1984)

61. Joab, I., Radanyi, C., Renoir, M., Buchou, T., Catelli, M.G., Binart, N., Mester, J., Baulieu, E-T: Nature 308, 850-853 (1984)

CELLULAR ANTAGONISTS OF GLUCOCORTICOID RECEPTOR BINDING

Mohammed Kalimi, Tanvir A. Shirwany, Dix P. Poppas

Department of Physiology and Biophysics
Medical College of Virginia
Virginia Commonwealth University
Richmond, Virginia 23298

John R. Hubbard

Dept. of Orthopedic Surgery (Biochemistry)
Brigham and Women's Hospital
75 Francis Street,
Boston, Mass. 02115

Introduction

Glucocorticoid hormones have many important physiological, metabolic and pharmacological roles. For example, glucocorticoids are known to (a) increase resistance to stress, (b) enhance gluconeogenesis, (c) relieve many symptoms of arthritis, and (d) depress inflammation (1-3). Regulation of glucocorticoid action by natural and synthetic antagonists could therefore be very important for both clinical and scientific reasons.

Glucocorticoids appear to exert many of their effects via a receptor mechanism. After diffusing through the cell membrane of a target cell, glucocorticoids are believed to bind to specific, high affinity cytoslic receptor proteins. The glucocorticoid-receptor complex, thus formed then undergoes a "transformation" or "activation" such that it can enter the cell nucleus and enhance the expression of specific genes. Thus, factors which hinder cytosolic glucocorticoid receptor binding are potential antagonists of glucocorticoid action.

Adrenal Steroid Antagonism

Many known glucocorticoid antagonists such as dexamethasone 21-mesylate (4), $\Delta^{1,9(11)}$-11-deoxycortisol (5), medroxyprogesterone (6), clotrimazole (7), 5'-deoxypyridoxal (8), and 17-α-methyltestosterone (9) are synthetic derivatives which are not normally present in animal tissue. However, there are also many factors which antagonize glucocorticoid receptor binding _in vitro_ and can be found in living cells.

In this chapter we will review observations (both past and current) on naturally occuring antagonists to cytosolic glucocorticoid receptor binding. Our knowledge of cellular glucocorticoid receptor binding antagonists is for the most part preliminary. Most studies have utilized _in vitro_ systems in which natural substances are exogenously added to cytosolic receptors. While these investigations are quite valuable, many future experiments will be needed to determine the mechanism of their action and the actual influence of these substances on glucocorticoid receptor binding _in vivo_. However, because of the potential influence of these factors on the regulation of glucocorticoid hormone action and in understanding the biochemical nature of steriod receptor interaction, past and future investigation in this area may prove very important.

Macromolecular cytosolic binding inhibitors. Recently, Fishman (10) reported that dextran-coated charcoal (DCC) treatment (2 h at 4°C) of uterine cytosol removed or inhibited components(s) which normally promote estradiol receptor inactivation. We therefore initially attempted to use DCC-treatment of rat liver cytosol in order to enhance and protect glucocorticoid receptor binding. Interestingly however, DCC-treatment of rat cytosol had quite different results on the liver glucocorticoid receptor as shown in Figure 1. Thus, treatment of cytosol with DCC for 30-40 min at 4° C resulted in approximately 50% reduction in subsequent

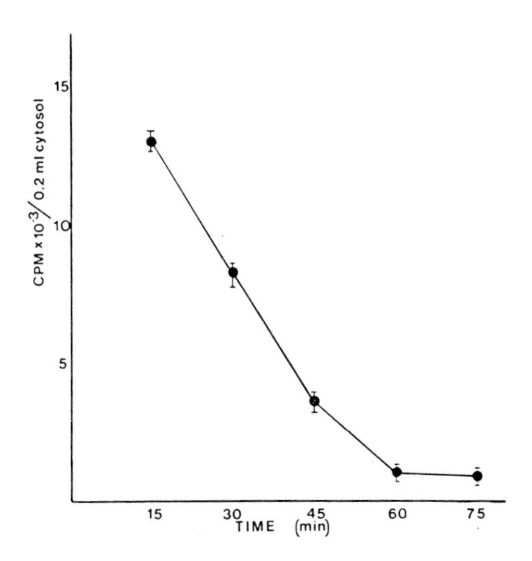

Fig. 1 Effects of DCC treatment on Glucocorticoid Receptor Inactivation - Cytosol was incubated at 4°C in the presence of DCC. 1 ml of DCC suspension (3.75 g of activated charcoal and 0.375 g of dextran T-500 in 100 ml of 10 mM Tris-HCl buffer, pH 7.5) was centrifuged for 5 min at 3,000 xg and the supernatant was discarded. To each DCC pellet, 1 ml of cytosol was added. Samples were mixed and allowed to stand at 4° C for 60 min (2 h for more concentrated cytosol) and centrifuged at 3,000 x g for 10 min. The supernatants were carefully removed, assayed for specific binding. Total protein remained about 10 mg per ml of cytosol after DCC treatment as determined by Lowry et al. (11). Control experiments were conducted in which homogenization buffer was treated with DCC (as described above). Specific binding was determined by incubation of 200 μl samples (in triplicate) with 3 x 10^{-8} M [^{3}H]dexamethasone in the presence and absence of 500-1000 fold excess cold steroid. After binding for 4 h at 4° C, the specific macromolecular bound fraction of ^{3}H-labeled steroid was determined using the charcoal-dextran separation technique (12). Nonspecific binding was generally less than 5% of the total binding. In some experiments (data not shown) Sephadex column chromatography was used to separate bound and free [^{3}H]dexamethasone. Essentially the same results were obtained using chromatographic assays as with the DCC procedure, showing that the DCC separation technique was not producing artifactual results in the DCC-treated (1-2h at 4°C) samples. The DCC binding assay was therefore used in other experiments described. Results are the mean ± SE of 5 experiments.

receptor binding to [^{3}H]dexamethasone. Essentially complete loss of glucocorticoid receptor binding ability occurred after 60 min incubation with DCC at 4°c. DCC-treated control buffer had no apparent inhibitory effect on glucocorticoid receptor binding. Recently, similar effects with DCC on glucocorticoid receptor binding were reported by Grippo et al. (13). While Fishman's observations indicated that DCC treatment of uterine cytosol removed or inhibited component(s) which normally promote estradiol receptor inactivation and reduce binding levels, DCC treatment of hepatic cytosol appeared to remove or inactivate factor(s) required for glucocorticoid receptor binding. These observations therefore suggest that the cytosolic environment of the uterine estradiol receptor and hepatic glucocorticoid receptor are functionally quite different, or that the receptors themselves are chemically very different (at least with regards to DCC effects).

Previous studies have shown that glucocorticoid receptor binding is reduced by gel filtration (14) or Amicon filtration (15). At present, it is unclear if these observations are related to the present DCC results. Certainly, however, unlike these other methods of inactivation, DCC treatment (i) totally inactivates glucocorticoid receptor binding, (ii) is simpler and possibly more rapid than the other procedures, and (iii) results in little or no dilution of the sample. DCC treatment of cytosol may therefore prove to be a particularly useful method for eliminating glucocorticoid receptor binding in cytosol for many other studies.

In order to test the possibility that a factor(s) in normal cytosol might restore receptor binding in DCC-treated cytosol, experiments were performed in which untreated cytosol and DCC-treated cytosol were mixed, and subsequent receptor binding determined. Surprisingly, addition of increasing

TABLE I

Effect of DCC-Treated Cytosol on Untreated Cytosolic Glucocorticoid Receptor Binding

Experimental Group	Specific [^{3}H]dexamethasone binding (CPM)
Untreated cytosol (100 μl)+buffer (100 μl)	12,450 ± 690
DCC-treated cytosol (100 μl)+buffer (100 μl)	265 ± 55
Untreated cytosol (100 l)+DCC-treated cytosol	
" " " (25 μl)	11,105 ± 710
" " " (50 μl)	9,680 ± 680
" " " (75 μl)	8,165 ± 485
" " " (100 μl)	5,645 ± 325

Untreated cytosol (100 μl) was mixed with the indicated amounts of DCC-treated cytosol and the final volume was adjusted to 200 μl with either 10 mg BSA/ml in homogenization buffer or with homogenization buffer. Specific [^{3}H]dexamethasone binding was determined after 4 h incubation at 4°C as described in Figure 1. Results are the mean ± SE of 5 experiments, each done in triplicate.

amounts of DCC-treated (25 to 100 μl) cytosol to a fixed amount of untreated cytosol (100 μl) resulted in inhibition of receptor binding in a dose-dependent manner (Table I).

Maximum binding inhibition (60-70%) was obtained at a ratio of 1:1. Similar results were obtained when control samples were adjusted with either homogenization buffer or a BSA solution (to keep the volume and total protein content constant to the experimental group). Thus, changes in protein content did not appear to influence receptor binding results. DCC-treated buffer did not inhibit receptor binding in untreated cytosol (data not shown), showing that the inhibitor appears to originate from cytosol (not from DCC).

TABLE II

Effect of DCC-Treated Cytosol on Prebound Cytosolic [^{3}H]Dexamethasone-Receptor

Experimental Group	Specific [^{3}H]dexamethasone binding (CPM)
Untreated cytosol (100 μl)+buffer (100 μl)	12,680 ± 615
DCC-treated cytosol (100 μl)+buffer (100 μl)	345 ± 60
Untreated cytosol (100 l)+DCC-treated cytosol	
" " " (25 μl)	10,605 ± 480
" " " (50 μl)	8,600 ± 385
" " " (75 μl)	7,460 ± 485
" " " (100 μl)	6,285 ± 340

Cytosol was labled with 3 x 10^{-8}M [^{3}H]dexamethasone for 3h at 4°C. 100 l cytosol was mixed with indicated amounts of DCC-treated cytosol and the final volume was adjusted to 200 μl with either 10 mg BSA/ml in homogenization buffer or with homogenization buffer. After incubation for 4h at 4°C, 100 μl DCC was added, incubated for 10 min at 4°C, centrifuged for 10 min at 3000 xg and supernatant (200 μl) counted. The results are the mean ± SE of five determinations, each in triplicate.

Next, we examined the effect of DCC-treated cytosol on steroid prebound receptors (Table II). For these studies a fixed amount (100 μl) of cytosol, prelabeled with [^{3}H]dexamethasone, was mixed with increasing amounts of DCC-treated cytosol (25-100 μl).

The addition of DCC-treated cytosol resulted in dissociation of steroid from the prebound steroid-receptor complexes. 40-50% dissociation was obtained at a ratio of 1:1 untreated to DCC-treated cytosol (Table II).

It may be concluded from the studies that DCC removes some factor(s) from the cytosol resulting in (i) loss of

TABLE III

Effect of Dialysis on Glucocorticoid Receptor Binding Inhibitor

Experimental Group	Specific [^{3}H]dexamethasone Binding (CPM)
Control (untreated) cytosol	13,285 ± 685
DCC-treated cytosol	325 ± 70
Control cytosol + DCC-treated cytosol	6,345 ± 355
Control cytosol+dialysed DCC-treated cytosol	6,650 ± 425

For dialysis experiments - DCC-treated cytosol was dialyzed for 4 h at 4°C against 10 mM Tris-HCl buffer containing 0.25 M sucrose, pH 7.5 (1 liter of buffer was used per ml of cytosol). The molecular cut off of dialysis tubing was 12,000- 14,000. The contents from the dialysing bag were carefully removed and added to untreated cytosol. In each case 100 µl of untreated cytosol was mixed with 100 µl of DCC-treated cytosol (either dialyzed, or not) and specific [^{3}H]dexamethasone binding was determined after incubation at 4°C as described in Figure 1. Results are the mean ± SE of 4 experiments, each done in triplicate.

glucocorticoid receptor binding and (ii) expression of some glucocorticoid binding inhibitor(s) in the DCC-treated cytosol. This inhibitor, in turn, acts on both unbound and steroid prebound receptors in the untreated cytosol. In order to test this hypothesis we tried to characterize the steroid binding inhibitor(s) in the DCC-treated cytosol.

Results presented in Table III show that dialysis of the DCC-treated cytosol was unable to remove the receptor binding inhibitor.

We next investigated the heat stability of the receptor binding inhibitor as shown in Table IV. Heating DCC-treated cytosol for 10 min at 50° or 60° C greatly depressed the action of the inhibitor of receptor binding indicating that the factor is thermal labile.

TABLE IV

Stability of Glucocorticoid Binding Inhibitor at various Temperatures

Experimental Group	Specific [^{3}H]dexamethasone Binding (CPM)
Control cytosol	12,235 ± 945
DCC-treated cytosol	195 ± 55
Control cytosol + DCC-treated cytosol	5,085 ± 265
Control cytosol + DCC-treated cytosol heated at 37°C for 10 min	5,895 ± 305
Control cytosol + DCC-treated cytosol heated at 45°C for 10 min	8,620 ± 370
Control cytosol + DCC-treated cytosol heated at 50°C for 10 min	10,845 ± 515
Control cytosol + DCC-treated cytosol heated at 60°C for 10 min	10,885 ± 465

DCC-treated cytosol was incubated for 10 min at temperatures shown. Cytosols were then mixed as described in Table III. Specific binding was determined as described in Figure 1. Results are the mean ± SE of 3 experiments, each done in triplicate.

In order to further investigate this factor, we incubated DCC-treated cytosol with various enzymes such as trypsin, pronase, chymotrypsin and papain (Table V). There was no significant effect of these proteolytic enzymes on the activity of the binding inhibitor. However, DNase treatment slightly (40%) and RNase treatment greatly (80%) reduced the effect of the inhibitor. In addition, 5 mM dithiothreitol completely blocked the activity of the inhibitor (Table V). (5 mM dithiothreitol had no effect on receptor binding in untreated cytosol.) These results suggest that reducing equivalents are lost during DCC treatment of cytosol, and that removal of reducing equivalents results in expression of some glucocorticoid receptor binding inhibitor(s) (possibly possessing a nucleic acid portion) in the remaining cytosol.

TABLE V

Effect of Enzymatic Treatment and Dithiothreitol on the Glucocorticoid Receptor Binding Inhibitor

Experimental Group	Specific [^{3}H]dexamethasone Binding (CPM)
Control cytosol	12,845 ± 585
DCC-treated cytosol	285 ± 50
Control cytosol + DCC treated cytosol	5,890 ± 285
Control cytosol + trypsin	5,605 ± 180
Control cytosol + papain	5,315 ± 175
Control cytosol + pronase	6,280 ± 310
Control cytosol + chymsotrypsin	6,525 ± 280
Control cytosol + ribonuclease	10,805 ± 605
Control cytosol + deoxyribonuclease	9,180 ± 730
Control cytosol + 2 mM dithiothreital	10,840 ± 515
Control cytosol + 3 mM dithiothreital	13,015 ± 605
Control cytosol + 5 mM dithiothreital	12,980 ± 815

1 ml of DCC-treated cytosol was incubated for 1 h at 0-4°C with 0.1 ml of the appropriate enzyme solution (1 mg per ml) or with buffer. After incubation, the samples were centrifuged for 15 min at 3,000 xg and supernatants were used for the experiments. Cytosols were mixed as described in Table III and specific binding determined as described in Fig. 1. Results are the mean ± SE of 3 experiments, each done in triplicate.

We then characterized this endogenous inhibitor by passing DCC-treated cytosol on Sephadex G-100 columns. The fraction which eluted between 40-50 Å was able to inhibit receptor binding, indicating that the inhibitor is macromolecular in nature having a molecular size of 4-5 nm (data not shown).

The biochemical characteristics of this cytosolic factor(s) (and reducing equivalents) are still under investigation. The role of this inhibitor in regulation of receptor binding *in vivo* is unknown but may be of considerable future interest.

Influence of metal ions - Metal ions, such as Ca^{2+}, are ubiquitous in animal tissues and have been shown to have numerous roles in metabolism and cell physiology. Since intracellular Ca^{2+} levels fluctuate greatly and are closely regulated, Ca^{2+} serves as a useful cell regulator in many systems.

Many *in vitro* studies have shown that metal ions greatly influence glucocorticoid receptor properties. For example, Ca^{2+} (2-5 mM) has been shown to significantly promote glucocorticoid receptor inactivation in thymus (16), kidney (17), liver (19), and lymphoid tissues (18).

Thus while control unbound thymus receptors had a half-life of approximately 90 min at 25°C, in the presence of 2 mM $CaCl_2$ receptor binding decreased 50% in only 30 min (16). Likewise, hepatic unbound receptors were shown to have a half-life of about 60 min at 25°c, while in the presence of 5 mM Ca^{2+} over 90% inactivation was observed in 30 min and in the presence of 10 mM $MgCl_2$ 80% inactivation was found after 60 min incubation (23). Ca^{2+} also enhanced the inactivation rate of unbound hepatic receptors at 4°C (23). Interestingly, Kalimi et al. (23) showed that Ca^{2+}-promoted glucocorticoid receptor inactivation was greatly inhibited by addition of molybdate. This suggested that the mechanisms of Ca^{2+} inactivation and molybdate stabilization may be related. Kalimi et al. (23) also showed that receptor binding after Ca^{2+} treatment could not be reactivated by molybdate plus dithiothretol. Ca^{2+} inactivation therefore may involve an irreversible process.

The observation discussed above using exogenous metal ions is supported by the stabilizing effect of metal chelators, such as EDTA and EGTA, on cytosolic glucocorticoid receptors (19-22). For example, Hubbard and Kalimi (19) recently showed that 10 mM EDTA significantly stabilized hepatic glucocorticoid receptors at 4°C, while 10 mM EGTA was a more powerful stabilizer at 25°C. This observation suggests that different endogenous metal ions may influence glucocorticoid receptor inactivation at different temperatures. At high temperature, the stabilizing effect of EGTA suggests a possible role of Ca^{2+}. Likewise, the greater effect of EDTA at low temperature is consistant with the possibility that other ions, such as Mg^{2+}, may be involved in low temperature glucocorticoid receptor inactivation. Thus studies with chelators indicate that at least two separate mechanisms of inactivation (one at low and one at high temperature) may exist and that both mechanisms seem to involve endogenous metal ions.

Rousseau et al. (24) has reported that submicromolar levels of free calcium decreases glucocorticoid receptor affinity for dexamethasone. This decreased affinity was due to a depressed rate of association, with little or no apparant alteration in dissociation rate or receptor concentrations. These results are particularly interesting since half-maximal effect was obtained with 0-3 M free Ca^{2+}. This concentration of Ca^{2+} is well within intracellular Ca^{2+} levels and again supports a role of Ca^{2+} in glucocorticoid receptor function.

Ca^{2+} could also influence glucocorticoid receptor function by physicochemical alteration of the receptor. For example, Aranyi and Naray (16) reported that 4 mM Ca^{2+} reduced receptor size as measured by Sephacryl S-200 chromatography. Kalimi et al. (23,25) showed that large amounts of Ca^{2+} (10-50 mM) caused rat liver receptor aggregation as assessed by agarose chromatography and sucrose-density gradients. While these reports appear conflicting, both observations indicate that Ca^{2+} could influence glucocorticoid receptor action by alteration of its physical properties.

The antagonizing effect of Ca^{2+} on glucocorticoid receptor binding could also be due to activation of specific enzymes such as proteinases. Thus Aranyi and Naray (16) showed that the tripeptidealdehyde BOC-D-Phe-Pro-Ary-aldehyde (known to inhibit trypsin and thrombin) depressed Ca^{2+} effects on the thymus receptor. Kalimi et al. (23) found that the cysteine proteinase inhibitor leupeptin and the rather non-specific proteinase inhibitor α_2-macroglobulin had little or no effect on Ca^{2+}-dependent glucocorticoid receptor inactivation in hepatic cytosol. In addition, trasylol (an inhibitor of trypsin-like proteinases) had no influence on Ca^{2+} alteration of thymus receptors. Further study of the mechanism of Ca^{2+} (and other metal ions) antagonism of glucocorticoid receptor binding may prove very important. Studies using purified receptors and reconstituted systems seem essential to explore Ca^{2+} action in more detail.

Heparin - Heparin is a sulphated mucopolysaccharide which is found in many animal tissues (in addition to being a widely used pharmaceutical) (33,34). Recently, Hubbard and Kalimi (35) showed that heparin significantly increased the rate of unbound rat hepatic receptor inactivation at 4°C and 25°C. Thus while control hepatic receptor binding decreased approximately 25% in 6h at 4°C, heparin (40 μg per ml cytosol) treated glucocorticoid receptor binding decreased about 75%. Likewise, at 25°C control receptors had a half-life of about 60 min, while heparin-treated receptor binding had a half-life of only 15 min. In addition, heparin was shown to have a dramatic effect on the physicochemical nature of the steroid-receptor complex. Thus, heparin reduced the sedimentation coefficients of prebound glucocorticoid-receptor complexes from 7-8 S to about 3-4 S (35). These results indicate that heparin could hinder glucocorticoid receptor function. Interestingly, 10 mM molybdate was able to significantly block heparin action on the receptor at 4° and 25° C (28).

This observation suggests that the mechanisms of heparin and molybdate action may be related. For example, they may compete for the same binding site on the receptor. Thus elucidation of molybdate action may shed light on heparin effects, and vice versa.

It is possible that heparin could act by regulation of a receptor-modulating enzyme. For example, heparin has been shown to alter the activity of proteases (36), protein kinases (37), and phospholipid lipases (38). The possibility of direct interaction of heparin with the receptor is supported by its high affinity to other steroid hormone receptors such as the estrogen receptor (39). Since heparin is made in living tissue and is also given as a drug, heparin effects on glucocorticoid receptors _in vivo_ is currently being investigated in our laboratory. The influence of other naturally occurring mucopolysaccharides on glucocorticoid receptor properties may also be of significant interest.

Enzymatic inactivation - It has become increasingly apparent that cellular enzymes could act as glucocorticoid antagonists by inactivation (loss of receptor binding) or degradation (reduction of receptor size by proteolysis, subunit dissociation, etc.) of the glucocorticoid receptor.

The glucocorticoid receptor is a protein, and therefore is expected to be degraded, like other cell proteins, by intracellular proteinases. It is possible that these proteinases could also serve a control function on receptor binding _in vitro_ and _in vivo_. Studies using proteinase inhibitors have therefore been conducted to identify and inhibit potential receptor modulating proteinases. Many proteinase inhibitors appear to be ineffective in stabilizing glucocorticoid receptor preparations. For example, the proteinase inhibitors D-tryptophan methyl ester (40), chymotrypsin inhibitor

(40), trypsin inhibitor (40), sodium tetrathionate (40), benzamidine (40), phenylmethane sulfonylfluoride (40-42), diisopropylfluorophosphate (40), TPCK (40-42), TLCK (43), $NaHSO_3$ (43) and E-aminocaproyl-p chlorobenzylamide (42) all appear to have little or no influence on receptor lability *in vitro*.

However, several other studies indicate that cellular proteinases could play a role in glucocorticoid receptor inactivation and degradation. For example, Naray (44) showed that Ca^{2+} enhances lymphoid glucocorticoid receptor inactivation at 25°C and that this Ca^{2+}-dependent process was significantly blocked by the proteinase inhibitor carbobenzoxy-D-Phe-Pro-Arg-aldehyde. These results are therefore consistent with the possibility that Ca^{2+} enhances the activity of cytosolic proteinases which inactivate glucocorticoid receptor binding. Proteinase inhibitors have also been shown to block cytosolic glucocorticoid receptor degradation *in vitro*. For example, Sherman et al. (45,46) showed that rat kidney and mouse mammary tumor glucocorticoid receptors were degraded to small (Rs = 23 Å) "meroreceptor" forms and that this process could be inhibited by the cysteine proteinase inhibitors antipain and leupeptin. In addition, Hazato and Murayama (47) reported that two separate enzymes degrade rat liver glucocorticoid receptors. One cellular enzyme was inhibited by phosphoramidon (which inhibits thermolysin-like proteinases) and the other was significantly blocked by the cysteine proteinase inhibitors antipain or leupeptin.

Finally, it is very interesting to note that Sherman et al. (48) recently found that molybdate, one of the most widely used glucocorticoid receptor stabilizing agents (49-51), has the capacity to act as an inhibitor of cytosolic tyrosine-specific and lysine-specific proteinases. The possibility that molybdate stabilizing effects are due to its ability to inhibit proteolytic enzymes is still largely unexplored but may, with further study, provide new understanding of glucocorticoid receptor inactivation and degradation.

In addition to proteolytic enzymes, it is clear that other cellular enzymes could act as glucocorticoid antagonists by inactivation of the glucocorticoid receptor. Of particular interest has been the role of phosphatases in glucocorticoid receptor lability. For example, in 1977 Nielsen et al. (52) showed that highly purified alkaline phosphatase inactivated both mouse fibroblast and rat liver unbound glucocorticoid receptors. Similar results with alkaline phosphatase were obtained by Leach et al. (53) using mouse L cells. In addition, Housley et al. (54) showed that both alkaline phosphatase and phosphoprotein phosphatase inactivated rat liver cytosolic glucocorticoid receptors. In the study by Housley et al. (54) cytosol was initially filtered (removing components of Mr less than about 10,000) to eliminate phosphorylated components and ions (like Mn^{+2}) which could interfere with phosphatase activity. Samples were then incubated with highly purified calf intestine alkaline phosphatase in the presence of 10 mM molybdate at 20°C. While no enhancement of inactivation rate was noticeable with 2 μg/ml of enzyme, a significant increase in inactivation rate was apparent at 5 g/ml alkaline phosphatase. Thus while control binding was stable for 4h at 20°C, in the presence of 20μg/ml of alkaline phosphatase receptor binding was essentially completely lost within 2h (54).

Together these results suggest that phosphorylase enzymes could inactivate glucocorticoid receptors and therefore hinder glucocorticoid hormone action. Further investigation by Housley et al. (54) indicated that dephosphorylation may cause oxidation of a chemical group (such as -SH) on the glucocorticoid receptor which leads to the loss of receptor binding. The possibility that the glucocorticoid receptor itself is a phosphoprotein (and therefore conceivably regulated by dephosphorylation) was supported by Housley and Pratt (55) who labeled cultured L-cell proteins with [^{32}P]orthophosphate for 18h and found ^{32}P labeling coincided with receptor binding protein.

Possibly related to phosphorylase results is the observation by Towle and Sye (56) that Na^+, K^+ - ATPase inactivates rat brain and liver unbound glucocorticoid receptors at 30°C.

Inactivation was significantly blocked by addition of Ouabain or omission of Mg^{+2}, Na^+, and K^+. These results suggest that cellular ATPases could act as an antagonist of glucocorticoid hormones by receptor inactivation. It is possible for example that ATP interaction with the receptor is necessary for hormone binding and that ATP dephosphorylation by ATPase (or other enzymes) could lead to loss of receptor binding ability.

Enzymatic studies clearly show that a number of enzymes such as proteinases and phosphatases have the ability to inactivate glucocorticoid receptors and therefore could act as cellular antagonists of glucocorticoid hormone action. Enzymatic inactivation of glucocorticoid receptors by other enzymes such as RNases (57) or phospholipases (58) is also possible. While these *in vitro* studies stimulate speculation of possible enzymatic regulation of glucocorticoid receptors *in vitro* further work is clearly needed to establish the role of these enzymes in living tissue.

In addition, studies with purified receptors and reconstituted systems are needed to determine if these enzymes (and other cellular modulators) act directly on the receptor or on an endogenous receptor modulating factor. For example, receptor inactivation could occur if a receptor stabilizing factor (59) is inactivated.

Conclusions and future perspectives - A number of investigations have shown that many naturally occuring substances hinder glucocorticoid receptor binding and/or alter the physicochemical nature of the receptor. Since hormone-receptor interaction is believed to be a vital step in hormone mechanisms, it is possible that these receptor antagonists act as glucocorticoid antagonists *in vivo*. For this reason continued study of these cellular receptor binding inhibitors and receptor modifying substances may prove very fruitful. In addition, even those substances which act *in vitro* but not *in vivo* may lend insight into the biochemical nature of the receptor, steroid-receptor interaction or receptor modulating factors. Clearly, in most instances studies with purified receptor(s) and studies using reconstituted *in vitro* systems will greatly enhance our understanding of the mechanism of these receptor modulating factors. In addition, *in vivo* models are needed where the concentrations of these substances can be controlled so that their effects on receptor function and glucocorticoid action *in vivo* can be examined.

Acknowledgements

M. Kalimi is a recepient of Research Career Development Award AM 00731.

References

1. Baxter, J.C., Rousseau, G.G.: Glucocorticoid Hormone Action, Springer-Verlag, New York (1979).

2. Kalimi, M., Hubbard, J.: Principles of Recepterology (Agarwal, M.K., ed) Walter de Gruyter, New York, 141-206 (1983).

3. Bethune, J.E.: The Adrenal Cortex, a Scope Monograph, The Upjohn Co., Michigan (1975).

4. Simons, Jr., S.S., Thompson, E.B.: Proc. Natl. Acad. Sci. USA 78, 8541-8545 (1981).

5. Chrousos, G.P., Barnes, K.M., Sauer, M.A., Loriaux, D.L., Cutler, Jr., G.B.: Endocrinology 107, 472-477 (1980).

6. Svec, F., Rudis, M.: J. Steroid Biochem. 16, 135-140 (1982).

7. Loose, D.S., Stover, E.P., Feldman, D.: J. Clin. Invent. 72, 404-408 (1983).

8. O'Brien, J.M., Cidlowski, J.A.: Biochemistry 21, 5644-5650 (1982).

9. Boucheix, C., Agarwal, M.K.: Antihormones, Agarwal, M.K. (ed) Elsevier/ North-Holland Biomedical Press, N.Y., 75-94 (1979).

10. Fishman, J.: Biochem. Biophys. Res. Commun. 110, 713-718 (1983).

11. Lowry, O., Rosenbrough, N., Farr, A. and Randall, R.: J. Biol. Chem. 193, 265-275 (1951).

12. Beato, M. and Feigelson, P.: J. Biol. Chem. 247, 7890-7896 (1972).

13. Grippo, J.F., Tienrungroj, W., Dahmer, M.K., Housley, P.R., Pratt, W.B.: J. Biol. Chem. 258, 13658-13664 (1983).

14. Cake, M., Goidl, J., Parchman, G. and Litwack, G.: Biochem. Biophys.(1976) Res. Commun. 71, 45-52.

15. Leach, K.L., Grippo, J.F., Housley, P.R., Dahmer, M.K., Salive, M..E. and Pratt, W.B.: J. Biol. Chem. 257, 381-388 (1982).

16. Aranyi, P., Naray, A.: J. Steroid Biochem. 12, 267-272 (1980).

17. Rafestin-Oblin, M.C., Michaud, A., Claire, M., Corrol, P.: J. Steroid Biochem. 8, 19 (1977).

18. Naray, A.: J. Steroid Biochem. 14, 71 (1981).

19. Hubbard, J., Kalimi, M.: Biochim. Biophys. Acta 755, 178-185 (1983).

20. Bell, P.A., Munck, A.: Biochem., J. 136, 97-107 (1973).

21. Hubbard, J.R., Kalimi, M.: J. Steroid Biochem. 19, 1163-1167 (1983).

22. Schaumberg, B.P.: Biochim. Biophys. Acta 214, 520-532 (1970).

23. Kalimi, M., Hubbard, J., Ray, A.: J. Steroid Biochem. 18, 665-672 (1984).

24. Rousseau, G.G., Baxter, J.D., Higgins, S.J., Tomkins, G.M.: J. Mol. Biol. 79, 539-554 (1973).

25. Kalimi, M., Colman, P., Feigelson, P.: J. Biol. Chem. 250, 1080-1086 (1975).

26. Munck, A., Brinck-Johnson, T.: J. Biol. Chem. 243, 5556-5565 (1968).

27. Ishii, D., Pratt, W.B., Aronow, L.: Biochemistry 11, 3896-3904 (1972).

28. Munck, A., Wira, C., Young, D.A., Mosher, K.M., Hallahan, C., Bell, P.A.: J. Steriod Biochem. 3, 567-578 (1972).

29. Wheeler, R.H., Leach, K.L., LaForest, A.C., O'Toole, T.E., Wagner, R., Pratt, W.B.: J. Biol. Chem. 256, 434-441 (1981).

30. Sando, J.J., LaForest, A.C., Pratt, W.B.: J. Biol. Chem. 254, 4772-4778 (1979).

31. Sando, J.J., Hammond, N.D., Stratford, C.A., Pratt, W.B.: J. Biol. Chem. 254, 4779-4789 (1979).

32. Barnett, C.A., Schmidt, T.J., Litwack, G.: Biochemistry 19, 3426-3435 (1980).

33. Kiss, J.: Heparin: Chemistry and Clinical Useage (Kakkar, V.V., Thomas, D.P., eds) Academic Press, New York, P. 3-20 (1978).

34. Nader, H.B., Straus, A.H., Takahashi, H.K., Dietrich, C.P.: Biochim. Biophys. Acta 714, 292-297 (1982).

35. Hubbard, J.R., Kalimi, M.: Biochim. Biophys. Acta 755, 363-368 (1983).

36. Gallus, A. Engel, G.: Heparin: The Society of Hospital Pharmacists of Australia, Victoia, P. 1-115 (1978).

37. Rose, K.M., Bell, L.E., Siefken, D.A. and Jacob, S.Y.: J. Biol. Chem. 256, 7468-7477 (1981).

38. Engelberg, H.: Monographs on Atherosclerosis 8, 1-69 (1978) .

39. Molinari, A.M., Medici, N., Moncharmont, B., Puca, G.A: Proc. Natl Acad. Sci. U.S.A. 74, 4886-4890 (1977).

40. Rafestin-Oblin, M.E., Michaud, A., Claire, M., Corvol, P.: J. Steriod Biochem. 8, 19-23 (1977).

41. Nielsen, C.J., Vogel, W.M., Pratt, W.B.: Cancer Research 37, 3420-3426 (1977).

42. Agarwal, M.K., Philippe, M.: Biochemical Medicine 26, 263-276 (1981).

43. Schmid, H., Grote, H., Sekeris. C. E.: Mol. Cell. Endo. 5, 223-241 (1976).

44. Naray, A.: J. Steroid Biochem. 14, 71-76 (1981).

45. Sherman, M.R., Pickering, L.A., Rollwagen, F.M., Miller, L.K.: Federation Proc. 37, 107-173 (1978).

46. Sherman, M.R., Barzilai, D. ,Pine, P., Tuazon, F. B.: Adv. Exp. Med. Bio. 117, 357-375 (1979).

47. Hazato, T., Murayama, A.: Biochem. Biophys. Res. Commun. 98, 488-493 (1981).

48. Sherman, M.R., Moran, M.C., Tuozon, F.B., Stevens, Y.W.: J. Biol. Chem. 258, 10366-10377 (1983).

49. Nielsen, C.J., Sando, J.J., Vogel, U.M., Pratt, W.B.: J. Biol. Chem. 252, 7568-7578 (1977).

50. Hubbard, J., Kalimi, M.: J. Biol. Chem. 257, 14263-14270 (1982).

51. Kalimi, M., Gupta,S., Hubbard, J., Greene, K.: Endocrinology 112, 341-347 (1983).

52. Nielsen, C.J., Sando, J.J,, Pratt, W.B.: Proc. Natl. Acad. Sci. USA 74,1598-1402 (1977).

53. Leach, K.L., Dahmer, M.K., Pratt, W.B.: J. Steroid Biochem. 18, 105-107 (1983).

54. Housley, P.R., Dahmer, M.K., Pratt, W.B.: J. Biol. Chem. 257, 8615-8618 (1982).

55. Housley, P.R., Pratt, W.B.: J. Biol. Chem. 258, 4630-4635 (1983).

56. Towle, A.C., Sye, P.Y.: Mol. Cell., Biochem. 52,145-151 (1983).

57. Rossini, G.P., Barbiroli, B.: Biochem. Biophys. Res. Commun. 113, 876-882 (1983).

58. Leach, K.L. Dahmer, M.K., Pratt, W.B.: J. Steroid Biochem. 18, 105-107 (1983).

59. Leach, K.L., Grippo, J.F., Housley, P.R., Dahmer, M.K., Salive, M.E., Pratt, W.B.: J. Biol. Chem. 257, 381-388 (1982).

GLUCOCORTICOID RECEPTOR DOMAINS AND ANTIGLUCOCORTICOIDS

Ulrich Gehring
Institut für Biologische Chemie der Universität Heidelberg
D-6900 Heidelberg

Introduction

Cell culture lines of mouse lymphomas have been used amongst other cell types to study the mechanism of glucocorticoid action. Several such cell lines respond to glucocorticoids by growth inhibition which is followed by lysis in some lines. This type of cellular response lends itself to the selection of unresponsive cell variants which then can be compared to the wild-type in an attempt to elucidate the mechanism of hormone action. In the case of S49.1 mouse lymphoma cells all the resistant variants that were characterized biochemically in some detail turned out to have defects of one type or another in the hormone-specific receptors (1-4).

Most abundant amongst the resistant sublines of S49.1 is the "receptorless" (r^-) phenotype which is characterized by negligible steroid binding activity. This phenotype could result from one of two kinds of defects: the receptor protein as such might be missing or the receptor molecule might have a defect in the steroid binding site. In addition, two types of variants with normal hormone binding have been found in which the interaction of the receptor-glucocorticoid complexes with cell nuclei, chromatin, or DNA is affected. Nuclear binding is decreased in the nt^- phenotype ("nuclear transfer deficient") while it is increased in the nt^i variant type ("increased nuclear transfer"). The isolation of these resistant variants with receptor defects in hormone and nuclear binding points to the qualitative importance of receptors for the physiological

Adrenal Steroid Antagonism

hormone response and to the significance of correct interaction of receptors with nuclear sites. The quantitative role of receptors for cellular responsiveness has been emphasized by the isolation of cell variants of decreased glucocorticoid sensitivity which have been shown to contain lower than normal receptor numbers (5). Receptor levels and hormone affinities may directly determine glucocorticoid sensitivity in mouse lymphoma cell lines (6).

The present paper summarizes experiments with wild-type and variant receptor-glucocorticoid and receptor-antiglucocorticoid complexes of S49.1 lymphoma cells which were carried out in order to obtain a better understanding of the function and structure of these receptors. Photoaffinity labeling of receptors with radiolabeled hormone ligands was used to analyze the receptor polypeptides under denaturing conditions in polyacrylamide gels. Chromatography of receptor-glucocorticoid and receptor-antiglucocorticoid complexes on DNA-cellulose as well as immunochemical analysis using monoclonal antibodies allowed to distinguish between some of the receptor types. The data are discussed in view of a domain model for glucocorticoid receptors.

Molecular Weights of Wild-type and Variant Receptors

In recent years affinity labeling of hormone receptors has become a valuable analytical method because it allows to determine the molecular weights of receptors in crude preparations (7). Since steroids containing α, β-unsaturated ketone structures can be excited by UV light to form a covalent bond with protein in the immediate molecular vicinity (8) we used the high-affinity glucocorticoid triamcinolone acetonide in tritium labeled form to investigate receptors of lymphoma cells (9 , 10). Gel electrophoresis in sodium dodecylsulfate followed by fluorography revealed a labeled polypeptide band of molecular weight 94 000 $\pm$ 5000 for wild-type and nt^- receptors of S49.1 cells (Figure 1). The same molecular weight was also

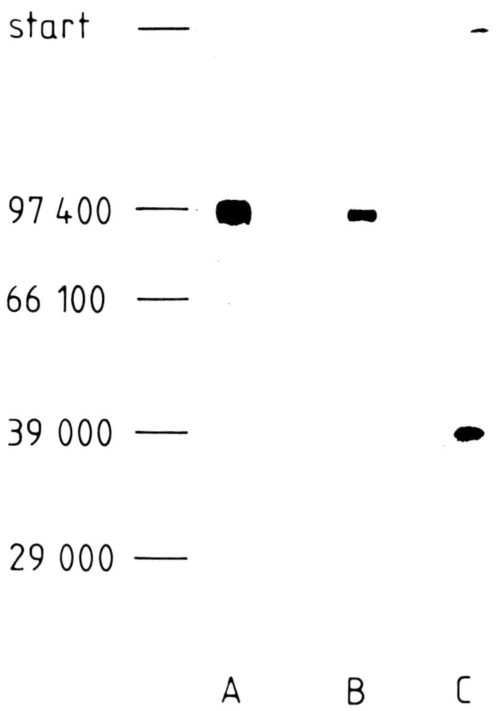

start —

97 400 —

66 100 —

39 000 —

29 000 —

A B C

FIG. 1: SDS gel electrophoresis of photoaffinity labeled receptors

Receptor complexes with [^{3}H]triamcinolone acetonide were subjected to photolabeling and subsequent gel electrophoresis. Fluorography was used to detect radiolabeled bands. (A) S49.1 wild-type; (B) S49.1 nt^{-}; (C) S49.1 nt^{i}. (Data modified from ref. 10).

found for wild-type receptors of other mouse lymphomas and human lymphoblastic leukemia cells (10). Variant receptors of the nt^{i} type, however, gave a major labeled band of molecular weigth 40 000 $\pm$ 2000 (Figure 1) and, in addition, in some experiments a minor band of about 37 000. The data are summarized in Table 1. It is noteworthy that the same steroid-labeled polypeptide molecular weight was obtained whether the S49.1

wild-type receptor was in the activated or in the non-activated form (9).

Table 1: Molecular weights of receptor types

Receptor type	Molecular weight of steroid-labeled receptor polypeptide		
	native	after chymotrypsin	after trypsin
wild-type	94 700	38 000	29 000 and 27 000
nt^- type (clone 22R)	94 000	37 400	29 000 and 27 300
nt^- type (clone 83R)	94 000	37 400	
nt^i type (clone 55R)	40 300	39 000	29 200 and 27 400
nt^i type (clone 143R)	40 800	41 000	

S49.1 receptor complexes with [^{3}H]triamcinolone acetonide were subjected to photolabeling and subsequent SDS gel electrophoresis. Treatment with chymotrypsin (10 μg/ml, 5 min) or trypsin (20 μg/ml, 30 min.) was after photolabeling. (Data from ref. 10).

Photoaffinity labeled wild-type and variant S49.1 receptors were subjected to mild treatment with various proteases in order to see whether receptor forms of the size of nt^i receptors could be produced (9, 10). Wild-type and nt^- receptors were indeed split by α-chymotrypsin to steroid-labeled polypeptides of molecular weight 38 000 while there was no change in the size of nt^i receptors (Table 1). Partial proteolysis with trypsin yielded steroid-labeled polypeptides with molecular weights of about 29 000 and 27 000 (Table 1). Under very mild conditions, however, trypsin cleaved the wild-type receptor to a 38 000 fragment that was readily further degraded to the smaller fragments (10). Fragments of about the same sizes were also generated by a lysine-specific protease but not by an arginine-specific enzyme (10).

Binding of Receptor-Glucocorticoid Complexes to DNA

The nt^- and nt^i variants of glucocorticoid resistant S49.1

cells were first detected because of abnormal receptor distributions in crude cell fractionation studies (1). Similar differences in nuclear binding were also seen in cell-free incubations of isolated nuclei with steroid-treated cytosols in which the origin of the cytosol rather than that of the nuclei determined the extent of nuclear binding (3). Decreased and increased nuclear binding in nt^- and nt^i variants, respectively, are reflected by decreased and increased binding to deproteinized DNA. This is particularly obvious when unfractionated DNA is adsorbed to cellulose and receptor complexes are chromatographed on this type of affinity matrix (1). As summarized in Table 2, wild-type receptor-triamcinolone acetonide complexes eluted with about 175 mM KCl while nt^- and nt^i receptor complexes required lower (70 to 90 mM) and higher (210 to 230 mM) salt concentrations, respectively. These differences in chromatographic behaviour reflect differences in the affinities to DNA of the variant receptor types (1, 11).

Table 2: DNA-cellulose chromatography of receptor-glucocorticoid complexes

Receptor type	KCl concentration required for elution (mM)	
	native	after chymotrypsin
wild-type	175	236
nt^- type (clone 22R)	75	129
nt^- type (clone 83R)	86	87
nt^i type (clone 55R)	225	229
nt^i type (clone 143R)	210	209

S49.1 receptor complexes with [^{3}H]triamcinolone acetonide were activated and chromatographed on DNA-cellulose (prepared from unfractionated calf thymus DNA) either prior to or after treatment with α-chymotrypsin. (Data from ref. 10).

Partial proteolysis with α-chymotrypsin of wild-type receptors of mouse lymphomas (11, 12) and rat liver (13) has previously been shown to produce receptor forms with abnormal DNA binding properties. This is shown in Figure 2 for wild-type S49.1 receptor-glucocorticoid complexes.

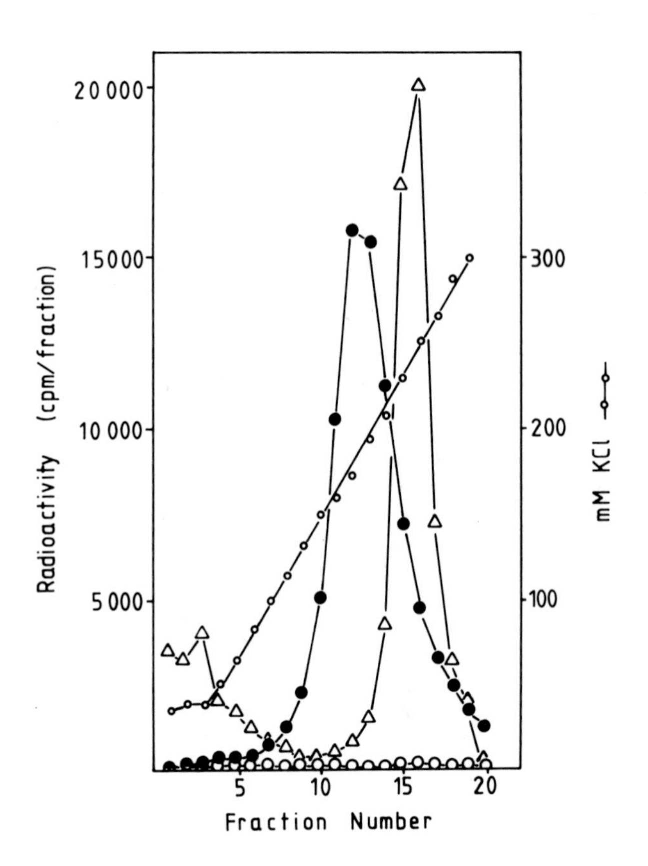

FIG. 2: DNA-cellulose chromatography of receptor-glucocorticoid complexes

S49.1 wild-type receptor complexes with [^{3}H]triamcinolone acetonide were activated at 20° and chromatographed on DNA-cellulose either in the native state (●) or following a treatment with α-chymotrypsin (Δ) or trypsin (o). (Data modified from ref. 9, 10).

The chymotrypsin treated receptors required about 230 mM KCl for elution from DNA-cellulose suggesting a DNA affinity similar to that of native nt^i receptors (Table 2). Despite the fact that nt^- receptors were cleaved by chymotrypsin (Table 1) they showed either no change or only a slight increase in the affinity for DNA, depending on the cell clone (Table 2).

Tryptic digestion of wild-type and variant receptor-glucocorticoid complexes produced receptor fragments mainly of molecular weight 29 000 which still carry the steroid label but did not bind to DNA-cellulose (Figure 2). This suggests that trypsin separates the domains for hormone binding and for nuclear interaction in wild-type and variant receptors.

Binding of Receptor-Antiglucocorticoid Complexes to DNA

The compound RU 38486 *) is a potent glucocorticoid antagonist (14). For more detailed information about this steroid the article by D.Philibert in this book should be consulted. In S49.1 wild-type lymphoma cells RU 38486 partially or completely overcomes the growth inhibitory and cytolytic effect of triamcinolone acetonide depending on the steroid concentrations used (data not shown). Triamcinolone acetonide and RU 38486 compete for the same binding sites in extracts of S49.1 cells. For investigating the interaction of receptor-antiglucocorticoid complexes with DNA we again used chromatography on DNA-cellulose. Similar to receptor-glucocorticoid complexes, the complexes with RU 38486 need to be activated in order to be able to bind to DNA; this is shown in Figure 3 for wild-type S49.1 complexes. Activation may be brought about by warming complexes to 20° or by exposure to high ionic strength as in the experiment of Figure 3. As summarized in Table 3, wild-type receptor-RU 38486 complexes eluted from DNA-cellulose with about 150 mM KCl while nt^- and nt^i receptor complexes required lower (73 mM) and higher (200 mM) salt concentrations, respectively.

*) 17ß-hydroxy-11ß-(4-dimethylaminophenyl)-17α-(1-propinyl)-estra-4,9-dien-3-one.

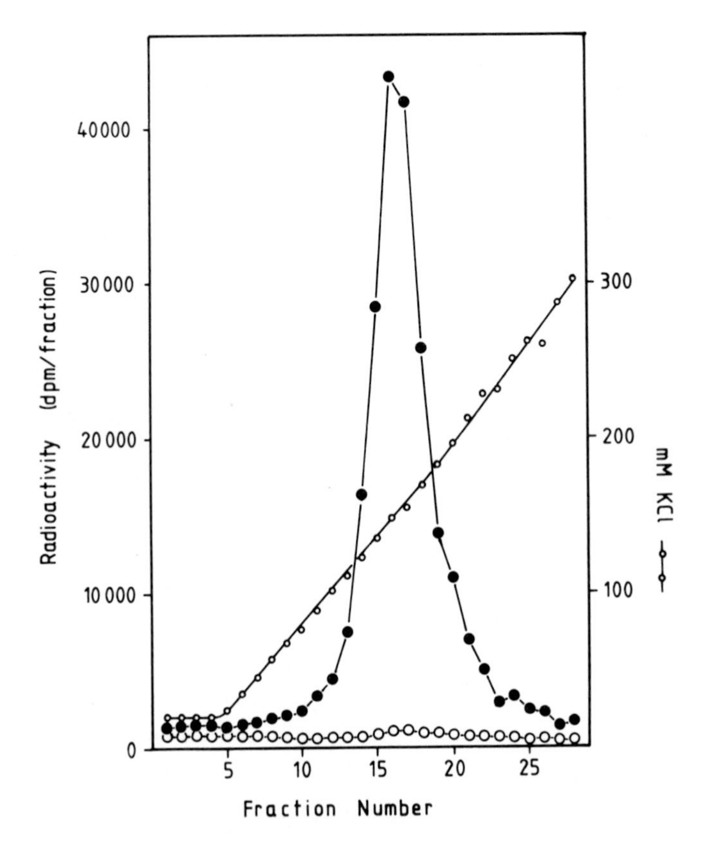

FIG. 3: DNA-cellulose chromatography of receptor-antiglucocorticoid complexes

S49.1 wild-type receptor complexes with [^{3}H]RU 38486 were chromatographed on DNA-cellulose either before (o) or after activation by incubation with 400 mM KCl for 60 min in the cold (●).

Table 3: DNA-cellulose chromatography of receptor-antiglucocorticoid complexes

Receptor type	KCl concentration required for elution (mM)
wild-type	152
nt^- type	73
nt^i type	201

S49.1 receptor complexes with [^{3}H]RU 38486 were activated and chromatographed on DNA-cellulose.

Immunochemical Properties of Wild-type and Variant Receptors

Monoclonal antibodies directed against the rat liver glucocorticoid receptor have recently been described (15, 16). Some of these were used with wild-type and variant S49.1 receptors (17); this is shown in Table 4.

Table 4: Reaction with monoclonal antibodies

Receptor type	Binding of receptors to antibodies (%)	
	mab 49 (IgG)	mab 57 (IgM)
wild-type	45	19
nt^- type (clone 22R)	55	8
nt^- type (clone 83R)	55	9
nt^i type (clone 55R)	< 1	< 1
nt^i type (clone 143R)	< 1	< 1

S49.1 receptor complexes with [^{3}H]triamcinolone acetonide were incubated at 0° over night with excess monoclonal antibodies. Subsequently complexes were incubated with rabbit anti-mouse immunoglobulins coupled to Sepharose, extensively washed, and the amount of bound steroid determined. (Data from ref. 17).

The antibodies reacted with wild-type and nt^- variant receptors but not with nt^i receptors. Using indirect immunochemical assays it was possible to demonstrate the presence of cross-reacting material in cell extracts of S49.1 variants of the r^- phenotype (17). This immunoreactive material had a molecular weight of 94 000 (17), i.e. it had the same size as wild-type and nt^- receptors under denaturing conditions (see above).

Functional Domains in Glucocorticoid Receptors

From the above mentioned experimental data a molecular model of the wild-type glucocorticoid receptor has evolved that is depicted schematically in Figure 4. Within a polypeptide chain of molecular weight 94 000 the receptor contains three functionally distinct domains: one for steroid binding, one for nuclear interaction, and a third domain that functions in modulating nuclear interaction or DNA binding of the receptor complex. The impact of the modulation domain on the rest of the receptor molecule is such that the affinity for nuclear sites is limited and biologically relevant acceptor sites can be recognized. If the modulation domain is missing from the molecule as is the case in nt^i variants the receptor-glucocorticoid complex might bind too tightly to chromatin such that it has no chance to find the appropriate gene loci which are normally under hormonal control.

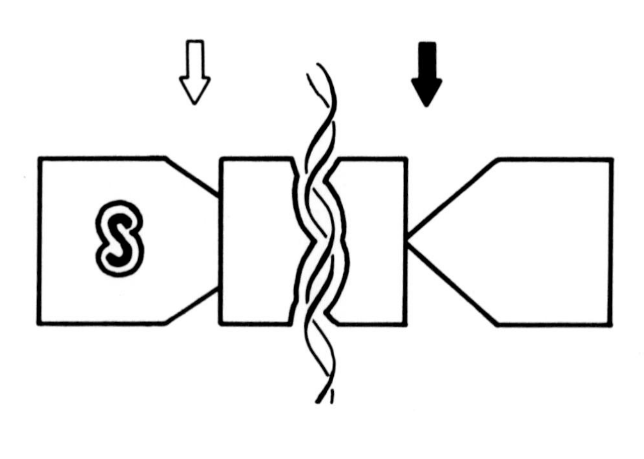

FIG. 4: Functional domains in glucocorticoid receptors

The open areas symbolize the functional domains of wild-type receptors. The filled arrow points to a linker area in the receptor structure that is particularly sensitive to proteolysis. The open arrow indicates another linker areathat is apparently less accessible to proteases but can be split, for example, by trypsin.

The types of receptor variants observed in the S49.1 mouse lymphoma cell system can be explained in view of the domain model of Figure 4. In the r^- phenotype the steroid binding site of the receptor molecule may be defective such that the hormone ligand is not being recognized; such a case has recently been identified by immunochemical methods (17). In the nt^- receptor variant a mutation in the nuclear interaction domain causes a significant decrease in the affinity for DNA. Lastly, nt^i variant receptors are lacking the modulation domain. This may either be due to a deletion mutation in the receptor gene or a nonsense mutation might cause premature protein chain termination. The immunochemical experiments of Table 4 using monoclonal antibodies show that the modulation domain of the receptor molecule contains the main antigenic determinants of wild-type receptors. A similar conclusion has also been drawn from experiments with polyclonal antibodies which did not react with nt^i variant receptors (18) and chymotryptic fragments of wild-type receptors (19, 20).

The DNA binding experiments with receptor-antiglucocorticoid complexes described here are of particular interest. The wild-type, nt^- and nt^i receptor complexes with RU 38486 clearle eluted from DNA-cellulose with lower salt concentrations than the respective complexes with triamcinolone acetonide (cf. Tables 2 and 3) suggesting decreased affinities for DNA. However, the wild-type receptor - RU 38486 complex did not behave like the nt^- receptor-glucocorticoid complex upon DNA-cellulose chromatography, likewise the nt^i receptor-RU 38486 complex is not equivalent to the wild-type receptor-glucocorticoid complex. This is in contrast to the conclusion recently drawn by Bourgeois et al. (21) using a slightly different binding assay. Nevertheless, the observed decrease in DNA binding affinity of the wild-type receptor complexed with an antiglucocorticoid rather than a glucocorticoid may be sufficient to account for the biological effects of the antihormone RU 38486. It should be pointed out that the experiments described here were done with unfractionated DNA; if cloned DNA sequences containing specific receptor binding sites were used instead (21) a more pronounced difference between receptor-glucocorticoid and receptor-antiglucocorticoid complexes might be observed.

In the schematic presentation of the receptor model (Figure 4) the functional domains are shown as areas of about equal size. In reality, however, these domains appear to be quite different in size. The modulation domain has a molecular weight of about 55 000 and the steroid and nuclear binding domains taken together of about 40 000. Even though trypsin and other proteases cleave the 40 000 molecular weight polypeptide to steroid binding fragments of 27 000 to 29 000 it might not be justified to conclude that the nuclear interaction domain is strictly localized within a polypeptide region of 10 000 - 13 000 molecular weight. The proteases may cleave off or destroy some of the essential parts of the nuclear binding domain. Therefore, the functional domains of the receptor molecule may partially overlap in ways that are not obvious from Figure 4. Likewise, the sequential order of the domains along the wild-type receptor polypeptide is not known at present except that the modulation domain is located distally to the remainder of the molecule. It is to be expected, however, that the same kind of domain structure as outlined in Figure 4 is also present in receptors for other steroid hormones.

Acknowledgements

I would like to thank Roussel-Uclaf for kindly providing RU 38486 and its [^{3}H]derivative. This work was supported by the Deutsche Forschungsgemeinschaft and Fonds der Chemischen Industrie.

References

1. Yamamoto, K.R., Gehring, U., Stampfer, M.R., Sibley, C.H.: Recent Progr. Horm. Res. 32, 3-32 (1976).
2. Pfahl, M., Kelleher, R.J., Bourgeois, S.: Mol. Cell. Endocrinol. 10, 193-207 (1978).
3. Gehring, U.: Biochemical Actions of Hormones (G.Litwack, ed.) Vol. VII, 205-232, Academic Press, New York, 1980.
4. Stevens, J., Stevens, Y.-W., Haubenstock, H.: Biochemical Actions of Hormones (G.Litwack, ed.) Vol. X, 383-446, Academic Press, New York, 1983.
5. Gehring, U., Ulrich, J., Segnitz, B.: Mol. Cell. Endocrinol. 28, 605-611 (1982).
6. Gehring, U., Mugele, K., Ulrich, J.: Mol. Cell. Endocrinol. 35, in press (1984).
7. Simons, S.S., Thompson, E.B.: Biochemical Actions of Hormones (G.Litwack, ed.) Vol. IX, 221-254, Academic Press, New York, 1982.
8. Benisek, W.F.: Methods Enzymol. 46, 469-479 (1977).
9. Dellweg, H.-G., Hotz, A., Mugele, K., Gehring, U.: EMBO J. 1, 285-289 (1982).
10. Gehring, U., Hotz, A.: Biochemistry 22, 4013-4018 (1983).
11. Andreasen, P.A., Gehring, U.: Eur. J. Biochem. 120, 443-449 (1981).
12. Stevens, J., Stevens, Y.-W.: Cancer Res. 41, 125-133 (1981).
13. Wrange, Ö., Gustafsson, J.-Å.: J. Biol. Chem. 253, 856-865 (1978).
14. Philibert, D., Deraedt, R., Teutsch, G.: 8th International Congress of Pharmacology, Tokyo, Abstract No. 1463 (1981).
15. Westphal, H.M., Moldenhauer, G., Beato, M.: EMBO J. 1, 1467-1471 (1982).
16. Gametchu, B., Harrison, R.W.: Endocrinology 114, 274-279 (1984).
17. Westphal, H.M., Mugele, K., Beato, M., Gehring, U.: EMBO J. 3, in press (1984).
18. Stevens, J., Eisen, H.J., Stevens, Y.-W., Haubenstock, H., Rosenthal, R., Artishevsky, A.: Cancer Res. 41, 134-137 (1981).
19. Carlstedt-Duke, J., Okret, S., Wrange, Ö., Gustafsson, J.-Å.: Proc. Natl. Acad. Sci. USA 79, 4260-4264 (1982).

20. Eisen, H.J.: Biochemical Actions of Hormones (G.Litwack, ed.) Vol. IX, 255-270, Academic Press, New York, 1982.

21. Bourgeois, S., Pfahl, M., Baulieu, E.-E.: EMBO J. 3, 751-755 (1984).

STEROID HORMONE AGONIST AND ANTAGONIST ACTION IN RELATION TO RECEPTOR CONFORMATION

G. Lazar[+] and M. K. Agarwal

Centre National de la Recherche Scientifique
15 rue de l'Ecole de Médecine, 75270 Paris Cedex 06, France
[+]University Medical School, Szeged, Hungary

Receptor heterogeneity is now a well established fact. A workshop devoted to this theme in 1977 clearly revealed by consensus that the receptor for all five classes of steroid hormones consists of polymorphic, heterogenuos, multiple peaks, that become variously evident depending upon the method of biochemical separation (1-3).

Heterogeneity in the composition of the rat glucocorticoid receptor (GR) was evident in our laboratory as early as 1970 when we were among the first to study GR, and to establish original techniques for GR separation and characterisation by double labelled chromatography (4-6). In later studies, we were able to establish a number of hydrodynamic properties of the glucocorticoid receptor separated into GR_1, GR_2, GR_3 and GR_4 entities (3). Similarily, the mineralocorticoid receptor (MR) was separated into MR_1, MR_2 and MR_4 moities (3). Other authors, too, have shown multiple GR binding systems in the brain, the adrenal cortex, the chicken liver, and the hepatoma cells (reviews in 1), using natural and various synthetic glucocorticoid agonists (7,8).

The purpose of this chapter is to review some of our work, using various steroid agonists and antagonists, in an effort to assign an eventual physiological or functional role to receptor heterogeneity.

Adrenal Steroid Antagonism

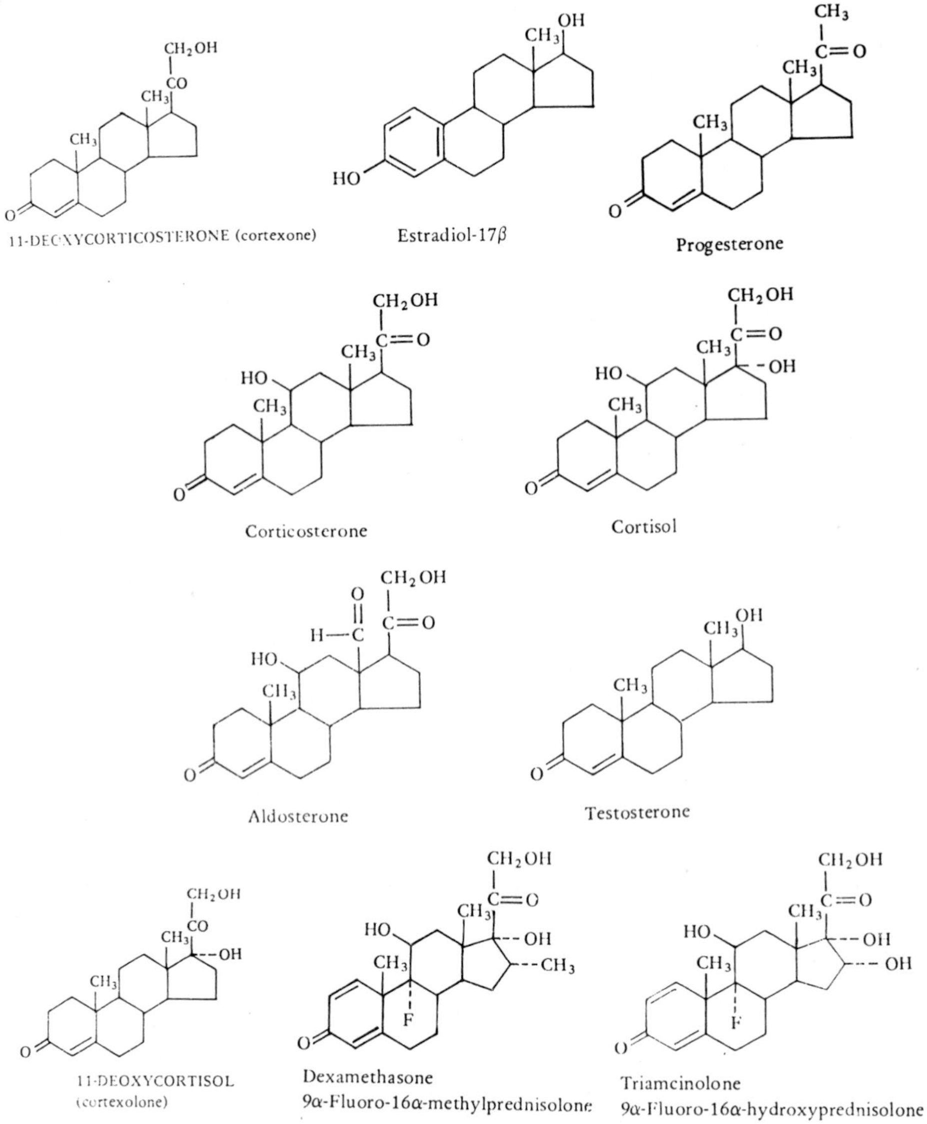

11-DEOXYCORTICOSTERONE (cortexone)

Estradiol-17β

Progesterone

Corticosterone

Cortisol

Aldosterone

Testosterone

11-DEOXYCORTISOL (cortexolone)

Dexamethasone
9α-Fluoro-16α-methylprednisolone

Triamcinolone
9α-Fluoro-16α-hydroxyprednisolone

Figure 1. Structural formulae of selected biologically active steroid hormones.

Table 1. STEROID HORMONE CONSTANTS IN THE HUMAN.

Hormone	Plasma concentration (µg/l)	Hypersecretion Syndrome	Output/24 h (Normal)
Cortisol	100	>30 mg/l (Cushing)	15-30 mg
Cortocosterone	10		2-4 mg
Progesterone	0.2-2	25 mg/l (ovulation)	2 mg
Estradiol	0.3	>1000x/day (ovulation)	0.03 mg ♂ 0.2 mg ♀
Testosterone	7 ♂ 0.7 ♀	>100x (cancer)	6 mg
Aldosterone	0.1	1 mg or > (Conn)	0.1-0.2 mg
Deoxycorticosterone (DOC)	Trace	ibid	
18-hydroxy-DOC	Trace	ibid	
Cortexolone (11-Deoxycortisol)	>1 µ		

For details see (14-19).

In Fig.1 are presented the structural formulae of some of the agonist and antagonists employed in the studies described here. Cortisol, corticosterone, dexamethasone and triamcinolone acetonide (TA), all induce gluconeogenesis and several liver enzymes (anabolic action) but cause lympholysis (catablic action). Endogeneous liver glycogen levels and TA induced gluconeogenesis can be antagonized by estradiol and testosterone analogues very effectively (9, 10). Under these conditions, liver tryptophan pyrrolase (TP) and tyrosine transaminase (TT) induction remains unaffected. On the other hand, testosterone analogues inhibit TT induction in hepatoma cells _in vitro_ (11), but had no influence in rat liver _in vivo_ (9,10). Progesterone analogues induced TT _in vivo_ in some studies (12) but were only a suboptimal inducer in hepatoma cells in vitro (11).

The contradictory nature of this problem is very evident from studies with cortexolone (11-deoxycortisol). This steroid is a suboptimal inducer of TT activity _in vitro_ (11) but not in rat liver _in vivo_ (9,10). It reverses the action of glucocorticoids in thymocytes (12) and He La cells (13). All these data were interpreted (13) in terms of a single binding site on the glucocorticoid receptor that was variously occupied by the agonist and/or antagonist in competition binding studies where large excess of one steroid over the other are commonly employed to gain an idea of the relative affinities of different steroids for the one and the same binding site. No effort is usually made to study these parameters in hormone ranges that are close to physiological concentrations _in vivo_.

As far as the mineralocorticoids are concerned, a suitable system for _in vitro_ or _in vivo_ action is still wanting. Nevertheless, it is well established that aldosterone is the major mineralocorticoid in man whose excess in hyperaldosteronism (Conn's syndrome) is well documented (15); 11-hydroxy--deoxycorticosterone (DOC) has only 1/50 the activity of aldosterone. Excess DOC (15) and 18-hydroxy-DOC (17) secretion

Table 2. INFLUENCE OF SEX STEROIDS IN LIVER.

Treatment	Glycogen (mg%)		3×10^{-6} M/ml	3×10^{-8} M/ml
	- TA	+ TA	competing steroid	^{3}H-TA CPM/mg protein
Control	9.03 ± 0.85	18.60 ± 3.50	—	1887
Estradiol (20 mg ip)	2.37 ± 0.31	9.42 ± 1.19	Estradiol	1195
Testosterone (20 mg ip)	2.1 ± 0.55	13.36 ± 1.11	Testosterone	1420
Tamoxifen (20 mg ip)	9.6 ± 2.0	23.9 ± 2.8	TA	61
Tamoxifen (20 mg ip) + Estradiol (2 mg ip)	7.6 ± 2.1	18.8 ± 1.8		

Triamcinolone acetonide was used to saturate the receptor in all cases. Further details in (9,10,20,21) syndromes, too, have been reported. Progesterone antagonizes mineralocorticoid action in vivo as evidenced by its clinical effect on the prevention of aldosterone induced sodium loss (7,8,14). Here again, all effort has centered on the manner in which various molecules compete for the one and the same binding site.

In table 1 are shown a representative example of some hormone concentrations in the plasma. This table is only indicative since wide fluctuations are a common occurrence due to diurnal rhythms, individual variations, assay methods, sex and so forth. It is obvious that the plasma concentration of estradiol (female) is 1/30 of cortisol and that of testosterone (male) 1/10. Data in table 2 show that even 100 fold excess of either gonadal steroid can not displace TA from its receptor in vitro but these same in 20 fold excess inhibit liver glycogen in response to 1 mg TA in vivo. It therefore follows that, even in syndromes of sex hormones hypersecretion (table 1), it is inconceivable that enough gonadal steroids should become available to displace the glucocorticoid from its receptor in vivo. In other words, sex steroids may be binding to GR sites distinct from those that bind the glucocorticoid. On the other hand, gonadal steroids may be influencing gluconeogenesis by modulation of the liver estrogen or the androgen receptors (7,8) although no evidence for this has as yet been obtained.

Data in Fig. 2 show that ^{3}H-estradiol and ^{3}H-testosterone labelled the ER_4 and the AR_4 components, respectively, that eluted in the same position as ^{14}C-corticosterone bound transcortin from DEAE-52 columns. Two other, minor AR_1 and ER_1 entities were also observed. In contrast, in similar concentration ranges, GR_2 and GR_3 moieties were obtained when corticosterone and TA, respectively, were used in place of the sex steroids. In other words, the physiological action of sex steroids would appear to proceed via saturation of those components that have little or no affinity for glucocorticoids.

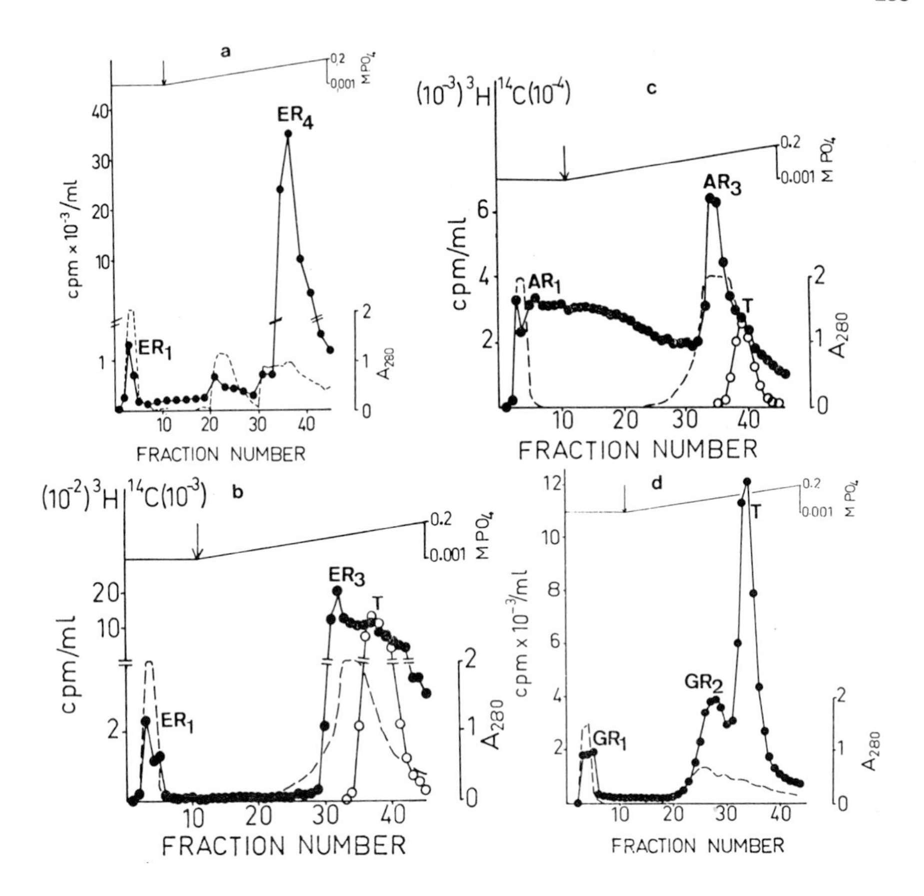

Figure 2. ION EXCHANGE SEPARATION OF VARIOUS RECEPTORS IN RAT LIVER.

4 ml cytosol in 0.001 M phosphate, pH 7.5, was incubated with 10^{-7} M of either ^{3}H-estradiol (a), ^{3}H-tamoxifen (b), ^{3}H-testosterone (c), or ^{3}H-corticosterone (d). 2 ml serum was similarly incubated with 0.5 µCi of ^{14}C-corticosterone. All samples were charcoal treated and loaded onto a column (1.0 x 25 cm) of DEAE-cellulose-25 (Whatman) equilibrated with 0.001 M phosphate. After an initial prewash, protein was eluted by a linear gradient (begun at the arrow in figures) between 0.001 M and 0.2 M phosphate, pH 7.5 (60 ml each). 1 ml samples were processed for radioactivity and the absorbance was recorded manually. For Figs b,c, serum was mixed with liver cytosol just before chromatography.

------------ A_{280}; ●——————● ^{3}H; o——————o ^{14}C

This was further confirmed by the action of tamoxifen under these conditions. This non steroidal anti-estrogen antagonizes estradiol receptor binding in the breast tissue and forms the basis of specific medical therapy of estrogen-dependent tumors (reviewed in 7,8). Tamoxifen was largely without effect on liver gluconeogenesis (table 2) and it was bound to an ER_3 component clearly different from estradiol bound ER_4 that was preferentially labelled with estradiol which exerts a physiological action in the liver.

Similar sorts of conclusions were obtained when the mineralocorticoid receptor (MR) system was analysed. Data in Fig. 3 show that, in equimolar concentrations, aldosterone, DOC and 18-hydroxy DOC labelled the MR_2, the MR and the MR components, respectively. Although renal MR_4 eluted in the same position as blood corticosteroid binding globulin (CBG), no radioactivity in this region was found when liver was used. In the other words, this region of DE-52 column represents the elution position of components that have organ specificity and are not found in tissues (eg. liver) which are not targets for the physiological action of the hormone (22-24). Going back to table 2, it becomes obvious that this sort of receptor polymorphism is really a physiological necessity. Since both DOC and 18-hydroxy-DOC are present only in trace amounts in human plasma, and possess weaker affinities than aldosterone for the same binding sites in classical competion studies (see 7,8 for reviews), they can never be expected to displace aldosterone from its receptor. Even in selected syndromes of hypermineralocorticoidism (15-17), plasma levels of these two do not exceed that of aldosterone yet both are potent in inducing typical Conn syndrome, presumably by saturating a mineralocorticoid specific receptor system.

Further proof for this sort of reasoning comes from studies with progesterone which is known to antagonize aldosterone action, physiologically (14). Data in Fig. 4 show that progesterone, but not the synthetic progestogen R-5020, was

3 a

DEOXYCORTICOSTERONE

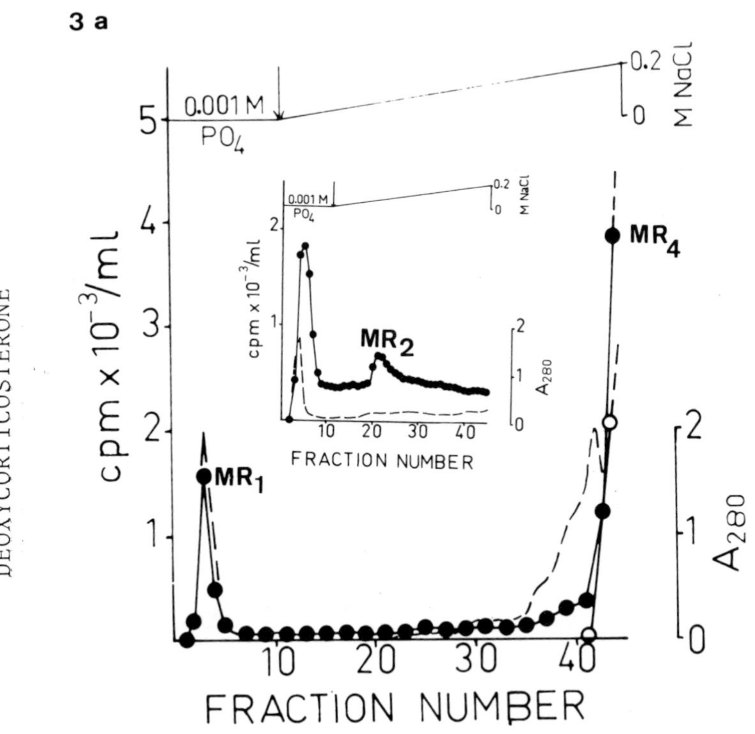
0.001 M
PO4
0.2
0
M NaCl
cpm x 10⁻³/ml
MR1
MR2
MR4
A280
FRACTION NUMBER

18-HYDROXY-DEOXYCORTICOSTERONE

b

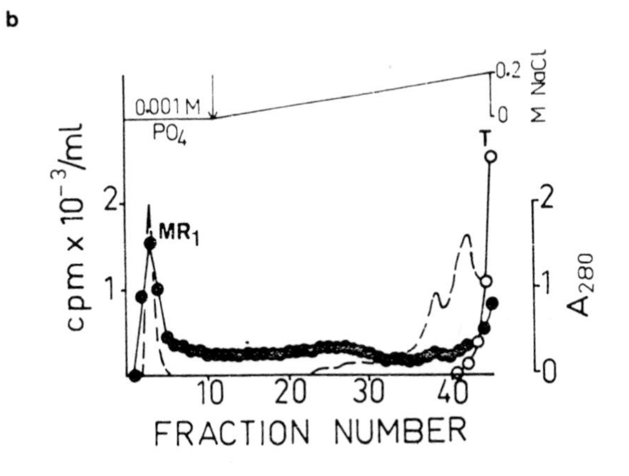
0.001 M
PO4
0.2
0
M NaCl
T
cpm x 10⁻³/ml
MR1
A280
FRACTION NUMBER

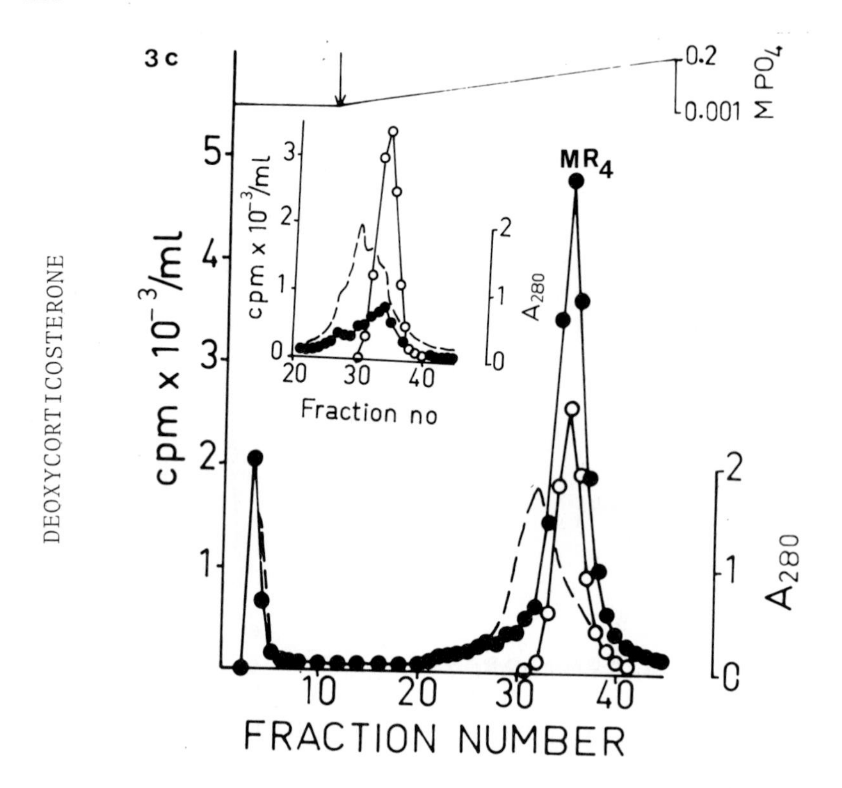

Figure 3. Ion exchange separation of rat kidney mineralocorticoid binding proteins. Blood serum (2 ml) was incubated with 0.2 µCi of ^{14}C-corticosterone for 60 min at 4°C. The 105 000 g supernatant fraction (5 ml) of rat kidney was equilibrated for 60 min at 4°C with 10^{-7} M of either ^{3}H-deoxycorticosterone (DOC) or ^{3}H-18-hydroxycorticosterone (18-OH-DOC) or 10^{-8} M ^{3}H-aldosterone. The cytosol and serum were finally mixed and layered on DEAE-52 (1 x 25 cm) column. After passage of 60-70 ml of the initial buffer (fraction volume 6-7 ml) elution was begun (at arrow) by a linear gradient between 60 ml each of 0.001 M PO_4, pH 7.5 (initial buffer) and this buffer containing 0.2 M NaCl, at a flow rate of 60 ml/h at 4°C (fraction vol. approximately 3 ml). For 3c, kidney cytosol (5 ml) and serum (2 ml) were equilibrated with 10^{-7} M ^{3}H-deoxycorticosterone and 0.2 µCi of 4-^{14}C-corticosterone, respectively, and eluted with a gradient between 0.001 M and 0.2 M PO_4, pH 7.5. For liver cytosol (insert), under exactly the same conditions as kidney, only the relevant portion is shown. Aliquots of 1 ml were mixed with 10 ml Unisolve (Kochlight) and counted in a Packard Tricarb scintillation spectrometer with corrections for quenching, spilling and background; A_{280} values were recorded manually.

as good as aldosterone in displacing the mineralocorticoid from its receptor in rat kidney both in classical competition studies and in aldosterone binding to the MR_1 and MR_2 components of MR during chromatography on DEAE-52 columns. However, in equimolar (10^{-8} M) concentrations, both gestogens were preferentially bound to the MR_4 entity which coeluted with ^{14}C-corticosterone-serum complex, which could not be labelled with aldosterone in the kidney, and which could not be detected in the liver and the serum under any condition.

In other words, an agonist and an antagonist may occupy different components of MR, simultaneously. Since plasma levels of aldosterone and progesterone are normally quite comparable (table 1), although subject to fluctuations, these two steroids may be concurrently involved in regulating the hydro-mineral balance in the human.

Technical considerations for this sort of analysis have been described in detail elsewhere (4,25-27). It has furthermore been demonstrated that this heterogeneity does not appear to be an expression of endogenous proteolysis limited in time and space (28,29). Tissue dependent specificity in binding of a hormone, too, has been shown (30,31) and criteria for receptor like nature of various peaks have been established (32). In other studies, species specific differences in the qualitative abundance of various components have been noted (32,34). In these and other experiments, it has been possible to differentiate receptor multiplicity from blood serum transcortin peaks (35,36). With these considerations in mind, table 3 summarizes selected hydrodynamic properties of various receptor systems and their components.

In order to understand the in vivo role and genesis of this sort of heterogeneity, one would like to purify the individual population and test its action in defined systems in vitro. Since required technical sophistication is still wanting, all efforts have hitherto attempted correlation of in vivo physiological effects with qualitative abundance of different

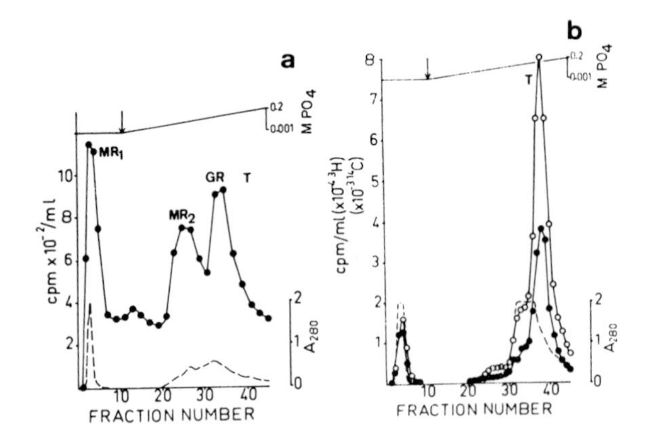

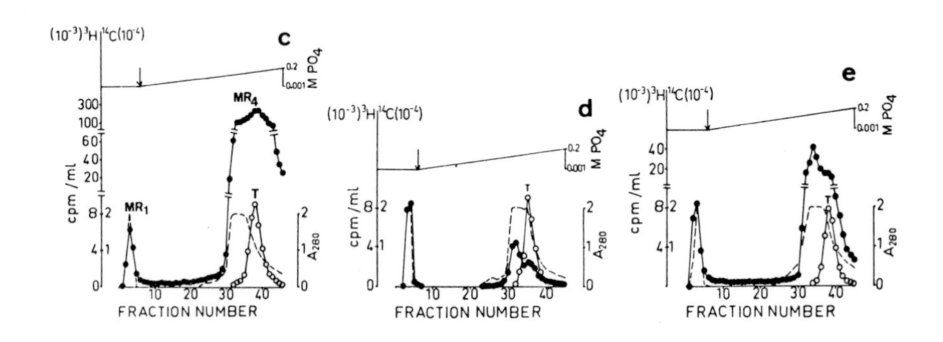

Figure 4. Ion exchange separation of steroid binders on DEAE-52 columns. 5 ml renal cytosol was incubated with 10^{-8} M ^{3}H-aldosterone (a); 4 ml of kidney (b) cytosol was incubated with 10^{-7} M ^{3}H-progesterone; 4 ml renal cytosol (c), undiluted serum (d), or liver cytosol (e) was incubated with 10^{-7} M ^{3}H-R-5020. For Fig. 4b, 2 ml serum was incubated in presence of 0.5 µCi of ^{14}C-progesterone; 2 ml fresh serum with 0.5 µCi ^{14}C-corticosterone was used for Figures 4c,4d,4e. All samples were charcoal treated (100 mg/ml mixture), separately, mixed immediately and loaded onto DE-52 columns (1 x 25 cm). After a low ionic prewash (0.001 M phosphate, pH 7.5), elution was begun (at arrow) with a gradient between 60 ml of 0.001 M and 0.2 M of this buffer, pH 7.5. 1 ml samples were processed for radioactivity and the absorbance was recorded manually.

---------A_{280}; ●———● ^{3}H; O———O ^{14}C

Table 3. HYDRODYNAMIC PROPERTIES OF RECEPTOR SUBPOPULATIONS

Component	Steroid specificity	Tissue Distribution	Elution Molarity	Conductance (miliSiemens)
GR_1	Synthetic > Natural	All tested	0.001 M	0.18 0.2
GR_2	Cortico-sterone	Liver only	0.02 M	5.0
GR_3	Synthetic only	All tested	0.04 M	5.8
GR_4	Natural only	All tested	0.06 M	9.2
MR_1	18-OH-DOC	Kidney	0.001 M	0.2
MR_2	Aldosterone > TA	Kidney Heart ?	0.006 M	2.3
MR_4	DOC Progesterone R-5020	Kidney	0.06 M	9.2
AR_1	Testosterone	Liver	0.001 M	
AR_3	R-1881	Liver	0.04 M	
AR_4	Testosterone	Liver	0.06 M	
ER_1	Estradiol	Liver	0.001 M	
ER_3	Tamoxifen	Liver	0.04 M	
ER_4	Estradiol	Liver	0.06 M	
Transc-ortin	Natural & Synthetic	Serum Tissue ???	0.06 M	9.8
Albumin	All tested	Plasma	0.06 M	8.9

AR = Androgen Receptor; ER = Estrogen Receptor;

GR = Glucocorticoid Receptor; MR = Mineralocorticoid Receptor

receptor peaks. One such instance has already been described above in the antagonism of glucocorticoid action in rat liver *in vivo*, using various gonadal steroids (20, 37).

Data in table 4 show that bacterial endotoxins can very effectively antagonize liver gluconeogenesis and tryptophan pyrrolase, but not tyrosine transaminase, induction *in vivo* (7,8,38). Since such conclusions were also supported by experiments in the isolated, perfused rat liver (38), it can safely be assumed that endotoxins react directly on the liver. However, the effect may be mediated via substances released by the reticuloendothelial system in the liver since parenchymal uptake of endotoxin is only minimal (38).

Table 4. INFLUENCE OF ENDOTOXIN ON LIVER ENZYME INDUCTION.

Treatment	TP activity	TT activity	Liver Glycogen
Control	43.9 ± 2.9	0.64 ± 0.04	12.63 ± 3.18
Endotoxin	31.3 ± 2.6	2.08 ± 0.18	2.04 ± 0.90
Cortisol	143.9 ± 4.4	4.41 ± 0.21	28.37 ± 6.58
Cortisol + Endotoxin	46.6 ± 1.7	5.20 ± 0.37	3.25 ± 2.25

500 µg endotoxin, 1 mg cortisol, either alone or concurrently, were given intraperitoneally 4 h prior to assay.

For details see (7-8).

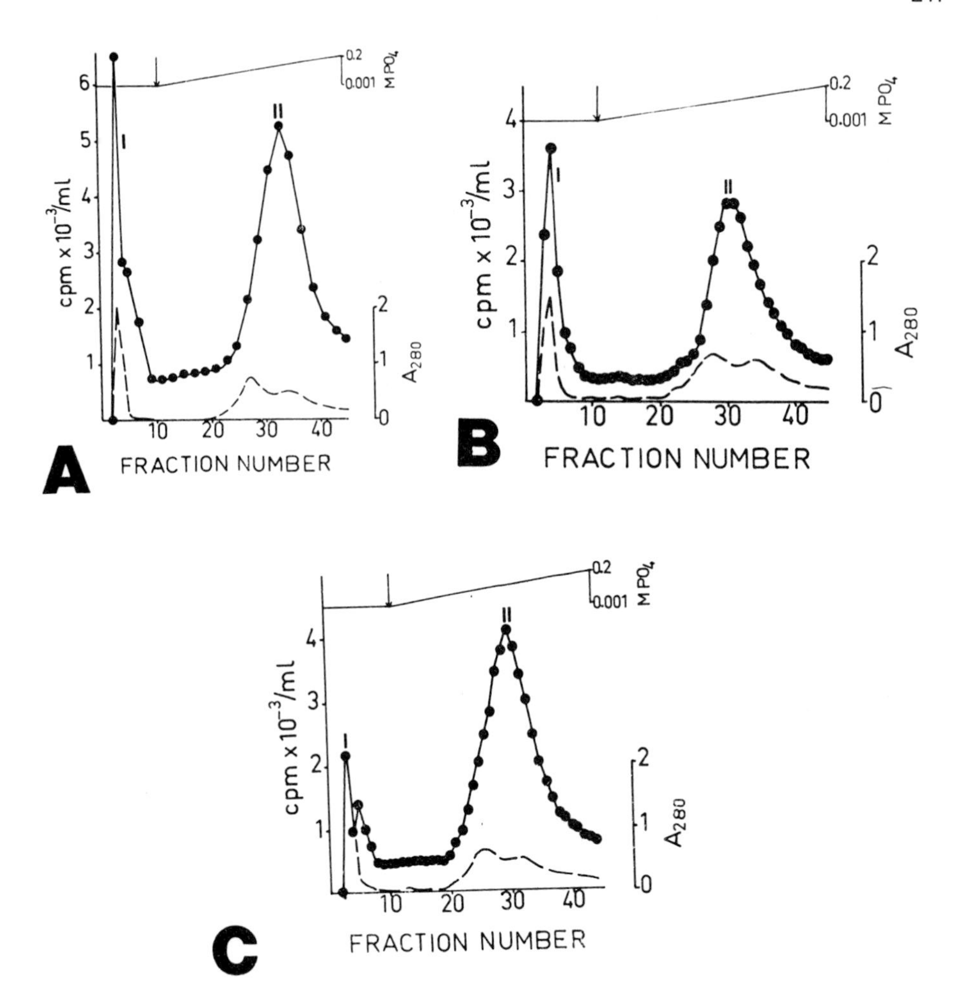

Figure 5. Qualitative variations in liver corticoid-receptor subpopulations following administration of RES-active materials in rats. Control (A) and endotoxin injected rats were sacrified 2 hr (B) or 16 hr (C) after treatment. Cytosol was processed by chromatography on DE-52 columns. After an initial prewash, the gradient elution was begun at the arrow in all cases. Radioactivity (•———•); A_{280} (-------).

For further details see (40).

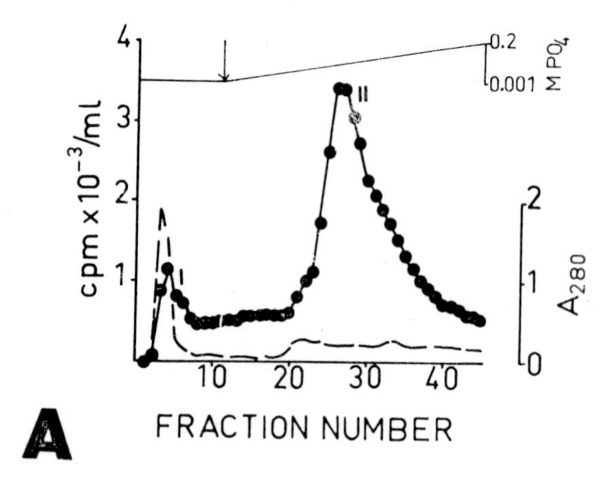

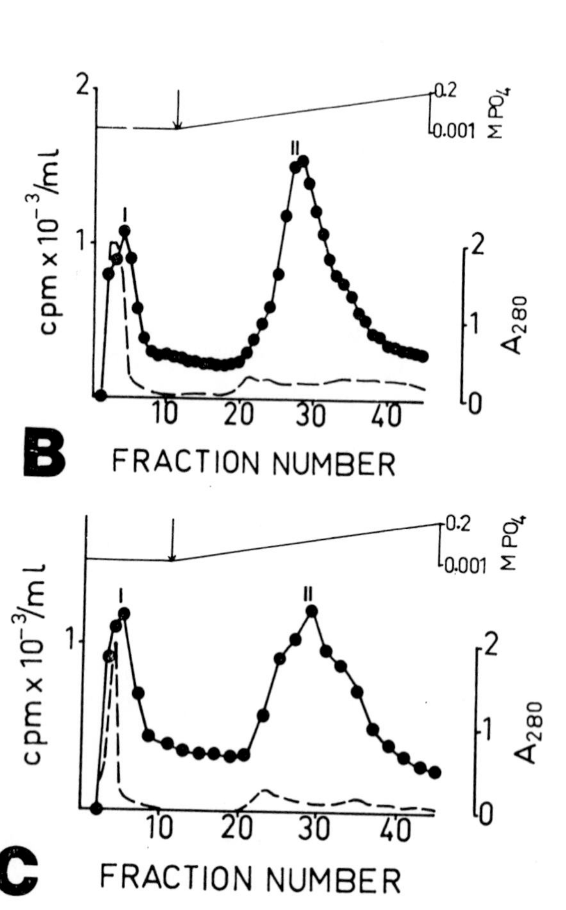

Figure 6. Influence of RE agents on corticoid-receptor subpopulations in rat spleen. Control (A) and endotoxin injected rats were sacrificed 2 hr (B) or 16 hr (C) after treatment. Spleens were processed for DE-52 chromatography. For other details see legend to Fig. 5.

These bacterial endotoxins interact with the glucocorticoid receptor and disturb the nucleo-cytoplasmic ratio of GR (39). In Fig. 5 it is shown that endotoxins lowered GR_1 and GR_3 within 2 h and GR_3 was replenished before GR_1 in rat liver (40). GR_3 was also lowered in the spleen but GR_1 was unaffected by the toxin (Fig. 6). Thus, a change in the qualitative abundance of various subpopulations of GR may explain the selective action of endotoxins. In addition, organ specificity (liver vs spleen) may reside in the saturation of different subspecies of GR. It is important to note that the receptor subpopulations were altered before the ensuing physiological process in the temporal sequence.

Liver TT can be induced just after birth but TP and glycogen induction are possible only one week or more post partum (25,26). Data in Fig. 7 show that the GR_2 component was minimum just after birth and increased progressively upto 22 days post partum. On the other hand, GR_1 exhibited a biphasic rise; after an initial maximum by the 8th day, it declined progressively until the 16th day and then rose again to reach adult levels by 22 days after birth. Biphasic evolution was also observed with GR_4 which was maximal after birth, declined in the first 8 days of extra uterine life, and then exhibited a second burst between 16-22 days.

TA bound GR_1 and GR_3 increased progressively between 1-22 days of extra uterine life although GR_3 was formed in greater quantities than GR_1 after the first 10 days (20,25,26).This time dependent difference in the genesis of various GR peaks, thus, may be related to the onset of various glucocorticoid responsive liver processes as a function of age (41). In another study (42), a receptor component not present at birth but appearing within a few days thereafter, was said to be specifically responsible for TP induction. In contrast, saturation binding studies had shown a gradual and progressive increase in the overall receptor concentration as a function of age (43).

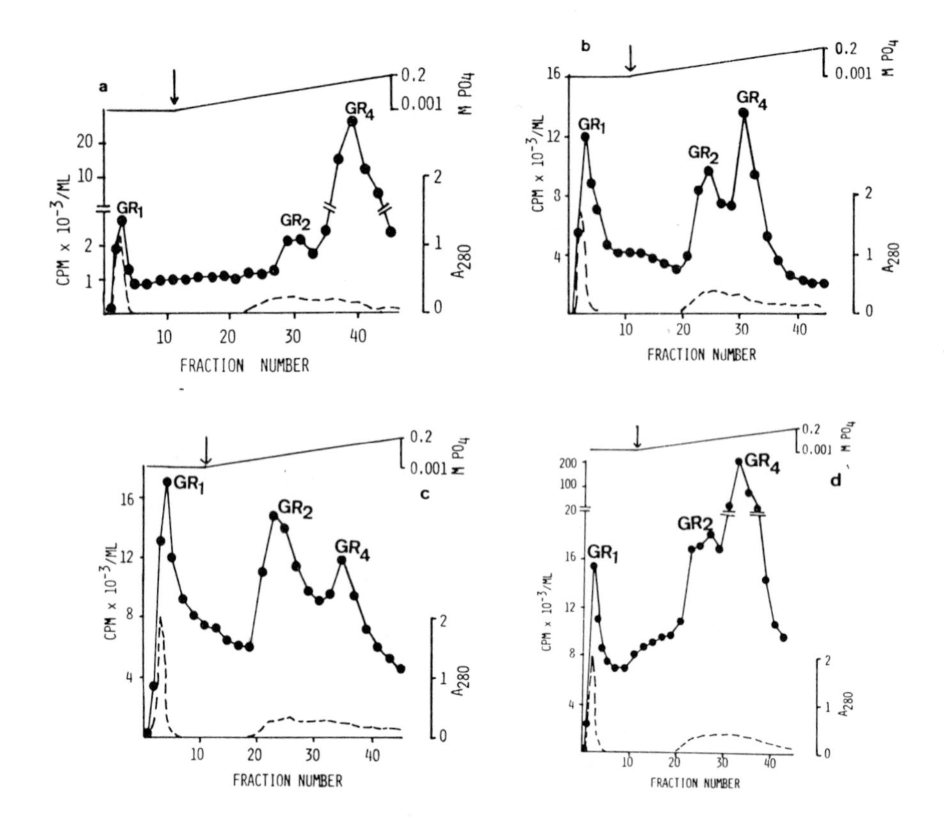

Figure 7. Ion exchange separation of corticosterone binders.

Livers were obtained from weanling rats 1 (a), 3 (b), 8 (c) or 22 (d) days after birth. 2 ml liver cytosol (105, 000 x g) in 0.001 M phosphate, pH 7.5, was incubated with 10^{-7} M ^{3}H-corticosterone and charcoal treated as before. After passage through glass wool, the cell sap was loaded onto DEAE-cellulose-52 columns (0.5 x 25 cm) equilibrated with 0.001 M phosphate. After a prewash with 20 ml of this initial puffer, protein was eluted by a linear gradient between 30 ml each of the initial buffer and the 0.2 M sodium phosphate, pH 7.5. The gradient elution was begun at the arrow in the figures. Fractions of 0.5 ml were processed for radioactivity in 10 ml Scintix (Isotec, France) and for absorbance.

From the foregoing it seems obvious that receptor heterogeneity is required to distinguish sense from antisense or non-sense signals in the overall mechanism of action of steroid hormones. Finally, a model to visualize the scheme of things in vivo, has been presented in the first chapter of this volume. These are in keeping with current models of regulation of mammalian protein synthesis (44-49).

Perspectives

Mineralocorticoid antagonists have been in clinical use since several decades. Specific glucocorticoid antagonists are being synthesized only now. Ru 38486 is one such anti-glucocorticoid and appears to hold much promise in future experime tal and clinical research.

In our laboratory, the influence of this material was tested in relation to endotoxin lethality and to the protective effect of glucocorticoid hormone during endotoxicosis. Data in Table 5 show that Ru 38486 very effectively reversed the protective effect of dexamethasone against endotoxin lethality. The protection was dependent upon the dose of Ru 38486 and was more effective by the intraperitoneal route.

Next, Ru 38486 was tested in an effort to assess its influence on the innate susceptibility to endotoxin lethality. Data in Table 6 show that Ru 38486 was ineffective in this respect by the subcutaneous route even when 200 mg/kg of the material were given 2 h prior to endotoxin challenge. However, Ru 38486 effectively sensitized when intraperitoneal route was used at the same dose level.

Finally, data in Table 7 show convincingly that the sensitizing effect of Ru 38486 on lethality was evident even when this product was injected 6 h before endotoxin challenge. However, the drug was clearly more potent when it was given con-

Table 5. INFLUENCE OF Ru 38486 ON ENDOTOXIN LETHALITY IN SWISS, OF_1 MALE MICE.

Treatment	Alive/Total in 48 h	% Survival 48 h	Statistics	
1 LPS Control	15/43	35		
2 LPS + DEX Control	25/30	83	2 vs 1	P<0.001
3 Ru 1 mg sc	21/21	100	3 vs 1	P<0.001
			3 vs 2	N.S.
4 Ru 5 mg sc	16/20	80	4 vs 1	P<0.001
			4 vs 2	N.S.
			4 vs 3	P<0.05
5 Ru 1 mg ip	8/15	53	5 vs 1	N.S.
			5 vs 2	P<0.05
			5 vs 3	P<0.001
6 Ru 5 mg ip	3/15	20	6 vs 1	N.S.
			6 vs 2	P<0.001
			6 vs 4	P<0.001
			6 vs 5	N.S.

S. typhymurium LPS (500 μg ip) and Dexamethasone (100 μg ip) were given concurrently in 0.2 ml 2 hours after the indicated dose of Ru 38486.

The Chi square test was used to determine significance between treatments.

currently with endotoxin challenge, or even 6 h after the toxin. The possibility must therefore be kept open that endotoxin may be sensitizing to Ru 38486. However, animals injected even with 200 mg/kg Ru 38486 show no sign of toxicity and the dose of 40 mg/kg in Table 7 is 5 times smaller.

Further studies are required to analyse these considerations and to assess the influence of Ru 38486 on various parameters altered by endotoxins. Clearly, Ru 38486 holds much promise as a potential antiglucocorticoid although its lack of specificity may become a limiting factor. Other chapters in this volume should be consulted for the effect of this novel antiglucocorticoid in different cell and organ systems. In the clinic, the antifertility activity of this material has already been exploited though its weak anti-androgen activity may pose long term problems. Finally, the affinity of RU 38486 for different receptor peaks has yet to be attempted.

Acknowledgements

Thanks are due to Dr. D. Philibert, Roussel-Uclaf, for a generous sample of RU 38486 and for sharing some of the unpublished data with this material. The vehicle used for the administration of RU 38486, too, was kindly supplied by his laboratory.
INSERM International Fellowship programme provided the financial assistance to Dr. Lazar for his visit to Paris.
This work was aided by grants from UER Broussais Hôtel Dieu.

Table 6. ROUTE DEPENDENT SENSITIZATION TO ENDOTOXIN LETHALITY BY Ru 38486.

Treatment	Alive/Total in 48 h	% Survival 48 h	Statistics
1 LPS Control	15/43	35	
2 Ru 1 mg sc	6/21	29	2 vs 1 N.S.
3 Ru 5 mg sc	7/20	35	3 vs 1 N.S. 3 vs 2 N.S.
4 Ru 1 mg ip	3/15	20	4 vs 1 N.S. 4 vs 2 N.S.
5 Ru 5 mg ip	0/15	0	5 vs 1 $P<0.01$ 5 vs 3 $P<0.02$

S. typhymurium LPS (500 μg ip) was given 2 hours after the Ru 38486.
The Chi square test was used to assess significance between different treatments.

Table 7. TIME DEPENDENT SENSITIZATION TO ENDOTOXIN LETHALITY BY Ru 38486.

Treatment	Alive/Total in 48 h	% Survival 48 h	Statistics
1 LPS Control	18/21	86	
2 Ru - 6 h	9/21	43	2 vs 1 P<0.005
3 Ru - 2 h	5/21	24	3 vs 1 P<0.01 3 vs 2 N.S.
4 Ru 0 h	2/20	10	4 vs 1 P<0.001 4 vs 2 P<0.02 4 vs 3 N.S.
5 Ru + 2 h	1/21	<5	5 vs 1 P<0.001 5 vs 2 P<0.005 5 vs 3 N.S.
6 Ru + 6 h	2/21	<10	6 vs 1 P<0.001 6 vs 2 P<0.02 6 vs 3 N.S.

Ru 38486 (1 mg ip) wasgiven at the indicated time points after LPS (250 μg ip).
The Chi square test was used for statistics.

References

1. Agarwal, M.K. (ed) Multiple Molecular Forms of Steroid Hormone Receptors, Elsevier/North Holland, Amsterdam, Oxford, New York, 1977.
2. Agarwal, M.K., Rossier, B.C.: FEBS Letters 82, 165 (1977).
3. Agarwal, M.K.: FEBS Letters 85, 1 (1978).
4. Agarwal, M.K.: FEBS Letters 62, 25 (1976).
5. Snart, R.S., Sanyal, N.N., Agarwal, M.K.: J. Endocrinol. 47, 149 (1970).
6. Snart, R.S., Shepherd, R.E., Agarwal, M.K.: Hormones 3, 293 (1972).
7. Agarwal, M.K. (ed) Antihormones, Elsevier/North Holland, Amsterdam, Oxford, New York, 1979.
8. Agarwal, M.K. (ed) Hormone Antagonists, Walter de Gruyter, Berlin, New York, 1980.
9. Agarwal, M.K., Lazar, G., Sekiya, S.: Biochem. Biophys. Res. Comm. 79, 499 (1977).
10. Agarwal, M.K., Coupry, F.: FEBS Letters 82, 172 (1977).
11. Sammuels, H.H., Tomkins, G.M.: J. Mol. Biol. 52, 57 (1970).
12. Kaiser, N., Milholland, R.J., Turnell, R.W., Rosen, F.: Biochem. Biophys. Res. Comm. 49, 2516 (1972).
13. Melnykovych, G., Bishop, C.F.: Endocrinology 88, 450 (1971).
14. Landau, R.L.: J. Clin. Endocrinol. 15, 1194 (1955).
15. Conn, J.W.: J. Lab. Clin. Med. 45, 3 (1955).
16. Crane, M.G., Harris, J.J.: J. clin. Endocr. Metab. 26, 1135 (1966).
17. Melby, J.C., Dale, S.L., Wilson, T.E.: Circulation Res. 28, 29 Suppl. 11, 143 (1971).
18. Harrison's Principles of Internal Medicine, Seventh Edition, Ed. by McGraw-Hill Book Company, New York, 1974.
19. Guyton, A.C.: Textbook of Medical Physiology, Third Edition, Ed. by W.B. Saunders Company, Philadelphia, London, 1966.
20. Agarwal, M.K.: Biochem. Biophys. Res, Comm. 109, 291 (1982).
21. Agarwal, M.K.: Biochem. Biophys. Res. Comm. 73, 767 (1976).
22. Agarwal, M.K.: FEBS Letters 67, 260 (1976).
23. Agarwal, M.K.: Nature 254, 623 (1975).
24. Agarwal, M.K.: Biochem. Biophys. Res. Comm. 89, 77 (1979).

25. Agarwal, M.K., Sekeris, C.E.(ed); Handbook of Receptor Research, vol. IV , Gluco- and Mineralocorticoids, Field Italia International, 1984.

26. Agarwal, M.K. (ed) Principles of Receptology, Walter de Gruyter, Berlin, New York, 1983.

27. Agarwal, M.K.: Experientia 25, 73 (1976).

28. Agarwal, M.K. (ed) Proteases and Hormones, Elsevier/North Holland, Amsterdam, Oxford, New York, 1977.

29. Agarwal, M.K.: FEBS Letters 106, 1 (1979).

30. Agarwal, M.K., Philippe, M., Coupry, F.: Biochem. Biophys. Res. Comm. 83, 1 (1978).

31. Agarwal, M.K.: Int. J. Biochem. 8, 7 (1977).

32. Agarwal, M.K.: Biochem. J. 154, 567 (1976).

33. Agarwal, M.K.: Biochem. Med. 24, 201 (1980).

34. Agarwal, M.K.: Die Naturwissenschaften 63, 50 (1976).

35. Agarwal, M.K.: Arch. Biochem. Biophys. 180, 140 (1977).

36. Agarwal, M.K.: Int. J. Biol. Macromol. 1, 211 (1979).

37. Agarwal, M.K., Sekiya, S., Lazar, G.: Res. Exptl. Med. 176, 181 (1979).

38. Agarwal, M.K., Hoffman, W.W., Rosen, F.: Biochem. Biophys. Acta 177, 250 (1969).

39. Agarwal, M.K.: Int. J. Biochem. 3, 408 (1972).

40. Agarwal, M.K.: Biochem. Med. 17, 193 (1977).

41. Agarwal, M.K.: Biochem. Biophys. Res. Comm. 106, 1412 (1982).

42. Grote, H., Schmid, W., Sekeris, C.E.: FEBS Letters 82, 329 (1977).

43. Giannopaulos, G.: J. Biol. Chem. 250, 5847 (1975).

44. Tymoczko, J.L., Philipps, M.M.: Endocrinol. 112, 142 (1983).

45. Winneker, R.C., Clark, J.H.: Endocrinol. 112, 1910 (1983).

46. Chothia, C., Lesk, A.M., Dodson, G.G., Hodgkin, D.C.: Nature 302, 500 (1983).

47. Firestone, G.L., Paywar, F., Yamamoto, K.R.: Nature 300, 221 (1983).

48. Jensen, E.V., Greene, C.L., Closs, L.E., De Sombre, E.R.: Rec. Prgr. Horm. Res. 38, 1 (1982).

49. Eckert, R.L., Katzenellenbogen, B.S.: J. Biol. Chem. 257, 8840 (1982).

SYNTHETIC AND NATURALLY OCCURRING ANTAGONISTS OF ADRENAL STEROID ACTION ON BRAIN AND BEHAVIOUR

H. Dick Veldhuis, E. Ronald De Kloet

Rudolf Magnus Institute for Pharmacology, Medical Faculty, University of Utrecht, Vondellaan 6, 3521 GD Utrecht, The Netherlands.

Introduction

The rat brain contains receptor sites for adrenocortical hormones (1). The receptor sites are heterogeneous and differ in cellular localization and binding specificity for mineralocorticoids, naturally occurring glucocorticoids and their synthetic analogs. The anatomy, binding specificity and capacity of the receptor system have provided the criteria to study receptor mediated steroid effects on brain function, as judged from neurochemical and behavioural parameters (2-5). Via this approach agonist and antagonist properties of adrenal steroids and their synthetic analogs were defined. Antagonism could be explained in most cases on the basis of competitive binding to a particular adrenal steroid receptor system in the brain.

2. Binding properties of soluble adrenal steroid receptors

The binding properties of the soluble receptor sites for adrenal steroids have been characterized on basis of affinity constants and specificity of the receptor(s) for various steroid ligands. Brain tissue was obtained of rats that were previously adrenalectomized for depletion of endogenous adrenal steroids and extensively perfused with saline through

Adrenal Steroid Antagonism

the heart for removal of plasma transcortin. Binding assays described below were performed in a low ionic strenght buffer (5 mM Tris buffer, 1 mM 2- mercaptoethanol, 1 mM EDTA, pH = 7.4) without ammonium molybdate. Separation of bound from free ^{3}H-steroid occurred via filtration over small sephadex LH_{20} columns (6).

^{3}H-Corticosterone (^{3}H-CS) labeled sites were clearly distinct from plasma transcortin on the basis of physicochemical properties, steroid binding specificity and immunological characterization (7-9). Our immunocytochemical studies showed that transcortin-like molecules were present intracellularly in some anterior pituitary cells (10), but not detectable in the brain. The distribution of cytosolic ^{3}H-CS-labeled receptor sites parallels the regional differences, observed in cell nuclear uptake in vivo (see section 3, for anatomy figure 3). In limbic brain regions (especially hippocampus and septum) the highest amount of sites was found. Parts of the midbrain (e.g. raphe area) are practically devoid of such receptor sites (11), while substantial CS binding was found in motor-neurons of the spinal cord (11,12). Furthermore, in hypothalamus and pituitary a considerable amount of high affinity binding sites could be measured (7, 13-15).

Scatchard analysis of ^{3}H-CS binding to soluble macromolecules present in hippocampal cytosol, measured over a 50-fold concentration range (0.5 - 25 nM), provided a linear plot with the following binding characteristics: B_{max} = 424 fmoles/mg protein, K_d = 3.4 nM. When ^{3}H-dexamethasone (^{3}H-DEX) is used, similar binding characteristics could be obtained: B_{max} = 507 fmoles/mg protein, K_d = 5.0 (13). Further insight in the binding specificity of the two glucocorticoids could be gained after determination of the relative binding affinity (RBA) of several steroids for the sites labeled with either ^{3}H-CS or ^{3}H-DEX (Figure 1). Hippocampal cytosol was incubated with a fixed amount (chosen so as to be near the K_d-level, i.e. 5.0 nM) of each ^{3}H-steroid in the presence of

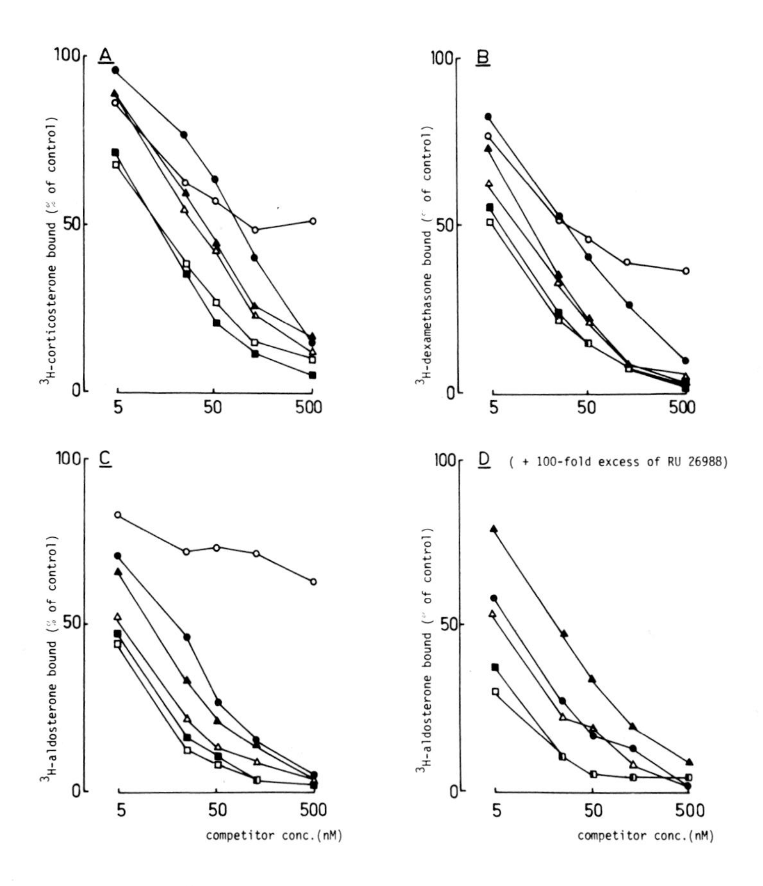

Figure 1: Competition of various steroids for specific binding of ^{3}H-corticosterone (A), ^{3}H-dexamethasone (B), ^{3}H-aldosterone in absence (C) or presence (D) of a 100 fold excess of unlabeled RU 26988 in hippocampal cytosol. Incubations were carried out at 4 °C.

The values shown are the means of three separate determinations; the SE values were less than 10% for each point.

0 —— 0: RU 26988; ● —— ●: aldosterone; ▲ —— ▲: dexamethasone

△ —— △: progesterone; □ —— □ : DOC; ■ —— ■ : corticosterone

increasing concentrations of unlabeled competing steroids. For ^{3}H-CS the RBA-ranking was as follows: B = DOC > PROG = > DEX > ALDO. The same order of potency was measured with ^{3}H-DEX as the binding ligand. The "pure" glucocorticoid RU 26988, defined on its exclusive glucocorticoid action in peripheral tissues (16) was used for further analysis of the nature of the sites labeled with ^{3}H-CS and ^{3}H-DEX. Inclusion of RU 26988 in a 100-fold excess displaced both ^{3}H-steroids to a maximum of 45%. The fraction occupied by RU 26988 then actually represents exclusively a population of glucocorticoid receptor sites. The remaining "non-glucocorticoid" receptor sites were further analyzed on binding affinity to the various steroid ligands, in an attempt to select ligands that had a comparable or even more pronounced affinity to the ^{3}H-CS labeled sites in rat hippocampus in the presence of a 100-fold excess of RU 26988. Since the glucocorticoid receptor sites were now occupied with RU 26988 the affinity of potent synthetic glucocorticoids such as dexamethasone decreased considerably to the remaining sites. In this study the same incubation conditions were used but separation of bound and free ^{3}H-steroid occurred with Dextran coated charcoal and on microtitration plates under these assay conditions we found that CS still displayed the highest affinity of 29 steroids, related to the pregnane structure (17). Aldosterone had a RBA of 11% and DEX of 6% compared to CS (100%).

Scatchard analysis of ^{3}H-ALDO binding in rat hippocampus gave a curvilinear plot, indicative for the presence of two populations of binding sites (Figure 2). The two constructed lines are respectively representing a high affinity site with K_d = 2.2 nM and B_{max} = 70 fmol/mg protein, and a low affinity binding site with K_d = 30.3 nM and B_{max} = 367 fmol/mg protein. Inclusion of a very small amount of CS (0.6 times excess) or a 100-fold excess of RU 26988 linearized the ALDO Scatchard plot, leaving only the high affinity site available

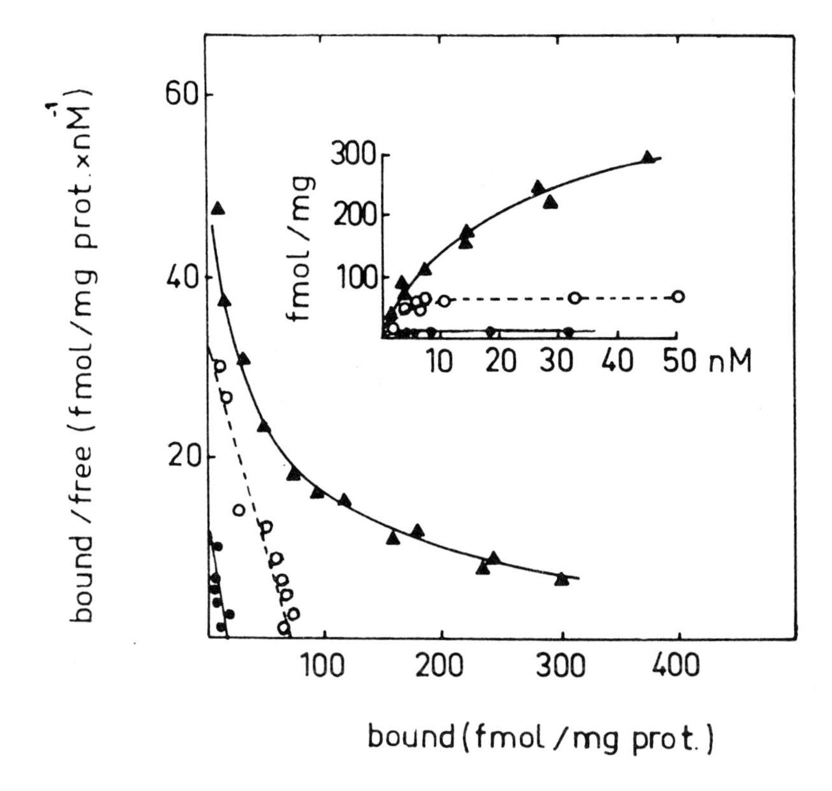

Figure 2: ^{3}H-Aldosterone binding in vitro to soluble macromolecules in hippocampal cytosol and competition with unlabeled corticosterone. Hippocampal cytosol was incubated with increasing concentrations of ^{3}H-aldosterone in the absence (▲ —— ▲) and presence of corticosterone in a concentration of 0.6 times (0 --- 0) or 10 times (● —— ●) that of ^{3}H-aldosterone. Nonspecific binding was assessed by inclusion of a 500-fold excess of unlabeled aldosterone. Each point represents the mean of triplicate determinations in a single experiment. Inset, saturation analysis in the absence (▲ ——▲) and presence of unlabeled corticosterone (0 --- 0, resp. ● —— ●).

(13). Increasing the amount of CS to ten-fold excess also depressed the high affinity ALDO binding sites. The RBA for ^{3}H-ALDO-labeld sites is in the order of: CS = DOC > PROG > DEX > ALDO. Inclusion of a 100-fold excess of the pure glucocorticoid RU 26988, which leaves only high affinity ALDO sites available, resulted in a decreased RBA of DEX; the order is then as follows: CS = DOC > PROG > ALDO > DEX. Nevertheless, CS and DOC still posses the highest RBA for these "mineralocorticoid" sites. Thus, based on Scatchard analyses of binding to soluble macromolecules in a particular brain region, the hippocampus, it appears that glucocorticoid, mineralocorticoid and/or "corticosterone preferring" receptor sites can be distinguished. This receptor heterogeneity is further discussed in section 5.

3. Cellular localization of adrenal steroid receptors.

When ^{3}H-CS is administered i.v. or s.c. to ADX rats, the labeled steroid is preferentially retained by cell nuclei of limbic structures, e.g. hippocampus, lateral septum, amygdala and parts of the cortex (18). Autoradiography revealed a principal localization in neurons of limbic brain regions (19-21). Most heavily labeled are the neurons of the hippocampus pyramidal cell layer and the dentate gyrus, the lateral septum, cortical and basal amygdala, entorhinal and cingular cortex. A more disperse distribution of labeled neurons is found in olfactory nucleus, habenular nucleus and red nucleus. Motor neurons of the cranial nerve nuclei of the spinal cord are heavily labeled. There exist only a few scattered labeled neurons in the hypothalamus. Apart from neuronal localization some glial cells also retain ^{3}H-CS (21). Thus, in general, this distribution pattern matches the regional distribution pattern of soluble ^{3}H-CS-labeled binding sites in the brain, as determined *in vitro* (see

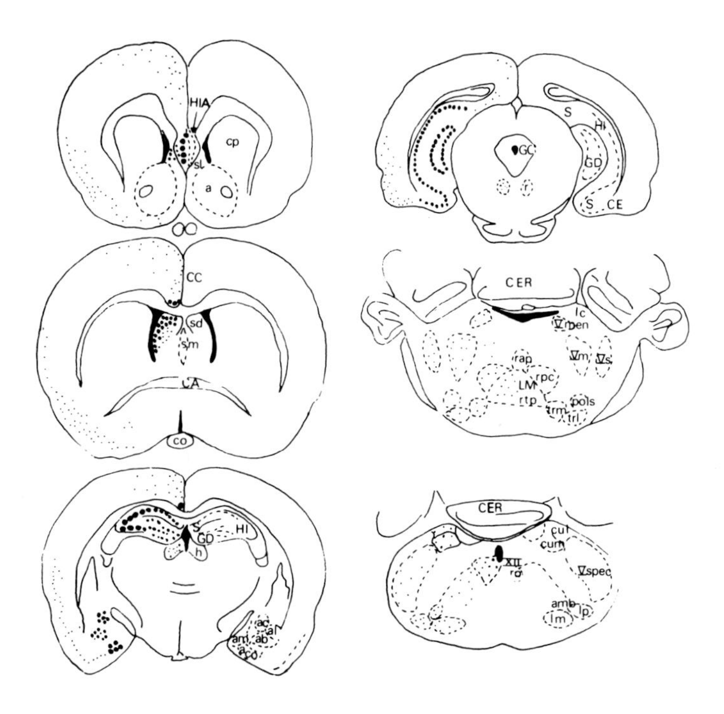

Figure 3: Schematic drawing of corticosterone retained by receptors in cell nuclei of the rat brain. The extent of cell nuclear uptake is represented by the intensity of the dots at the left hand part of the brain sections. Data taken from autoradiograms (2,3,4).

section 2 and figure 3). Based on radioimmunoassay it appeared that in non-adrenalectomized rats endogenous CS is localized in a nearly identical pattern over cell nuclei of the various brain cell regions. Highest concentrations are found in cell nuclei of the hippocampus, which implies that in intact rats in the presence of the various adrenocortical secretion products corticosterone is preferentially retained by the hippocampal receptor system. Circulating levels of ALDO, PROG and DOC do not displace CS in significant amounts.

^{3}H-Cortisol uptake reaches in the hippocampus about 5% of the amount of ^{3}H-CS; the neuroanatomical distribution however, is identical to that of CS (22).

In contrast to CS and cortisol, synthetic glucocorticoids are poorly retained in cell nuclei of neurons of the limbic brain structures (22-26). The retention of ^{3}H-DEX is evenly distributed over the brain regions and does not exceed 10% of the retained amount of ^{3}H-CS (23). Endothelial cells of blood vessels, epithelial cells lining the choroid plexus and ventricles show the most pronounced labeling, as revealed by autoradiography. Also cells of the circumventricular organs and the medial basal hypothalamus retain ^{3}H-DEX. Glial cells are labeled weakly throughout the whole brain (25,26).

In ADX rats, ^{3}H-ALDO displays a neuroanatomical distribution pattern almost identical to that of ^{3}H-CS (27,28). Extrahypothalamic neurons in limbic brain structures and the motor neurons of the cranial nerves, thus, show the highest uptake. The rate of uptake of ^{3}H-CS is different. Fifteen minutes after administration of ^{3}H-ALDO, its concentration in cell nuclei is maximal, but this amount is still half the amount accumulating after an equimolar dose of ^{3}H-CS as measured 60 min after administration (13). This difference may be due to the more rapid clearance of ^{3}H-ALDO from the plasma. The other naturally occurring mineralocorticoid, DOC, is in its tritiated form only poorly retained by cell nuclei in vivo; about 1% as compared to that of ^{3}H-CS (22). Finally,

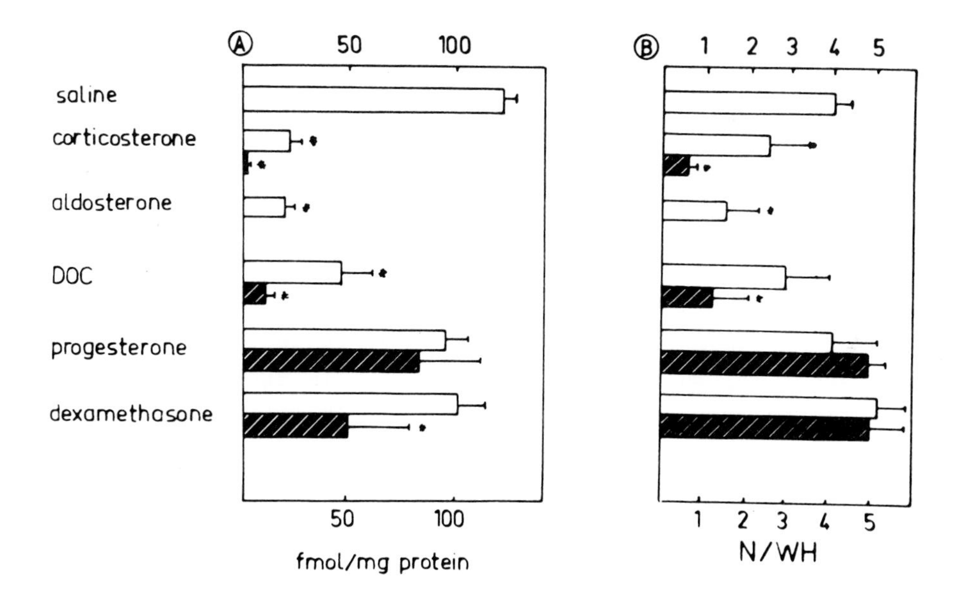

Figure 4: Blockade of hippocampal cell nuclear uptake of ^{3}H-corticosterone in vivo with prior administration of unlabeled steroids, administered in a dose of 30 µg (open bars) or 300 µg (hatched bars) per 100 g rat. The unlabeled steroids were administered 30 min prior to administration of a tracer dose of ^{3}H-corticosterone (50 µCi) to rats adrenalectomized three days previously. The animals were sacrificed by decapitation one hour after administration of the tracer dose.

A) cell nuclear retention of ^{3}H-corticosterone, expressed as fmol/mg nuclear protein. Saline: adrenalectomized rats receiving saline 30 min prior to the tracer.

B) ratio of cell nuclear uptake (see A) to tissue uptake of ^{3}H-corticosterone.

*) $p < 0.05$ vs saline (Newman Keuls multiple range test).

^{3}H-PROG uptake is negligible. However, the synthetic progestagen, ^{3}H-RU 5020, revealed a predominant localization in cell nuclei of neurons of hypothalamic and preoptic regions (29).

These differences in uptake of labeled steroids partly can be explained by binding to blood proteins, metabolism and clearance from the vascular compartment, and penetration of brain cells. Yet, the findings on receptor specificity and binding affinity of the various ligands suggest that the differences in cell nuclear localization in the brain reflect an intrinsic property of the brain steroid receptors (6).

4. Competition for cell nuclear localization of ^{3}H-corticosterone.

Endogenous CS in intact rats or pretreatment of ADX rats with CS (30 μg/100 g bodyweight) suppresses cell nuclear retention of a tracer dose of ^{3}H-CS in the hippocampus (7, 13,24). Pretreatment with the same dose of ALDO or DOC similarly suppresses the uptake of ^{3}H-CS. DEX and PROG do not compete (Figure 4). Only a ten times higher dose of DEX (300 μg/100 g b.w.) is partially effective. Some further insight in the mode the steroids interact with the cell nuclear retention mechanism of ^{3}H-CS is derived from a comparison of the amount of ^{3}H-CS, localized in the cell nuclear compartment and in the whole tissue. This ratio (N/WH) is depressed after pretreatment with CS and ALDO, indicating that the two steroids in particular block cell nuclear localization and that they do so via interaction with the same receptor. DEX as well as PROG both suppress the uptake in cell nuclei and in the tissue to the same content (unchanged N/WH-ratio), reinforcing the notion that these two steroids do not interact *in vivo* to a great extent with the receptor-mediated cell nuclear retention of CS.

5. Heterogeneity of the adrenal steroid receptors.

The retention of labeled ^{3}H-CS and ^{3}H-ALDO in cell nuclei of limbic brain regions in vivo indicate the presence of CS preferring receptor sites, that resemble mineralocorticoid receptors. The in vitro binding experiments in cytosol of brain thoroughly perfused to remove transcortin show, that CS has the highest affinity to these receptors provided that glucocorticoid receptors are occupied with RU 26988. ALDO and DOC have lower affinity to these CS preferring sites. Heterogeneity is clearly demonstrated with Scatchard analysis of ^{3}H-ALDO binding, that reveals a small population of high affinity sites and a large population of RU 26988 sensitive sites.

The fraction of labeled sites that can bind RU 26988 are glucocorticoid receptor sites, which are probably identical to the glucocorticoid receptors occurring in peripheral glucocorticoid target tissues, such as the corticotrophs of the pituitary, the thymus and others. In the brain these glucocorticoid receptors are present in endothelial and epithelial cells (25). Enucleated optic nerve does not contain neurons, since these degenerate after denervation, but only glial cells. Cell nuclei of this tissue retain ^{3}H-DEX better than ^{3}H-CS, which is in accordance with a preferential glial cell localization of glucocorticoid receptors (30,31). However, also certain neuronal cell groups may contain glucocorticoid receptors as appeared from receptor assays in microdissected brain regions. A large proportion of glucocorticoid receptors was found in the lateral septum and in the paraventricular nucleus, the site of synthesis of corticotrophin releasing hormone (CRH) (32).

A number of studies appeared in 1982 and 1983, that were devoted to the heterogeneity of adrenal steroid receptors in brain cytosol (see Table 1). Those studies performed without inclusion of RU 26988 stated the presence of only

TABLE I

STUDIES ON HETEROGENEITY OF ADRENAL STEROIDS IN BRAIN

Brain region	Removal CBG	Molybdate	Assay	RU 26988	GR	MR	CSR	Reference
hippocampus	perfusion	no	DCC	yes	+	+		29
hippocampus	perfusion	no	LH_{20}	yes	+	+	(+)	25
striatum	perfusion	no	LH_{20}	no	+			33
hippocampus	perfusion (cortisol-17β 2h)	yes	DCC	no	+			34
hippocampus	perfusion	(yes)	LH_{20}	yes	+	+		35
whole brain	perfusion	yes	LH_{20}	no	+			36
hippocampus	isoelectric focussing	no	DCC	yes	+	+		37
hippocampus	perfusion	yes	DCC	yes	+	+	(+)	38
hippocampus	hydroxyl apatite	yes	DCC	yes	+	+	(+)	39

glucocorticoid receptor sites (33,34,36). The use of RU 26988 permitted to differentiate glucocorticoid and mineralocorticoid receptor sites (25,29,35,37-39). Characterization of these sites occurred by kinetic analyses or by physicochemical properties revealed by isolectric focussing (37) or by sucrose density centrifugation (35,37). The mineralocorticoid and glucocorticoid receptor sites were separated by these two techniques. Wrånge and Gustafsson (37) identified a different pattern of labeled proteins after limited trypsin digestion and isoelectric focussing. The mineralocorticoid receptors of hippocampus and kidney showed an identical fragmentation pattern. In addition, an antiglucocorticoid receptor antibody showed no cross reactivity with the mineralocorticoid receptor.

The appearance of receptors in activated or unactivated form further complicates the issue of receptor heterogeneity. Molybdate has been advocated as the substance to stabilize receptors in the unactivated state. Activation of the glucocorticoid receptor has been studied in the brain (34,36,49). but there have been no systematic studies undertaken on the affinity changes to the various steroid ligands depending on or resulting from receptor activation.

The binding studies disclosed in most cases a high affinity of CS for the mineralocorticoid receptors provided CBG was removed from the cytosol (25,35,38,39). If CBG was present the affinity of CS dropped considerably in contrast to ALDO which does not bind to CBG and DOC which has only 4% of the affinity. Therefore, the presence of CBG is a likely explanation for the disparity in binding affinity of the steroids to the kidney and brain receptor systems. It is known where the actual localization of CBG is in the kidney, but the brain virtually blocks CBG or CBG-like molecules after perfusion (10,11).

The presence of CBG-like binders in tissue in larger concentrations than can be accounted for by blood contamination has been observed in a number of tissues (42). One of these tissues is the pituitary (40,41,43-45). CBG-like binding activity remained associated with isolated anterior pituitary cells (44) and antigens to a CBG antiserum were localized immunocytochemically intracellularly in some pituitary cells (10,46).

CBG or intracellular CBG-like binding molecules, thus, may modulate the interaction of CS with mineralo- or glucocorticoid receptor sites. This is the case in a glucocorticoid target tissue such as the pituitary and the mineralocorticoid target cells in the kidney. CBG-like molecules may serve as a buffer against rapidly changing steroid levels or otherwise constitute an intracellular pool of binding molecules to ensure availability of steroid in the cell (44). The lack of these molecules in the hippocampus implies that the receptor system responds rapidly to changes in CS concentration and is most likely the reason that the CS receptor system manifests itself under the appropriate experimental conditions as a receptor species with highest affinity to CS *in vitro* in cytosol and *in vivo* after administration of ^{3}H-CS. This line of reasoning has led to the term "CS preferring receptors" in rat hippocampus as a satisfactory term to describe the biochemical and pharmacological (see below) features. It is conceivable that mineralocorticoid activity is only expressed in cases where access to the receptor is blocked by CBG-like molecules. However, at present it is not yet possible to exclude a separate population of specific mineralocorticoid receptor sites. Perhaps the CS preferring population of receptor sites has its origin early in evolution (47). The glucocorticoid receptor blocked by RU 26988 and a separate population of high affinity sites may be later specializations in CS receptor phylogeny. Autoradiography and immunocytochemistry are the techniques to further resolve the cellular localization of such receptor molecules in the brain.

6. Adrenal steroid action

Environmental challenges (stress) evoked by psychological, traumatic, toxic or infectious events result in autonomic, behavioural and neuroendocrine responses aimed to maintain homeostasis. The neuroendocrine responses involve primarily activation of CRF's (CRH and vasopressin), that stimulate pituitary ACTH release and subsequently lead to enhance CS secretion of the adrenal cortex. CS feeds back on brain and pituitary. In the brain CS acts on neuronal processes involved in interpretation of environmental stimuli and expression of behavioural and neuroendocrine responses. In general, feedback action leads to blockade of pituitary-adrenal activity and elimination of behaviour that is of no more relevance. Autonomous responses are also modulated under the influence of circulating levels of CS. The concept of adrenocortical and medullary hormones in control of stress-responses stems from the pioneering work of Selye (50) and Cannon (51).

As behavioural paradigm for CS actions on the brain we studied aspects of fear motivated behaviour (conditioned avoidance behaviour, exploration of a novel environment and acquired immobility tests).

In order to resolve the physiological function of gluco- and mineralocoids in brain functioning, studies were designed based on the properties of the respective receptor systems. This implies that deficiencies occurring after bilateral removal of the adrenals should be specifically restored with either gluco- or mineralocorticoids in a dose range not exceeding the physiological level. Besides these requirements the implication of medullary catecholamines in disturbed behaviour and the high circulating ACTH levels after ADX should be considered. Removal of the adrenals results in both adrenocortical steroids and medullary hormones and in high levels of circulating ACTH. Sofar, only a few studies have

TABLE II

EFFECT OF ADRENALECTOMY AND SUBSEQUENT STEROID REPLACEMENT ON BEHAVIOUR

Behavioural paradigm	Effect of ADX	Agonist[a]	non effective[a]	Reference
Free-operant avoidance effect of pre-stimulation	absent	ALDO	CS	70
Forced extinction of passive avoidance	impaired	CS	PROG,DOC,DEX	58,59
Adrenaline action on passive avoidance	changed dose-response curve	CS	DEX	54,55
Extinction of food-reinforced response	facilitated	CS	DEX	60,61
Exploration novel environment	reduced	CS	PROG,ALDO,DEX	4,62,63
Retention of immobility during swimming	reduced	DEX,CS	DOC	68,69

[a] Steroid specificity in the normalization of disturbed behaviour after ADX; "agonist" defined on basis of restoration of deficit, "non effective" on being inactive.

been reported in which implication of all these criteria has been fulfilled. In Table 2 an overview is given of behavioural responses that are changed after bilateral ADX and restored after steroid substitution in the physiological dose range.

6.1. Agonist action of corticosterone on behaviour.

The behavioural paradigm was a one-trial learning passive avoidance test (52). Rats, like other rodents, prefer dark to light. If placed on a brightly lit, elevated platform, they readily enter into a dark chamber to which the platform is attached. Following adaptation to the dark compartment and a pretraining to enter the dark, the rats receive an unescapable scrambled footshock through the grid floor of the dark compartment (learning trial). Passive avoidance behaviour manifests itself in a long latency to re-enter the dark compartment from the elevated platform 24 h after the learning trial.

It was found that bilateral removal of the adrenals up to 5 days prior to a single learning trial of passive avoidance behaviour induces a severe deficit in the retention of this behaviour (53). Replacement therapy with CS chronically or with a single injection failed to normalize, while adrenaline treatment (dose range 0.005 to 5 μg/100 g b.w.) resulted in a dose-dependent recovery of the behavioural deficit of ADX-rats. Although adrenomedullectomy (ADXM) resulted in a similar behavioural deficit as ADX, higher dose of adrenaline were required to correct the retention of passive avoidance of ADXM rats than that of ADX ones (54). Pretreatment with CS in a physiological dose (30 μg/100 g b.w.) of ADX rats decreased the efficiency of adrenaline to affect impaired inhibitory avoidance behaviour. This shift in efficacy of adrenaline was about tenthousand fold (Figure 5). DEX and

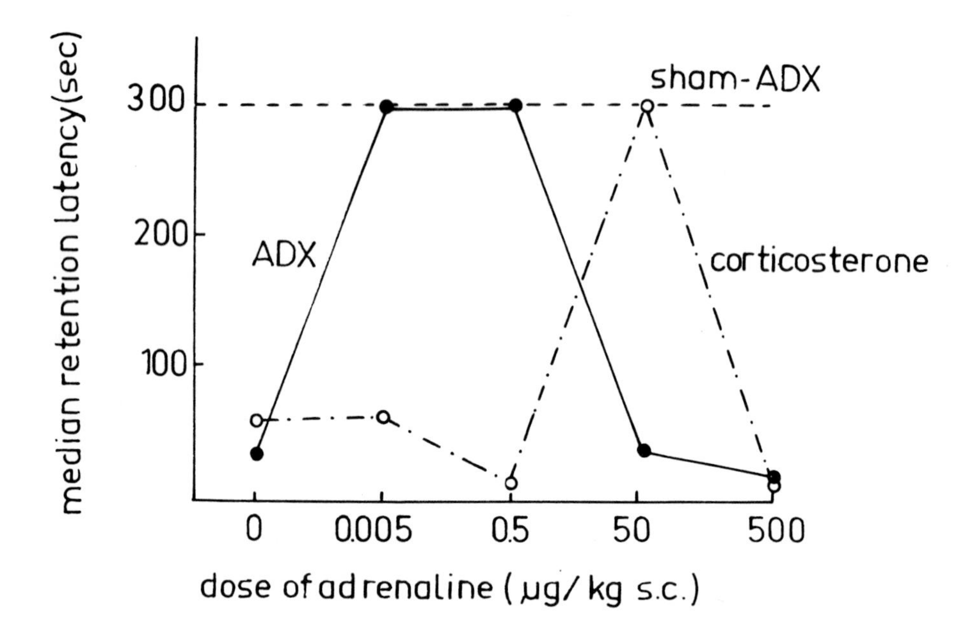

Figure 5: Effect of corticosterone-CS (30 μg/100 g bodyweight s.c.) administered one hour prior to the learning trial. Rats received adrenaline (A) in doses indicated at the horizontal axis immediately postlearning. Rats were adrenalectomized 48 h before learning trial and retention of the passive avoidance behaviour was tested 24 hours after learning. ——, ADX; —.—.—, ADX + corticosterone; — — sham-ADX. Data are median latencies (sec.). n = 12 animals per group.

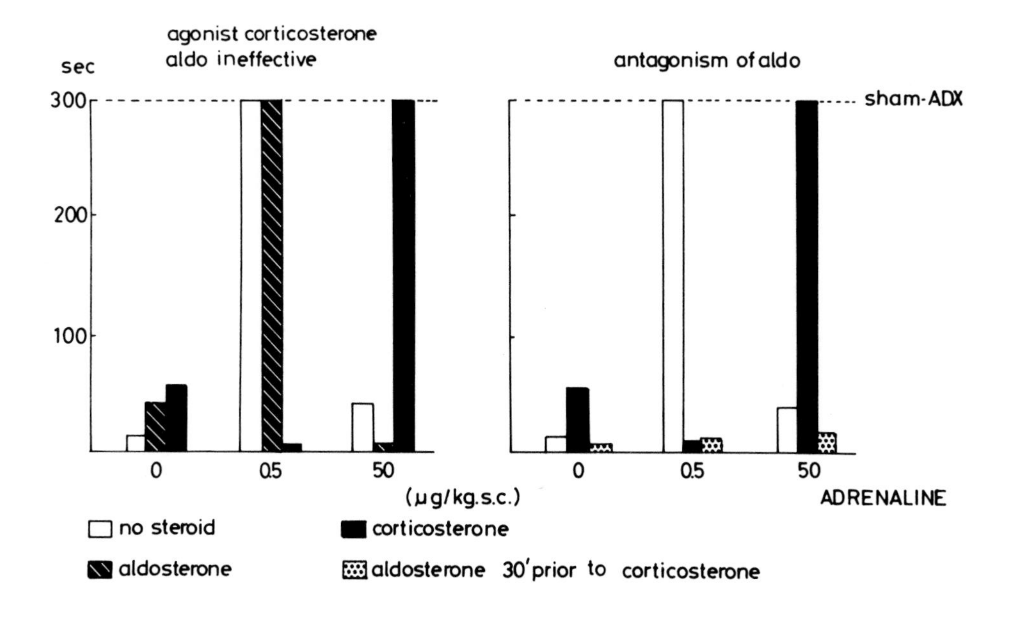

Figure 6: Effect of Aldosterone(ALDO) administered 90 min before learning and corticosterone(CS) 60 min before learning (left hand side) and of ALDO pretreatment (- 90 min) of CS replaced (- 60 min) rats (right hand side). Rats received either 0.5 µg/kg or 50 µg/kg of adrenaline (A) immediately post-learning.
Data are median latencies (sec). n = 8 animals.

ALDO (Figure 6) administered in an equimolar dose failed to modify the behavioural responsiveness to adrenaline (54,55). Accordingly, CS specifically decreases the efficacy by which adrenaline affects later passive avoidance retention behaviour of ADX rats, without having an own effect on behaviour. The specificity and the physiological conditions of this action suggest involvement of the hippocampal CS receptor system.

While CS had a modulatory effect on behavioural responses facilitated by adrenaline, it appeared that another aspect of the avoidance response was specifically under control of CS. Confinement of the rats to the dark compartment where they have experienced the aversive stimulus for 5 min shortly (3 h) after the learning trial, results in a complete extinction of the passive avoidance response ("forced extinction") (Figure 7). Apparently, the experience of the absence of footshock punishment in the dark compartment was sufficient to eliminate a behavioural response that is no longer relevant. Removal of the adrenals one hour before the forced extinction training led to an almost complete absence of extinction, as studied by the retention test 24 h later, i.e. forced extinction was ineffective in ADX rats. A low dose of CS (30 μg/100 g b.w.) administered immediately after ADX normalized extinction behaviour (58, 59). For experimental scheme see Figure 5.

Progesterone, DOC and the synthetic glucocorticoid DEX were not effective at the same or ten times higher dose as CS (Figure 8). The dose of CS (30 μg/100 g b.w.) was sufficient to maintain receptor occupation in the hippocampus up to the level of sham-operated controls (70 - 80% occupation) and produced plasma CS levels, that were within the physiological range (59). Furthermore, since DEX is ineffective, while the steroid is a potent suppressor of pituitary ACTH release, it seems unlikely that pituitary ACTH participates in the steroid effects on behaviour.

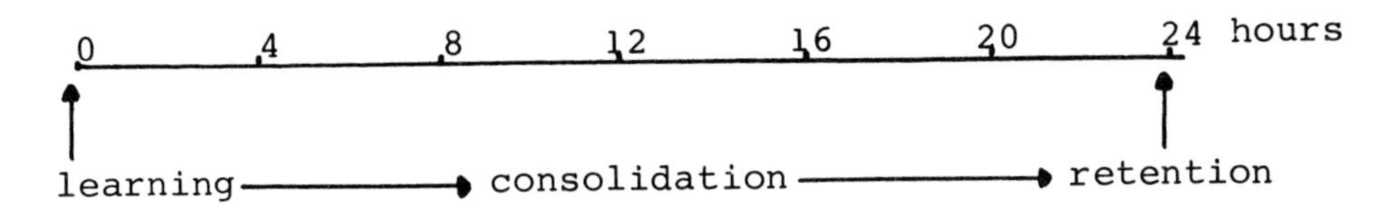

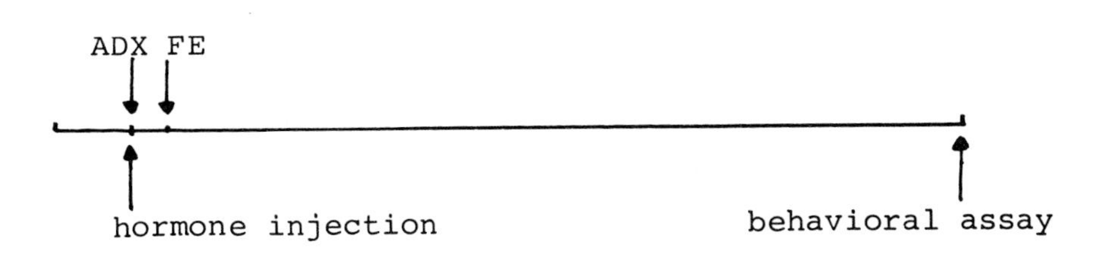

Other studies suggested that the specific involvement of CS in aversively-motivated also holds for extinction of appetitive tasks (60,61). A rapid extinction, using an appetitive runway of chronically ADX rats was observed. The extinction of the food-motivated behaviour was delayed when ADX rats received a subcutaneously implanted CS pellet, that provided physiological plasma levels of circulating CS. Implanted DEX pellets did not result in effects on this behaviour.

A novel environment evokes arousal and consequent activation of pituitary-adrenal activity. When rats are placed in a novel environment, they show a number of spontaneous behaviours. These are exploration (ambulation and rearing) and grooming responses. Since a novel environment enhances CS secretion it may be not surprising that the hormone also affects indices for exploratory activity, when the animal is placed in an open field test situation (4,62,63). Ten days after ADX decreased ambulatory and rearing activities are found in rats, if placed in an open field and tested for 5

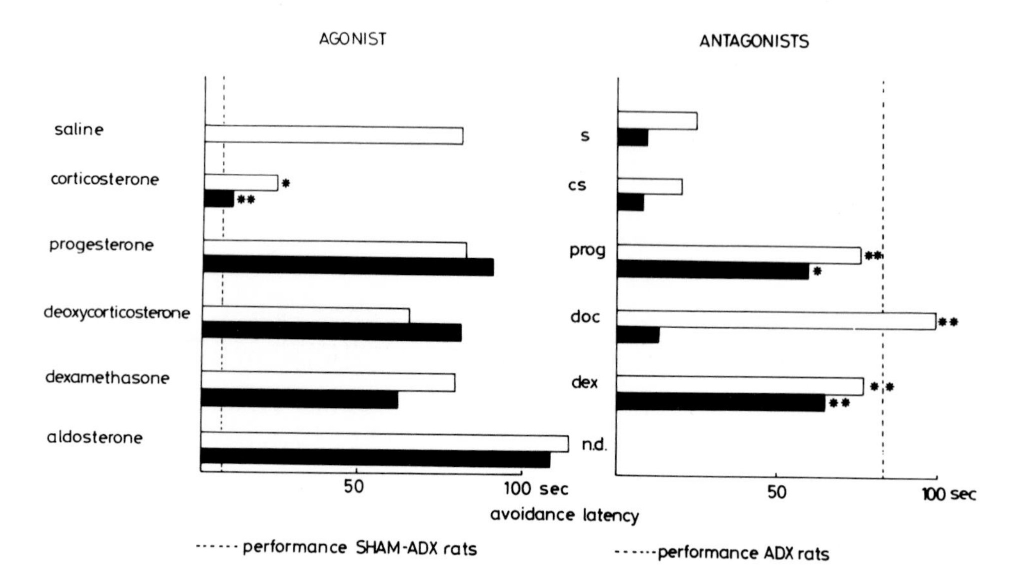

Figure 8: Agonist effect of corticosterone and no effect of progesterone deoxycorticosterone, dexamethasone and aldosterone on extinction behaviour of ADX rats. (left hand figure). Treatment was given s.c. 60 min before forced extinction, that is immediately after ADX in doses of 30 μg (open columns) or 300 μg/100 g b.w. (black columns).
Antagonistic properties of progesterone, deoxycorticosterone and dexamethasone on extinction behaviour of ADX rats substituted with corticosterone (right hand figure). The steroids were given at two hours before forced extinction in a dose of 30 μg/100 g b.w.. This dose (open columns) or 300 μg/100 g b.w. (hatched columns) corticosterone was administered to all rats 60 min before forced extinction.
Significance of differences between steroid vs saline treated (left) and saline-corticosterone vs steroid corticosterone treated rats is given (Mann-Whitney test; * $p < 0.05$, ** $p < 0.01$).

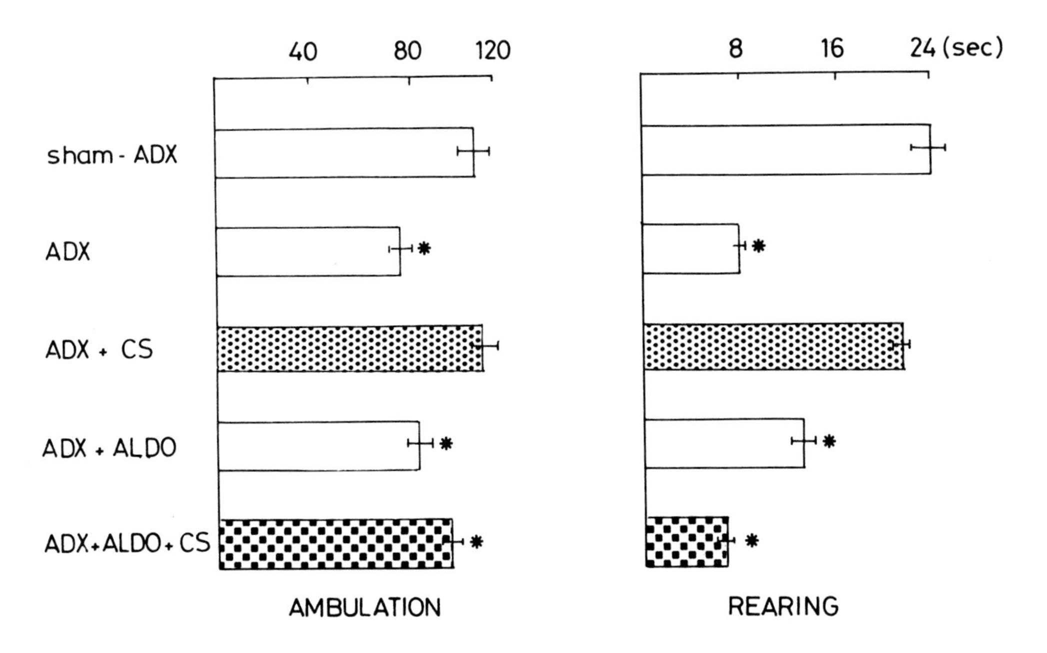

Figure 9: Five minute open field behaviour of ADX rats. Treatments were given to ADX rats (10 days) 1 h before the open field test (30 μg/100 g b.w. subcutaneously).
[dotted pattern] shows agonist effect of corticosterone.
[checkered pattern] shows antagonist effect of aldosterone on normalization of exploratory activity (ambulation and rearing) by corticosterone. Aldosterone was administered 90 min prior to open field test and 30 min prior to corticosterone (dose ALDO 30 μg/100 g b.w. subcutaneously).
* $p < 0.01$ vs SHAM-ADX and ADX + CS (Newman-Keuls multiple range test).

min (64). These two components of exploratory behaviour are restored to the level of sham-operated controls when ADX rats are treated with a single s.c. injection of CS (30 μg/ 100 g b.w.) one hour prior testing. DEX, PROG and ALDO were ineffective (Figure 9).

6.2. Agonist action of glucocorticoids on behaviour.

In contrast to the CS-specific behavioural actions, that are generally believed to require participation of the hippocampus, and that can be explained on basis of the involvement of "specific" CS receptors in that brain region, another type of behavioural test seems to display a different steroid specificity. Recently, an experimental mode, designed to screen potential antidepressants (Porsolt swimming test) has been described (65-67). Rats are individually placed in narrow (plexi)glass cylinders containing 15 cm of water at 25 °C. Initially, the rats will swim vigorously, but over a 15 min test period they spend increasing amounts of time relatively immobile, moving only sufficiently to keep their head above the water (68). When such rats are retested 24 hours later, they spend about 70% of a 5 min retest period immobile. ADX, performed 4 - 6 days previously, does not interfere with the initial 15 min swimming, but retested 24 hours later, ADX rats show a significantly reduced immobility (30%) compared to sham-operated controls (70%) (68,69). The effects of ADX can be reversed by the administration (15 min after the initial swimming test) of DEX (1 - 10 μg/100 g b.w.) or CS (1 - 5 mg/100 g b.w.) . DOC was ineffective (68). Thus, the steroid specificity in reversing the ADX effect on immobility appears to be regulated by a receptor system that requires glucocorticoids as agonists;DEX is already potent in low doses, while supraphysiological doses of CS are necessary.

6.3. Agonist action of mineralocorticoids on behaviour.

Concerning specific mineralocorticoid actions on fear-motivated behavioural responses, only a very few studies have been reported. An example is the enhancement of an avoidance response following prestimulation by a very weak stressful event (e.g. handling or an air blast). ADX completely prevents the effect of prestimulation. Replacement with ALDO restores this phenomenon, while CS is ineffective (70). Naturally, salt appetite is most potently affected by mineralocorticoids (71).

6.4. Antagonists of corticosterone action on behaviour

Given the steroid specificity in normalization of behavioural deficits, induced by bilateral removal of the adrenals, we have also designed experiments to reveal steroid antagonism in behaviour. Since ALDO and DOC blocked the uptake of ^{3}H-CS in neuronal cell nuclei of limbic brain regions, we assumed that these steroids may act as antagonist in appetitively and aversively motivated behaviour and exploratory behaviour. Indeed DOC and ALDO pretreatment of CS administration to ADX rats abolished normalization of behaviour. The action of CS was antagonized by DOC on forced extinction of passive avoidance behaviour and by ALDO on the facilitation of the avoidance response by adrenaline (4,55,58) (Figure 5,6,8). In the latter behavioural paradigm, ALDO blocked the dessensitizing action of CS on responsiveness to A. In addition treatment of ADX rats with ALDO at 30 min prior to CS administration prevented the normalization of the indices for exploratory activity (ambulation and grooming) by the naturally occurring glucocorticoid (63).

DEX pretreatment, however, also blocked normalization of the above mentioned behaviours by CS (58,59,63). DEX in the dosage used does not interfere with CS uptake in limbic cell nuclei and DEX shows only about 6% of the affinity of CS to the CS receptor system. These biochemical observations suggest that DEX displays its antagonistic action via an entirely different mechanism, which counteracts the CS effect, and as such DEX can be considered as a functional antagonist

6.5. Antagonists of glucocorticoid action on behaviour.

The second type of behavioural response (immobility during swimming test), studied in relation to steroid specific normalization after ADX, was shown to require gluco-corticoids, and especially synthetic ones rather than CS. For this reason we used a synthetic antiglucocorticoid, RU 38486, in order to antagonize the effect of the potent synthetic glucocorticoid, DEX. RU 38486 (17-hydroxy-11-(4-dimethylaminophenyl)17-(1-propynyl)estra-4,9-diene-30ne) has been reported to display a rather high RBA for rat thymic glucocorticoid receptors. In vivo, this antagonist has been shown to block the action of glucocorticoids on ACTH secretion and diuresis (see chapter by Philibert). As shown in figure 10 injection of the antiglucocorticoid in a dose of 2.5 mg/100 g b.w. directly after the initial swimming test, i.e. 15 min prior to DEX administration (10 μg/100 g b.w.) completely blocked the normalizing effect of DEX on the duration of immobility during the 5 min retest period 24 h later. Furthermore, figure 11 shows that, using intact animals, administration of RU 38486 directly before the initial swimming, significantly decreased the immobility, scored 24 h later (69). Since it has been reported earlier (68) that hypophysectomized animals behave like intact ones during the retest period, the action of the anti-gluco-corticoid could not be explained by changes in pituitary ACTH

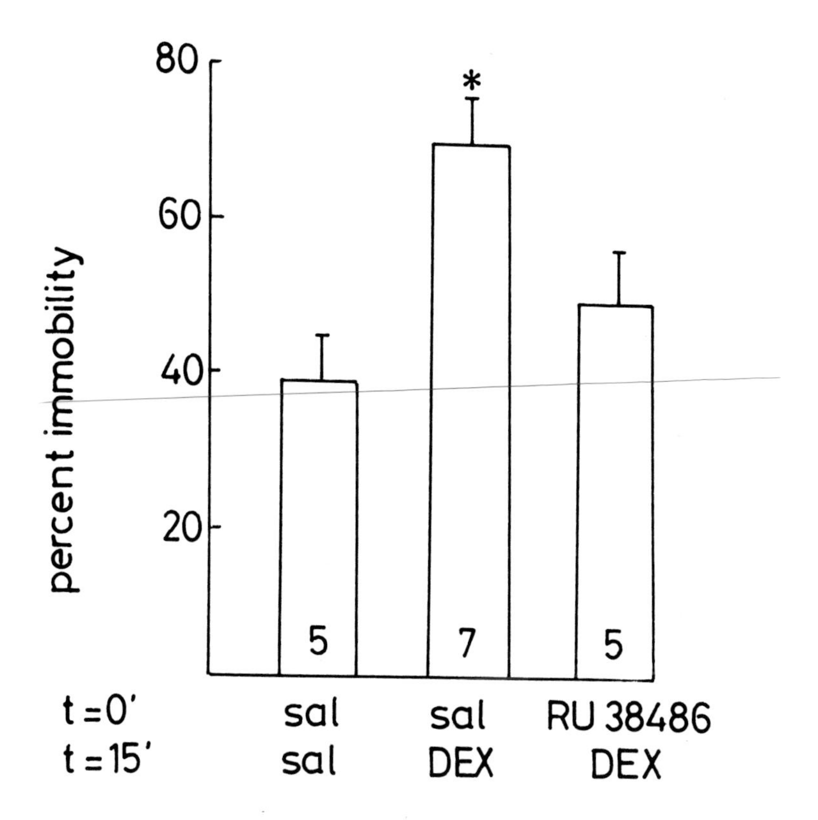

Figure 10: Antagonistic action of RU 38486 on immobility (during a 5 min retest period) of ADX rats substituted with DEX. The antiglucocorticoid (2.5 mg/100 g b.w.) or the vehicle (sal) was given s.c., directly after the initial swimming, i.e. 15 min prior to s.c. administration of DEX (10 μg/100 g b.w.) or vehicle.

* $p < 0.005$ vs (sal-sal) and (RU 38486-DEX) (Student-Newman-Keuls multiple range test).

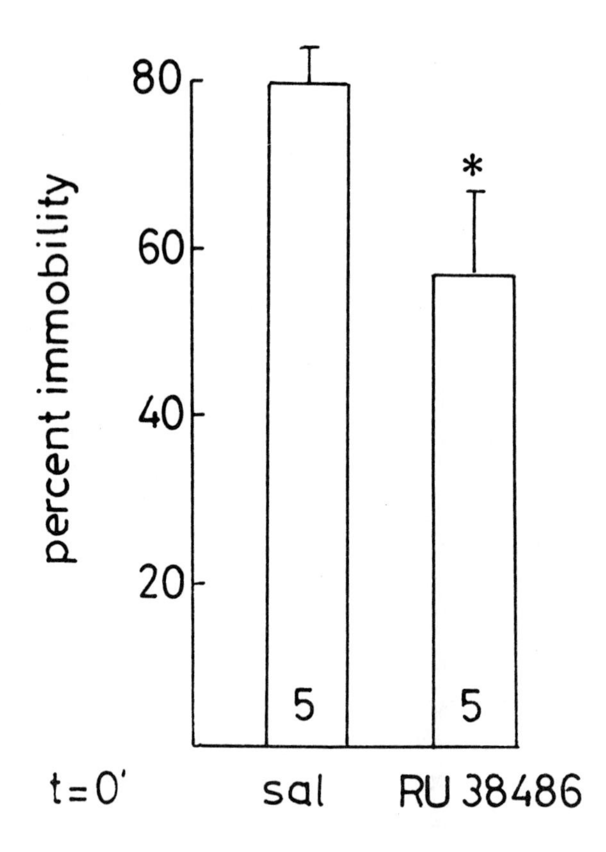

Figure 11: The effect of vehicle (sal) or RU 38486 (2.5 mg/ 100 g b.w.) s.c. administered directly prior to the initial swimming on immobility (during a 5 min retest period 24 h later) of intact rats.
* $p < 0.05$ vs sal injected rats (Student's t-test).

release. RU 38486 interferes with receptor mediated processes activated by synthetic glucocorticoids. Thus, while the previous fear-motivated and exploratory behaviours are specifically affected by CS suggesting mediation by hippocampal CS receptors, the swimming test represents a behavioural test situation, that is affected via glucocorticoid interaction with the glucocorticoid receptor system present in target tissues for potent synthetic glucocorticoids such as DEX.

6.6. Adrenal steroid agonist and antagonist action on brain biochemistry.

Although numerous effects of adrenal steroids on brain biochemistry have been reported, there have been not many studies designed to find neurochemical responses to adrenal steroid receptor interaction. Recently, some actions were reported that fulfill the criteria as prescribed by the receptor systems on localization, binding specificity and capacity. The strategy in this type of experiments is to define neurochemical disturbances after ADX, that are restored with doses of the receptor specific ligand in doses not exceeding the physiological range.

ADX reduces the amount of protein I in hippocampus, which is a constituent of neurotransmitter vesicles. CS induced protein I, while DEX was ineffective (72), which is in line with the specificity of the CS receptor system in the hippocampus. CS also affects GABA-reuptake (73) and the number of β-adrenergic receptor sites (74) in this tissue. The most striking specificity of CS action is, however, found in the 5-HT neuron, that projects from the midbrain raphe to the hippocampus. ADX reduces the activity of tryptophan hydroxylase (75,76) in midbrain and hippocampus, the uptake of the precursor tryptophan (77) and the release of 5-HT from

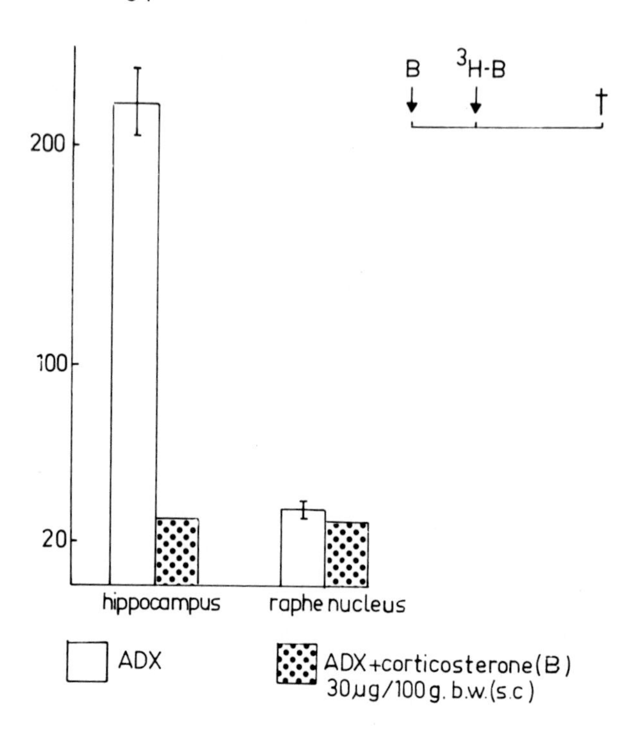

Figure 12: Cell nuclear uptake of ^{3}H-corticosterone in hippocampus and raphe area of ADX rats one hour after subcutaneous administration.

Pretreatment with corticosterone (30 µg/ CS/100 g b.w. s.c.) 30 min prior to administration of ^{3}H-corticosterone. Data are mean of 5 animals ± SEM and expressed as fmoles ^{3}H-corticosterone/mg cell nuclear protein.

Figure shows suppression of ^{3}H-CS uptake by unlabeled CS in hippocampus and very low cell nuclear uptake of the labeled steroid in the raphe area.

synaptosomes *in vitro* (78) and the synthesis rate of 5-HT in hippocampus and midbrain (79,80). Replacement of ADX rats with CS restores 5-HT synthesis (79,80). The action of CS on 5-HT synthesis occurs simultaneously in the raphe nucleus (cell body region) as well as in the hippocampus (terminal) region (80,81). The raphe neuron is devoid of CS receptors (81) (Figure 12). The 5-HT response is thus presumably triggered via receptor sites, which are localized post-synaptically in the hippocampal pyramidal neurons. Thus, it was concluded that CS induced a change in cellular metabolism to which the raphe neuron responds with increased 5-HT synthesis and release. The nature of the chemical processes in the hippocampal neurons that trigger the raphe 5-HT response is not known. It could be an excitatory projection of the hippocampus, since lesioning of the efferent pathways of the hippocampus prevents the CS induced 5-HT response (82).

CS is the only agonist, all other steroids including ALDO and DEX are ineffective. These steroids act as antagonist (80,81,83), when given prior to CS (Figure 13). Specificity of the CS action, thus, corresponds to the specificity of the receptor for CS in the hippocampus neurons and is the best biochemical correlate defined sofar with the CS action on fear-motivated and exploratory behaviour. Antagonism by ALDO may well be exerted via competition for the specific CS receptors in the hippocampus neurons. As was the case with the antagonist action of DEX on behaviour, it is unlikely that DEX exerts this effect via competitive antagonism. Regarding 5-HT, the finding could be explained by the reduced availability or blockade of the uptake process of the tryptophan precursor (77) after DEX, which counteracts the effect of a physiological dose of CS given to ADX rats.

DEX displays a more potent action than CS on glycerol-6-phosphate dehydrogenase (83) and ornithine decarboxylase (84). Brain damage also induces ornithine decarboxylase in

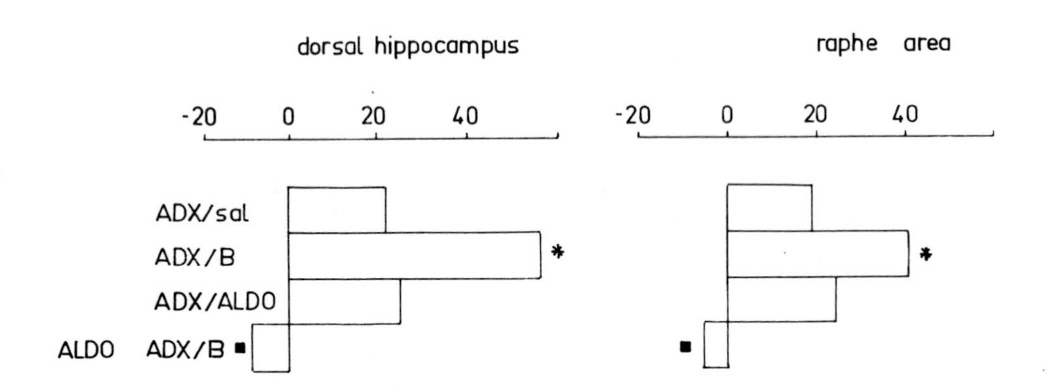

Figure 13: Effect of corticosterone and aldosterone on the pargyline induced accumulation of serotonin (5-HT) in the dorsal hippocampus and raphe area of the acutely 1 h ADX rat. Pargyline blocks 5-HT degradation and results in an increase in 5-HT concentration over time, which can be used as index for 5-HT synthesis rate. Given in the figure are Δ μg 5-HT/g tissue/0.5 h. B = corticosterone ALDO = aldosterone. Corticosterone is given at the time of ADX (dose 30 μg/100 g b.w., s.c.). Pargyline (150 mg/kg, i.p.) administered 30 min later. Sacrifice 60 min after corticosterone and 30 min after pargyline. Aldosterone is given (30 μg/100 g b.w., s.c.) 30 min prior to ADX and corticosterone replacement. Figure shows agonist action of corticosterone and antagonist action of aldosterone.
Data are mean of 16 animals ± SEM and analysed by one way ANOVA and subsequently by Student's t-test (two-tailed).

areas with nerve cell degeneration and local glial cell proliferation, but the magnitude of the response can be suppressed by low doses of glucocorticoid (85) with DEX being more potent than CS. The glucocorticoid effects on glial enzymes and the more potent action of DEX suggests that this action takes place via glucocorticoid receptors reminiscent to the ones in peripheral glucocorticoid target tissues.

CONCLUDING REMARKS

Uptake and binding studies revealed heterogeneity of adrenal steroid receptor systems in the brain. Glucocorticoid receptors are present in glial cell types and may be present in certain neurons such as those localized in the paraventricular nucleus involved in synthesis of corticotrophin releasing hormone and vasopressin. The glucocorticoid receptors are not different from the receptors in peripheral target tissues that mediate glucocorticoid action on cell growth, intermediary metabolism and expression of the pro-opiomelanocortin gene in pituitary corticotrophs. DEX is among the glucocorticoids with the highest affinity to these sites. Mineralocorticoid-like receptors are present in neurons of the extrahypothalamic limbic brain structures. Physicochemically these receptors are not different from the mineralocorticoid receptors in the kidney. CS is the steroid with highest affinity to these sites. Access of CS to glucocorticoid and mineralocorticoid-like receptors seems determined by the presence of CBG-like binding molecules that may compete with the receptor for binding of CS. This line of reasoning serves as explanation for the expression of ALDO effects in the kidney, since locally near or at the receptor containing cells high amounts of CBG have been detected. Similarly, CBG-like molecules determined immunocytochemically intracellularly in the pituitary are assumed to modulate CS access to the glucocorticoid receptors. Neurons are devoid of significant amounts of CBG as judged from immunocytochemistry. The thousand fold excess of circulating CS over ALDO would permit a unique responsiveness to CS.

Functional studies on receptor pharmacology have disclosed indeed a remarkable specificity of CS action on certain fear-motivated behaviours and neurochemical responses. Among the neurochemical responses the 5-HT synthesis rate is subject to the same stringent specificity as the behaviour. CS is the sole agonist, the parameters are associated with hippocampal function and doses are required that do not result in CS levels exceeding the physiological range. DEX and ALDO are ineffective, and act as antagonist. ALDO antagonism can be explained by competitive inhibition at the CS receptor, while DEX apparently interferes via a different as yet unknown mechanism. ALDO, however, also affects specific brain functions related to mineralocorticoid activity such as salt appetite. DEX effects seem primarily confined to glial cells.

In view of these observations on binding and pharmacology, it is advisable to define a CS preferring receptor system in limbic neurons and glucocorticoid receptor (DEX receptor) in glial cells and presumably in certain neurons, while the presence of a separate set of mineralocorticoid receptors cannot be excluded.

REFERENCES

1. McEwen, B.S., Weiss, J.M., Schwartz, L.S.: Brain Res. 16, 227-241 (1969).
2. McEwen, B.S.: Adrenal actions on the Brain (eds. D. Ganten, D.W. Pfaff), Springer-Verlag, Berlin 1982, pp. 1-22.
3. McEwen, B.S., Micco, D.J.: The Brain as an Endocrine Target Organ in Health and Disease (eds. P.A. van Keep, D. De Wied), MTP Press, Lancaster 1980, pp. 11-28.

4. Bohus, B., De Kloet, E.R., Veldhuis, H.D.: Adrenal Actions on the Brain (eds. D. Ganten, D.W. Pfaff), Springer Verlag Berlin 1982 pp. 108-148.

5. De Kloet, E.R.: J. Ster. Biochem. 20, 175-181 (1984).

6. De Kloet, E.R., Veldhuis, H.D.: Handbook of Neurochemistry vol. VIII (ed. A. Lajtha), Pergamon Press New York 1984 pp.

7. McEwen, B.S., Zigmond, R.E., Gerlach, J.L.: Structure and Function of Nervous Function, vol. 5 (ed. G.H. Bourne), Academic Press New York 1972 pp. 205-291.

8. McEwen, B.S., Wallach, G.: Brain Res. 57, 373-386 (1973).

9. De Kloet, E.R., Leunissen, J.L.M, Voorhuis, Th.D., Koch, B.: J. Ster. Biochem. 20, 367-371 (1984).

10. De Kloet, E.R., Voorhuis, Th.D., Leunissen, J.L.M., Koch, B.: in preparation.

11. De Kloet, E.R., Versteeg, D.H.G., Kovács, G.L.: Brain Res. 264, 323-327 (1983).

12. Towle, A.C., Sze, P.Y.: Brain Res. 253, 221-229 (1982).

13. Veldhuis, H.D., Van Koppen, C., Van Ittersum, M., De Kloet, E.R.: Endocrinology 110, 2044-2051 (1982).

14. De Kloet, E.R., Wallach, G., McEwen, B.S.: Endocrinology 96, 598-609 (1975).

15. Grosser, B.I., Stevens, W., Reed, D.J.: Brain Res. 57, 387-395 (1973).

16. Moguilevski, M., Raynaud, J.P.: J. Ster. Biochem. 12, 309-314 (1980).

17. De Kloet, E.R., Veldhuis, H.D., Wagenaars, J.L., Bergink, E.W.: J. Ster. Biochem. in press.

18. McEwen, B.S., Schwartz, L.S., Weiss, J.M.: Brain Res. 17, 471-482 (1970).

19. Gerlach, J.L., McEwen, B.S.: Science 175, 1133-1136 (1972).

20. Warembourg, M.Y.: Brain Res. 89, 61-70 (1975).

21. Stumpf, W., Sar, M.: J. Ster. Biochem. 11, 801-807 (1979).

22. McEwen, B.S., De Kloet, E.R., Wallach, G.: Brain Res. 105, 129-136 (1976).

23. De Kloet, E.R., McEwen, B.S.: Molecular and Functional Neurobiology (ed. W.H. Gispen), Elsevier Amsterdam 1976 pp. 257-307.

24. Rhees, R.W., Grosser, B.I., Stevens, W.: Brain Res. 83, 293-300 (1975).

25. Rees, H., Stumpf, W.E., Sar, M.: Anatomical Neuroendocrinology (eds. W.E. Stumpf, L.D. Grant), Karger Basel

1975 pp. 262-269.

26. Warembourg, M.Y.: Cell Tiss. Res. 161, 183-191 (1975).
27. Ermisch, H., Rühl, H.J.: Brain Res. 147, 154-158 (1978).
28. Birmingham, M.K, Stumpf, W.E., Sar, M.: Experientia 35, 1240-1241 (1979).
29. Moguilevsky, M., Raynaud, J.P.: Steroids 30, 99-109 (1977).
30. Meyer, J.S., McEwen, B.S.: J. Neurochem. 9, 435-442 (1982).
31. Meyer, J.S., Leveille, P.J., de Vellis, J., Gerlach, J.L., McEwen, B.S.: J. Neurochem. 39, 423-434 (1982).
32. Reul, H.J.M., De Kloet, E.R.: in preparation.
33. Defiore, C.H., Turner, B.B.: Brain Res. 278, 93-101 (1983).
34. Alexis, M.H., Stylianapoulos, F., Kitraki, E., Sekeris, C.E.: J. Biol. Chem. 258 , 4710-4714 (1983).
35. Coirini, H., Marusic, E.T., de Nicola, A.F., Rainbow, T.C., McEwen, B.S.: Neuroendocrinology 37, 354-360 (1983).
36. Luttge, W.E., Densmore, C.L.: J. Neurochem. 42, 242-247 (1983).
37. Wrånge, Ö, Yu, Z.Y.: Endocrinology 113, 243-250 (1983).
38. Beaumont, K., Fanestil, D.D.: Endocrinology 113, 2043-2051 (1983).
39. Krozowsky, Z.S., Funder, J.W.: Proc. Natl. Acad. Sci. USA 80, 6056-6060 (1983).
40. Koch, B., Sakly, M., Lutz-Bucher, B.: Mol. Cell. Endocr. 22, 169-178 (1981).
41. Siitiri, P.K., Murai, I.F., Hammond, G.L., Misker, J.A., Raymoure, W.J., Kuhn, R.W.; Rec. Progr. Horm. Res. 38, 457-510 (1982).
42. Milgrom, E., Atger, M., Baulieu, E.E.: Nature 228, 1205-1207 (1970).
43. De Kloet, E.R., McEwen, B.S.: Biochim. Biophys. Acta 421, 115-123 (1976).
44. De Kloet, E.R., Burbach, J.P.H., Mulder, G.H.: Mol. Cell. Endocr. 7, 261-273 (1977).
45. Koch, B., Lutz, B., Briaud, B., Mialhe, C.: Biochim Biophys. Acta 444, 497-507 (1976).
46. Perrol-Applanal, M., Milgrom, M.: J. Ster. Biochem., in press.
47. Loose, D.S., Feldman, D.: J. Biol. Chem. 257, 4925-4926 (1982).

48. Dahmer, M.K., Honskey, P.R., Pratt, W.B.: Ann. Rev. Physiol. 46, 67-81 (1984).

49. De Kloet, E.R., McEwen, B.S.: Biochem. Biophys. Acta 421, 124-132 (1976).

50. Selye, H.: Stress, the Physiology and Pathology of Exposure to Stress, Acta Medica Publ., Montréal 1950.

51. Cannon, W.B.: Bodily Changes in Pain, Hunger, Fear and Rage, Appleton, New York 1915.

52. Ader, R.J., Weijnen,J.A.W.M., Moleman, P.: Psychon. Sci. 26, 125-128 (1972).

53. Borrell, J., De Kloet, E.R., Versteeg, D.H.G., Bohus, B.: Behav. Neural Biol. 39, 241-258 (1983).

54. Borrell, J., De Kloet, E.R., Bohus, B.: Life Sci. 34, 99-105 (1984).

55. De Kloet, E.R., Borrell, J., Bohus, B.: Stress: The Role of Catecholamines and Other Neurotransmitters (ed. E. Usdin), Gordon and Breach Publ., New York 1984, pp. 775-785.

56. Robustelli, F., Geller, A., Jarvik, M.E.: J. Comp. Physiol. Psychol. 81, 472-482 (1972).

57. Bohus, B.: Brain Res. 66, 366-367 (1974).

58. Bohus, B., De Kloet, E.R.: J. Endocrinol. 72, 64P-65P (1977).

59. Bohus, B., De Kloet, E.R.: Life Sci. 28, 433-440 (1981).

60. Micco, D.J., McEwen, B.S.: J. Comp. Physiol. Psychol. 93, 323-329 (1979).

61. Micco, D.J., McEwen, B.S.: J. Comp. Physiol. Psychol. 94, 624-633 (1980).

62. Veldhuis, H.D., De Kloet, E.R., Van Zoest, I., Bohus, B.: Horm. Behav. 16, 191-198 (1982).

63. Veldhuis, H.D., De Kloet, E.R.: Horm. Behav. 17, 225-232, (1983).

64. Weijnen,J.A.W.M., Slangen, J.L.: Progr. Brain Res. 32, 221-235 (1970).

65. Porsolt, R.D., Le Pichon, M., Jalfre, M.: Nature 266, 730-732 (1977).

66. Porsolt, R.D., Anton, N., Blavel, N., Jalfre, M.: Eur. J. Pharmacol. 47, 379-383 (1978).

67. Porsolt, R.D.: Biomedicine 30, 139-140 (1979).

68. Jefferys, D., Copolov, D., Irby, D., Funder, J.W.: Eur. J. Pharmacol. 92, 99-103 (1983).

69. Veldhuis, H.D., De Kloet, E.R.: in preparation.

70. Gray, P.: J. Comp. Physiol. Psychol. 90, 1-17 (1976).

71. Forman, B.H., Mulrow, P.J.: Handbook of Physiology, section 7: Endocrinology vol. VI (eds. R.O. Greep,E.B. Astwood), Am. Physiol. Soc., Washington 1975, pp. 179-198.

72. Nestler, E.J., Rainbow, T.C., McEwen, B.S., Greengard, P.: Science 212, 1162-1164 (1982).

73. Miller, A.L., Chaptal, A.C., McEwen, B.S., Peck, E.J.: Psychoneuroendocrinology 3, 155-164 (1978).

74. Mobley, P.L., Sulser, F.: Nature 286, 608-609 (1980).

75. Azmitia, E.C., McEwen, B.S.: Science 166, 1274-1276 (1969).

76. Azmitia, E.C.: Handbook of Psychopharmacology, vol. 9 (eds. L.L. Iversen, S.D. Iversen, S.H. Snyder), Plenum Press, New York 1978, pp. 214-233.

77. Sze, P.Y., Neckers, C., Towle, A.C.: J. Neurochem. 26, 169-173 (1976).

78. Vermes, I., Smelik, P.G., Mulder, A.H.: Life Sci. 19, 1719-1726 (1976).

79. Van Loon, G.R., Shum, A., Sole, M.J.: Endocrinology 108, 1392-1402 (1981).

80. De Kloet, E.R., Kovács, G.L., Szabo, G., Telegdy, G., Bohus, B., Versteeg, D.H.G.: Brain Res. 239, 659-663 (1982).

81. De Kloet, E.R., Angelucci, L., Kovács, G.L., Versteeg, D.H.G., Bohus, B.: Integrative Neurohumoral Mechanisms (eds. E. Endcröczi, D. De Wied, L. Angelucci, U. Scapagnini), Elsevier,Amsterdam 1983, pp. 147-155.

82. Azmitia, E.C., Conrad, L.C.A.: Neuroendocrinology 21, 338-349 (1974).

83. DeVellis, J., McEwen, B.S., Cole, R., Inglish, D.: J. Ster. Biochem. 5, 392-393 (1974).

84. Cousin, M.A., Lando, D., Moguilevski, M.: J. Neurochem. 38, 1296-1304 (1982).

85. De Kloet, E.R., Cousin,M.A., Veldhuis, H.D., Voorhuis, Th.D., Lando, D.: Brain Res. 275, 91-98 (1983).

STRUCTURE-ACTIVITY RELATIONSHIP OF SPIRONOLACTONE DERIVATES

Correlation of the Affinity for Rat Renal Mineralocorticoid Receptors in vitro and the Antialdosterone Activity in the Adrenalectomized Rat in vivo

Gerhard Wambach

Medizinische Klinik Köln-Merheim und Medizinische Poliklinik, Lehrstuhl für Innere Medizin II der Universität Köln, W-Germany

Jorge Casals-Stenzel

Research Laboratories Schering Aktiengesellschaft, Berlin and Bergkamen, W-Germany

Introduction

Mineralocorticoid antagonists are widely used in clinical medicine in the treatment of disorders with primary or secondary aldosterone excess. Among a number of different compounds, spironolactone was the first competitive aldosterone antagonist used in clinical medicine and it remained the standard compound for almost 25 years (1).

Meanwhile, more insights were gained in the mechanism of action of spironolactone and other aldosterone antagonists as recently summarized by Corvol et al. (2). Spironolactone inhibits binding of ^{3}H-aldosterone at rat renal cytoplasmic receptor proteins (3). Moreover, ^{3}H-labelled spironolactone and labelled prorenone were found to bind at high affinity sites in rat renal cytoplasma (4, 5). However, the spirolactone-receptor-complex did not bind to nuclear structures as has been shown for the aldosterone-receptor-complex. Spironolactone interferes with the synthesis of aldosterone

Adrenal Steroid Antagonism

and other steroids in the adrenal gland. In vitro studies revealed the inhibition of aldosterone synthesis stimulated by ACTH, potassium or angiotensin II by spironolactone (6). Later experiments confirmed this additional action of spirolactones and demonstrated an inhibition of mitochondrial 11-β and 18-hydroxylase-activity (7). Although the effect on steroidgenesis in vitro is evident, the decrease in aldosterone plasma levels and urinary aldosterone excretion during spironolactone treatment is only transient (8, 9, 10).
This discrepancy can be explained at least in part by the high concentrations of spironolactone needed to inhibit aldosterone synthesis in vitro. From the experimental and clinical data, the natriuretic effect of spirolactones appears to be mainly due to its peripheral antagonism of mineralocorticoids at the target tissues.
The clinical value of spironolactone particularly during long term treatment is limited by its endocrine side effects, including gynecomastia and impotence in men and menstrual irregularity in women (11). These undesired side effects correlate with the action of spironolactone on other steroid hormone systems.
Spironolactone inhibits menstruation in rabbits and monkeys suggesting a progestational activity of the drug (12). More important, spironolactone interferes with the secretion and peripheral action of androgens. Administration of large doses of spironolactone (200 mg/kg) causes a marked reduction in the formation of testosterone in the rat testis (13). This effect is primarily due to a blockade of the 17-hydroxylase, a microsomal enzyme dependent upon cytochrome P-450. These experimental data are consistent with studies in normal man. Treatment with spironolactone (400 mg/day) led to a significant rise in serum progesterone and serum 17-OH-progesterone (14). These data suggest an inhibition of the 17-hydroxylase as well as the 17-20-desmolase-activity in the testis, or the adrenal, or both. Plasma levels of testosterone may remain constant or may decrease depending on the duration

and on the dose of spironolactone-therapy (15).Spironolactone also inhibits the action of androgens at the target tissue:

1) The action of exogenous testosterone is antagonized in the rat (16).
2) Spironolactone decreases specific binding of dihydrotestosterone in prostate cytosol fraction both in experimental animal and man (17, 18, 19, 20, 21).
3) Nuclear uptake of ^{3}H-dihydrotestosterone in prostate tissue is reduced by spironolactone (18).

It becomes obvious, that spironolactone is far from being an ideal mineralocorticoid antagonist. In search for a more specific aldosterone antagonist, we compared the affinity of 17 spironolactone-like compounds in vitro and their anti-aldosterone activity in vivo in order to get insights into the structure-activity-relationship of aldosterone antagonists.

Methods

a) In vitro experiments: binding to renal cytoplasmic mineralocorticoid receptors (MCR)

Male Sprague-Dawley rats (body weight 180-200 g) were adrenalectomized under phenobarbitone anaesthesia 4 days prior to the actual mineralocorticoid receptor test, and were given 1 % sodium chloride solution to drink and fed Purina rat diet. On the day of the test, 3 - 4 rats were anaesthetized with phenobarbitone (5 mg/100 g body weight, given i.p.) and exsanguinated by the carotid artery. The kidneys were perfused with ice-cold buffer (0.17 M NaCl, 0.25 M K_2HPO_4, at pH 7.4) via vena cava; the kidneys were then removed and homogenized in Tris buffer (0.1 M Tris, 3 mM $CaCl_2$, at pH 7.4). The homogenate was centrifuged for 45 min at 30 000 g. The supernatant (cytosol) was taken for further incubation immediately

afterwards.
^{3}H-aldosterone (2.5 nM) and a number of unlabelled steroids in increasing concentrations were incubated together with 500 µl of cytosol for 1 hr at 23°. This concentration of ^{3}H-aldosterone represents the range of the Scatchard analysis, in which aldosterone is predominantly bound to Type I binding sites (22).
Protein bound ^{3}H-aldosterone was separated from the free activity by passage through a Biogel P-10 column (0.5 x 10 cm) which had been previously equilibrated with Tris buffer. The void vols. were collected in scintillation vials. The activity in the protein fraction was measured in the scintillation counter after addition of 10 ml of aquasol (Tricarb Model 3385, Packard Co.).
^{3}H-aldosterone which was non-specifically protein bound was determined by incubating ^{3}H-aldosterone (2.5 nM) together with unlabelled aldosterone at a 1000-times higher concentration. This 'non-specifically bound fraction' of the total protein bound ^{3}H-aldosterone accounted for about 10-20 % and was subtracted from the total bound ^{3}H-aldosterone. The inhibitory effect of a non-labelled steroid on the binding of ^{3}H-aldosterone to the receptor sites was calculated from the following relation:

$$\frac{\text{specific bound } ^3\text{H-aldosterone in presence of the substance} \times 100}{\text{specific bound } ^3\text{H-aldosterone in absence of the substance}}.$$

The affinity for aldosterone receptors of the various substances investigated was compared with that of spironolactone in each incubation sample. The relative affinity, referred to that of spironolacton (which was taken to be 100) was established by the following relation:

$$\frac{\text{concentration of spironolactone at 50 \% displacement}}{\text{concentration of reference substance at 50 \% displacement}} \times 100 .$$

For each component and concentration, at least two determinations were performed in duplicate. The method used is similar to that described previously (23, 24).

b) In vivo experiments: antialdosterone activity in rats

The method used for evaluation of the antialdosterone activity in rats was described previously (25). Adrenalectomized Wistar rats with a body weight of 140-160 g were substituted with 1 mg fluocortolone caproate/kg at the day of surgery and 10 mg fluocortolone/kg s.c. 1 day before the diuresis expt. These glucocorticoid-substituted rats were infused intravenously with a saline-glucose solution (0.05 % NaCl; 5 % glucose) containing aldosterone (50 µg/l) at a rate of 3 ml/hr for 10 or 15 hr. The aldosterone antagonist was administered 1 hr before the start of the aldosterone infusion. Urine excretion was measured in fractions of 1 hr. Sodium and potassium concentrations in urine were determined by flame photometry. The antialdosterone activity was assessed by the ability of the compounds to reverse the aldosterone effect on the urinary Na/K ratio.
Substances which showed antialdosterone activity in a preliminary test were examined in the diuresis expt over a period of 10 or 15 hr, to determine their relative potency in comparison to spironolactone after oral administration. The various steroids and spironolactone were adminstered at oral doses of 6.7, 13.4 and 26.8 mg/kg. The dose-response relationship was tested for each fraction (hr) by regression analysis after logarithmic transformation of the doses. The relative potency of a steroid, compared to spironolactone

was established by comparing linear and parallel dose-response curves. The potency of the standard substance, spironolactone, was allocated the value of 100. The maximum relative potency (MRP) achieved by the investigated steroid vs spironolactone during long-term expt is specified in Fig. 1 - 6.

c) Substances

d-Aldosterone (Aldocorten ^{R}) was from Ciba-Geigy. [1,2,6,7 ^{3}H] Aldosterone (85 Ci/mmol) was purchased from New England Nuclear. All derivates of spironolactone were synthesized by Schering Aktiengesellschaft Berlin/Bergkamen.

Results

a) 17α-Hydroxy-derivates of spironolactone

Compared to unlabelled aldosterone, spironolactone in ten times higher concentration was needed to displace ^{3}H-aldosterone from the receptor binding sites. Thus, the relative affinity of spironolactone was 10 % compared to aldosterone. Replacement of the 17-spirolactone group of spironolactone by a 17α-Hydroxypropyl and a 17β-Hydroxyl group resulted in a dramatic reduction in affinity for the aldosterone binding sites (Fig. 1). Two additional compounds had modifications of the Hydroxypropyl side chain: 17α(3-Hydroxy-3,3-dimethylpropyl) and 17α-Hydroxybutyl instead of 17α-Hydroxypropyl. The relative affinity for these three compounds was found to be below 1 % compared to spironolactone (100 %). However, the in vivo activity was quite different. The derivative with a 17α-Hydroxypropyl was even 1.84 times more potent than spironolactone, whereas the two others did not show any antialdosterone activity in vivo.

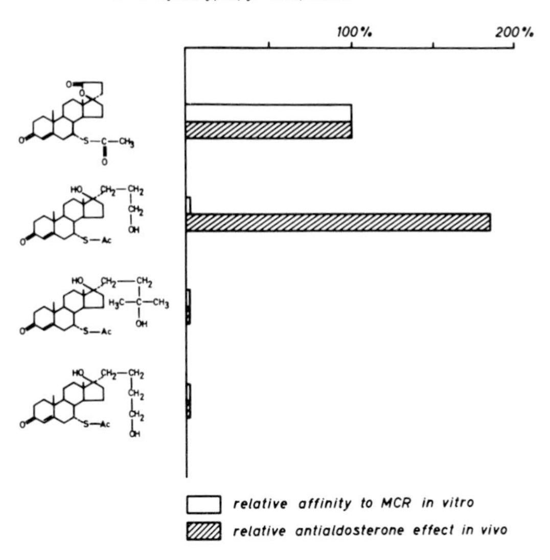

Fig. 1 Affinity for rat renal mineralocorticoid receptors in vitro and maximal relative antialdosterone potency in rats of 17α-Hydroxypropyl derivates of spironolactone.

b) Compounds unsaturated at C_6/C_7

Substitution of the 7α-thioacetyl group of spironolactone by a 6,7 double bond leads to canrenone, a compound generated in vivo (Fig. 2). The in vitro affinity of canrenone was found to be 24 %. A similar reduction in antimineralocorticoid activity of canrenone was observed in vivo (46 %). Modification of the lactone ring of canrenone resulted in a considerable loss of receptor affinity. Nevertheless, the in vivo action of the 17α-Hydroxypropyl-17β-OH derivate of canrenone was only slightly reduced. The Δ^{15}derivate of canrenone showed a similar reduction in receptor affinity (54 %) and in vivo activity (34 %) compared to spironolactone.

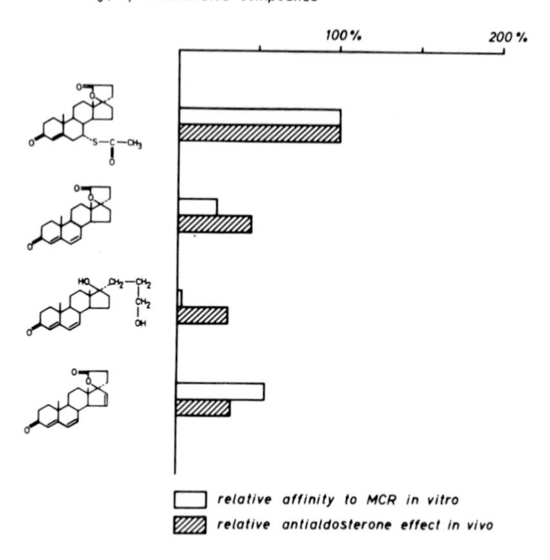

Fig. 2 Affinity for rat renal mineralocorticoid receptors in vitro and maximal relative antialdosterone potency in rats of C_6/C_7 unsaturated derivates of spironolactone.

c) 6β,7β-Methylene compounds

Hallmark of the prorenone molecule is a substitution of the 7α-thioacetyl group by a 6,7-methylene group in beta position (Fig. 3). This modification results in a rise in receptor affinity and in biological activity by 52 % and 41 % respectively in comparison with spironolactone. Additional modification of the 17-spirolactone ring, however, resulted in a substantial drop in affinity for the aldosterone binding sites in rat renal cytosol. The relative affinity for the 17α-hydroxypropyl, 17β-OH derivate of prorenone was found to be 4 % and for the compound with an additional double bond in ring A below 1 %. The biological activity of these two derivates was only slightly lower than that of spironolactone (71 % and 72 %). Potassium prorenoate was tested in both systems. Its antimineralocorticoid action was lower than that of prorenone and similar to spironolactone. The in vitro affinity (37.1 %) was considerably lower than for prorenone, however higher than for the two other 6β,7β-methylene derivates.

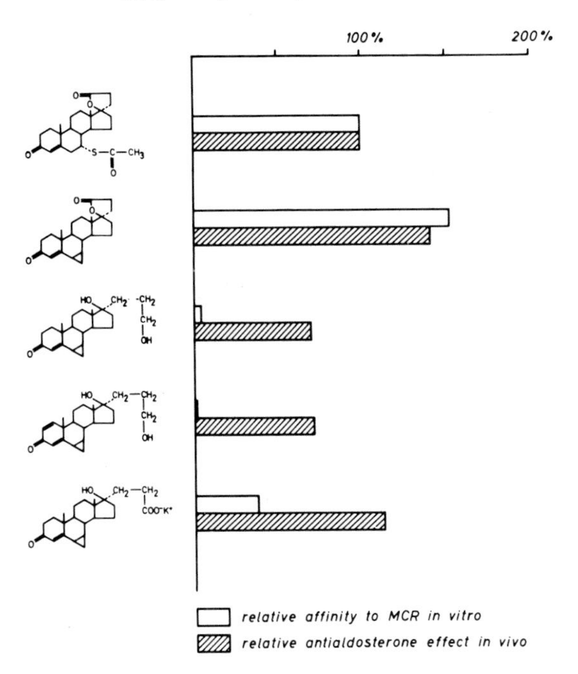

Fig. 3 Affinity for rat renal mineralocorticoid receptors in vitro and maximal relative antialdosterone potency in rats of 6β,7β-methylene derivates of spironolactone.

d) Methylation at the D-ring

The position of the methyl group at the D-ring with reference to the 17-spirolactone ring was of particular interest (Fig. 4). 15β-methyl-spironolactone had a relative affinity of 7.9 % and 16α-methyl-spironolactone had an affinity of less than 1 % compared to spironolactone. Similarly, the 15β-methyl compound showed less antimineralocorticoid activity in vivo than the mother compound while the 16α-methyl-derivate did not show antagonistic activity at all.

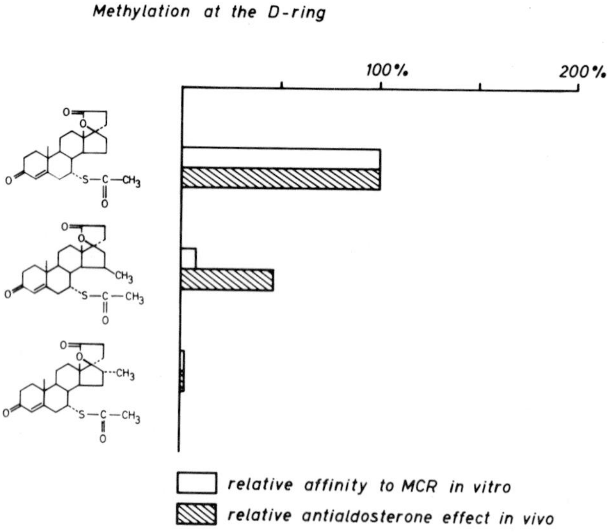

Methylation at the D-ring

100% 200%

relative affinity to MCR in vitro

relative antialdosterone effect in vivo

Fig. 4 Affinity for rat renal mineralocorticoid receptors in vitro and maximal relative antialdosterone potency in rats of spironolactone derivates methylated at ring D.

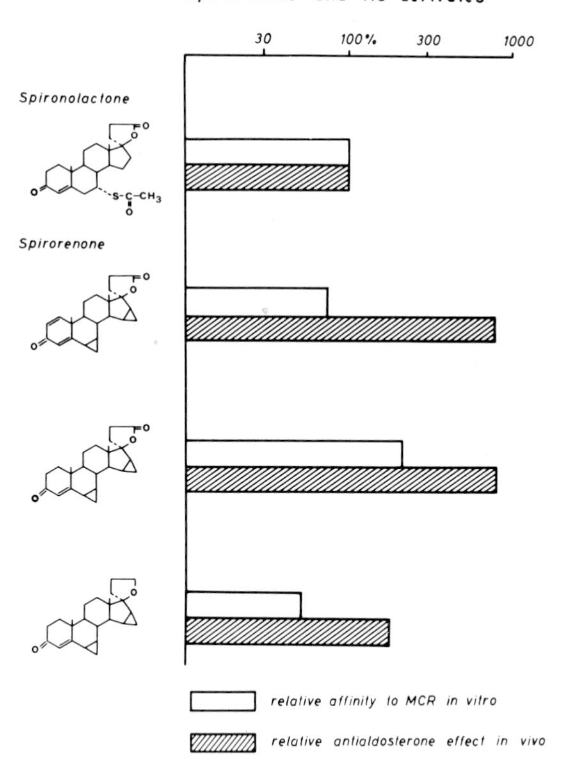

Fig. 5 Affinity for rat renal mineralocorticoid receptors in vitro and maximal relative antialdosterone potency in rats of spirorenone and related compounds in comparison to spironolactone.

e) Spirorenone and derivates

Spirorenones are a new group of potent aldosterone antagonists (Fig. 5, Fig. 6). The mean relative potency of spirorenone was 8.6 higher than that of spironolactone. This however was not reflected by an increase in receptor binding, the relative binding affinity of spirorenone being 73 %.
In contrast, the C_1,C_2 saturated derivate of spirorenone showed in vivo as well as in vitro higher effects than spiro-

nolactone. The in vivo activity averaged 7.5 times that of spironolactone and its affinity doubled that of the standard. The 17-spiroether derivate of spirorenone was still 1.7 times more potent than spironolactone its receptor affinity being only 50 % compared to spironolactone. Modification of the lactone formation (17α-hydroxy, 17β-hydroxypropyl-derivate of spirorenone) resulted in a loss of antialdosterone potency. The in vivo activity was still 1.3 times higher than that of spironolactone, the binding capacity in vitro however was

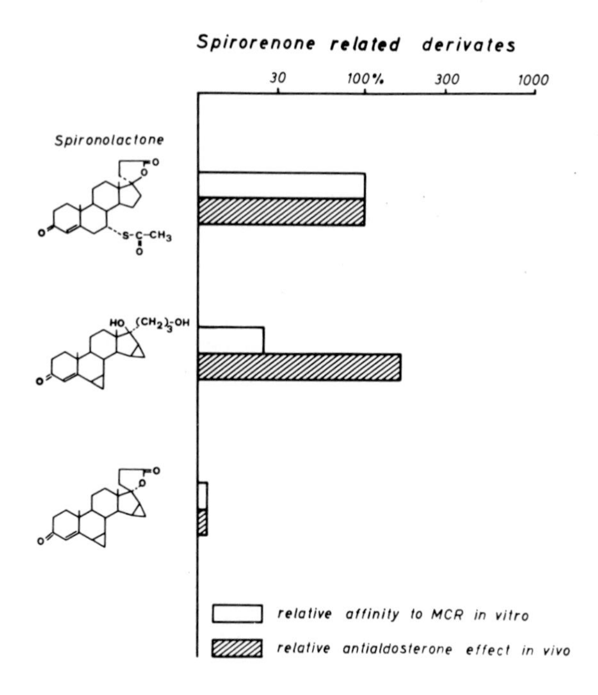

Fig. 6 Affinity for rat renal mineralocorticoid receptors in vitro and maximal relative antialdosterone potency in rats of spirorenone related compounds in comparison to spironolactone.

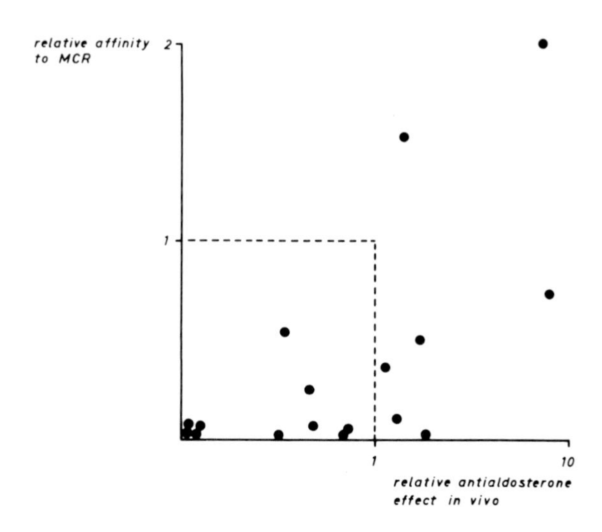

Fig. 7 Correlation between in vivo and in vitro activity of all spironolactone derivates tested.

very low. Finally, a spirorenone analogue with reversed spirolactone configuration at C_{17} was included. This compound did not show activity neither in vivo nor in vitro.

f) Correlation between in vitro affinity and in vivo activity

The relative affinity to rat renal mineralocorticoid receptors and the relative in vivo antialdosterone action of all spironolactone-like compounds is summarized in Fig. 7 and Table I. There is an overall significant correlation between the in vivo and in vitro activity ($r = 0.509$, $p < 0.025$). However, several compounds do show in vivo activities similar to spironolactone with only negligable receptor binding. On the other hand, there is no derivate with receptor binding and without in vivo antialdosterone activity.

Table 1. Affinity for renal mineralocorticoid receptors (MCR) of rats in vitro and maximal relative antialdosterone potency (MRP) in rats of spironolactone derivates in comparison to spironolactone

Compound	Relevant modifications compared to spironolactone	Affinity for MCR vs spironolactone (= 100 %) (%)	Maximal MRP vs spironolactone (= 100 %) (95 % confidence limits) (%)
17α-Hydroxypropyl compounds			
I	17α-Hydroxypropyl, 17β-OH	< 1	184 (108-387)
II	17α-(3-Hydroxy-3, 3-dimethylpropyl), 17β-OH	< 1	No activity
III	17α-Hydroxybutyl, 17β-OH	< 1	No activity
C_6/C_7 unsaturated compounds			
IV	Δ^6 (=canrenone)	24	46 (28-83)
V	Δ^6-17α-Hydroxypropyl, 17β-OH	< 1	32 (21-45)
VI	$\Delta^{6,15}$	54.4	34 (19-56)
6β,7β-Methylene compounds			
VII	6β,7β-Methylene (prorenone)	152	141 (98-221)
VIII	6β,7β-Methylene; 17α-hydroxypropyl, 17β-OH	4.0	71 (21-157)
IX	Δ^1; 6β,7β-methylene; 17α-hydroxypropyl, 17β-OH	< 1	72 (21-157)
X	6β,7β-Methylene-17α-propionate, K^+ (= K^+prorenoate)	37.1	114 (51-268)
Methylation at the D-ring			
XI	15β-Methyl	7.9	47 (19-153)
XII	16α-Methyl	< 1	No activity

Discussion

The observations, that the action of aldosterone at the target tissues is initiated by specific binding at cytoplasmic hormone receptors and that this initial step is inhibited by competitive aldosterone antagonists such as spironolactone offered a unique chance to study structure-affinity relationship of compounds with a potential agonistic or antagonistic activity. Funder et al. first investigated the affinity of 24 spironolactone analogues to rat renal cytoplasmic aldosterone receptors in vitro (26). They already demonstrated a reduction in receptor affinity of compounds with 1) unsaturation at C_6/C_7 position in ring B, 2) unsaturation of the γ-lactone and 3) opening of the γ-lactone ring in comparison to spironolactone. 19-nor-spironolactones show greater as well as less affinities in relation to their parent compounds. The significance of these in vitro affinities remained unclear since the data were not compared with the biological activities of these compounds. In further studies the relationship of agonistic, antagonistic and binding activity of seven of these compounds was investigated (27, 28).The in vivo activity was studied in the toad bladder by means of the short-circuit-current method allowing a differentiation between agonistic and antagonistic activity. Moreover, the system is independent from pharmacokinetic properties of the compounds under investigation. The data revealed a discrepancy between binding affinity and antagonistic activity whereas compounds with higher receptor binding manifested significant agonistic activity.

Recent studies confirmed the tremendous dissociation between the relative affinity for mineralocorticoid receptor in vitro and the in vivo pharmacological activity for certain groups of antimineralocorticoid compounds. Prorenone was found to be more potent than spironolactone both in vitro and in vivo (29). K^+-prorenoate however was almost without affinity for rat renal mineralocorticoid receptors despite an in vivo activity superior to spironolactone. Further studies were performed by the same group to support the hypothesis, that prorenoate and canrenoate are converted in vivo to prorenone and canrenone respectively, both compounds with γ-lactone ring and with a considerable affinity for aldosterone binding sites (30). 17-O-methyl-5,6-dihydrocanrenoic acid, a derivate which cannot be lactonized was inactive both in vivo and in vitro. K^+-canrenoate however, was active in the rat in vivo despite an extremely low affinity for aldosterone receptors.

Our studies again clearly demonstrate the importance of an intact structure of 17-spirolactone ring for the binding to the mineralocorticoid receptor. Opening of the spirolactone ring, e.g. in spironolactone, canrenone or prorenone, leads either to a partial or an almost complete loss of the binding capacity, depending on the remaining structure at C-17. Similar observations were also made by Funder et al. (26) and Peterfalvi et al. (30).

Among the compounds without the 17-spirolactone ring, the prototype of the series containig a 17α-hydroxypropyl side chain and a 17β-hydroxy group, showed a striking result: its pharmacological activty in vivo was higher than that of spironolactone while its affinity for the mineralocorticoid receptor appeared to be negligible. Assuming that a blockade of aldosterone binding sites in the cytoplasma is essential for competitive antagonism, this result implies that this compound is converted in vivo to one or more active metabolites which then have affinity for the mineralocorticoid receptors.

It is likely that the in vivo conversion to active substances takes place by oxidation of the hydroxyl group of the side chain at C-17 to a carboxyl group and further a closure to the spirolactone ring.

Peterfalvi et al. (30) already pointed to the importance of the lactonic ring at C-17 for the antialdosterone activity. They found that 17-O-methyl-5,6-dihydrocanrenoic acid, a derivative which cannot be lactonized, was inactive in vivo as well as in the receptor assay in vitro. This assumption for ring closure is supported in the present study by two additional compounds. A branched 17α side chain - the 17α-hydroxy-3,3-dimethylpropyl chain of one compound - or an extended 17α side chain - the 17α-hydroxybutyl chain of the other compound - seems to impede a ring formation by oxidation of the hydroxyl group and ring closure. Thus, not only the affinity to the mineralocorticoid receptor is very low but also their activity in vivo is abolished. Nevertheless, this mechanism can, however, only be answered on the basis of additional biochemical and pharmacological investigation.

Modifications of the structural featurs at the B-ring of spironolactone influence the affinity of its derivates for the aldosterone receptor sites to a variable extent. Canrenone presented a reasonable affinity for the receptors in vitro and a corresponding antialdosterone activity in vivo. This compound is one of the active metabolites of spironolactone (31). As in the case of spironolactone the 17α-hydroxypropyl derivate did not show affinity for the receptor sites but caused a clear antialdosterone effect in vivo.
The double bond between C-15 and C-16 increased slightly the action in vitro as well as in vivo when compared to canrenone.
The second group of 7α-dethioacetylated compounds is the 6β, 7β-methylene series. The 6β,7β-methylene moiety, as shown by prorenone, increases both the affinity for the receptor sites in vitro and the pharmacological activity in vivo, as shown extensively by Claire et al. (29). As in the previous group, the introduction of the 17α-hydroxypropyl and 17β-hydroxyl groups practically abolished the affinity for the receptor but reduced only to a minor extent the antialdosterone activity. In the case of potassium prorenoate, the open form of prorenone, a part of the binding capacity to the receptor was lost but the antialdosterone power in vivo was kept in comparison to prorenone.
The affinity for mineralocorticoid receptors was in accordance with the in vivo activity for the spironolactone derivates methylated at the D-ring.

Spirorenone and its C_1/C_2 saturated derivate showed a similar potency in vivo but their affinity for the mineralocorticoid receptor was markedly different. The binding activity of the C_1/C_2 saturated derivate (2 times the affinity of spironolactone) was higher than that of spirorenone (0.73 times the affinity of spironolactone). This finding indicates, that spirorenone must probably be transformed into a more active metabolite to achieve the strong antimineralocorticoid activity in vivo. It might be, that one of these active metabolites is the C_1/C_2 saturated compound. Another possible explanation for this result would be a slower inactivation and elimination of spirorenone than of the C_1/C_2 saturated compound.
It has been postulated that the spirolactone ring at C_{17} is essential for the development of aldosterone antagonistic activity (30, 32). The 17-spiroether derivate of spirorenone was less active than the congener, but still active in vivo and in vitro, and moreover, more potent than spironolactone. Thus, from that result it can be concluded, that not only the spirolactone form but also the spiroether ring lead to active aldosterone antagonistic compounds.
The spirorenone derivate containing a 17α-hydroxypropyl side chain and a 17β-hydroxy group, showed a striking result: its pharmacological activity in vivo was higher than that of spironolactone (markedly lower in comparison to spirorenone) while its binding capacity for the mineralocorticoid receptor in vitro was practically negligible. This result implies that this compound has to be converted in vivo to an active metabolite. It is likely, that this in vivo conversion to an active substance takes place by oxydation of the hydroxyl group of the side chain at C_{17} to a carboxyl group and further a closure to the 17-spirolactone ring. The latter step has been well demonstrated for potassium canrenoate (open form) which rapidly equilibrates with its lactonic form (33, 34). The importance of the lactonic ring at C_{17} for antialdosterone activity was also pointed by Peterfalvi et

al. (30). From our own experience with 17α-hydroxypropyl derivates of the spironolactone series, the branching, shortening or lengthening of the 17α-side chain results in a complete loss of the antialdosterone activity in vivo and of the receptor binding activity in vitro. This is probably due to a hindrance of the ring closure after the oxydation of the hydroxyl group at the end of the 17α-side chain (unpublished).

Finally, the reversion of the spirolactone configuration at C_{17} abolished completely the pharmacological activity in vivo as well as the affinity for the MCR in vitro. Similar results were found for other diastereomers spirolactone compounds (32). In conclusion, spirorenone is a potent aldosterone antagonist and represents a new series of compounds with aldosterone blocking properties which possess a higher pharmacological potency than the classical spironolactone. As has bee shown in preliminary studies (35), this new type of aldosterone antagonists is also active in healthy volunteers at lower doses than spironolactone. However, the clinical use of these compounds is limited by their affinity for the gestagen receptors.

In comparing the in vivo aldosterone antagonistic effect of various steroids with their relative affinities for the renal mineralocorticoid receptor, there are a number of aspects to be taken into consideration. In the receptor assay system, the investigation is confined to the competition between a steroid and ^{3}H-aldosterone for binding to cytoplasmatic receptor sites. However, in vitro affinity for the receptor of a given compound could induce an agonistic or antagonistic action (36, 37). In the bioassay, on the other hand, there are differences in intestinal absorption, distribution, metabolism and excretion which effect the aldosterone-agonistic or -antagonistic action of a substance in vivo.

For the principal reasons discussed above, one cannot expect a complete agreement between affinity for receptor sites and antimineralocorticoid activity in vivo of a given compound. However, the present studies demonstrate the usefulness of studying the affinity, for mineralocorticoid receptors in vitro, of compounds with expected mineralocorticoid or antimineralocorticoid activity in the search for more specific mineralocorticoid antagonistic steroids, and for a better understanding of the metabolism as well as the mechanism of action of new compounds.

Summary

We tested the ability of 17 steroids with structures similar to spironolactone to compete with ^{3}H-aldosterone for binding at rat renal cytoplasmic receptors in vitro and the antialdosterone activity in adrenalectomized rats in vivo in comparison with spironolactone.
Replacement of the 17-spirolactone ring by a 17α-hydroxypropyl group and a 17β-hydroxyl group resulted in a loss of receptor affinity without a reduction in antialdosterone action in vivo. Compared to spironolactone, C_6/C_7 unsaturated compounds showed a reduced activity both in vitro and in vivo. Substitution of the 7α-thioacetyl group in β-position (prorenone) increased the in vivo as well as the in vitro activity by 41 and 52 % respectively. Introduction of a methyl group at the D-ring resulted in a similar reduction in activity both in vivo and in vitro. Spirorenone and two of its derivates were 3-8 times more potent than spironolactone. Their receptor affinity was only slightly increased. Taken together, measuring the receptor affinity does not replace testing the in vivo antimineralocorticoid activity of new compounds. Comparison between affinity for mineralocorticoid receptors and biological activity however, provides insights into the metabolism of potential antimineralocorticoids.

References

1. Kagawa, C.M.: Endocrinology 67, 125-132 (1960)
2. Corvol, P., Claire, M., Oblin, M.E., Geering, K., Rossier, B.: Kidney Int. 20, 1-6 (1980)
3. Funder, J.W., Feldman, D., Highland, E., Edelman, I.S.: Biochem. Pharmacol. 23: 1493-1501 (1974)
4. Marver, D., Stewart, J., Funder, J.W., Feldman, D., Edelman, I.S.: Proc. Nat. Acad. Sci. (USA) 71, 1431-1435 (1974)
5. Claire, M., Rafestin-Oblin, M.E., Michaud, A., Roth-Meyer, C., Corvol, P.: Endocrinology 104, 1194-1200 (1979)
6. Erbler, H.C.: Naunyn-Schmiedeberg's Arch. Pharmacol. 273, 366-372 (1972)
7. Cheng, S.C., Suzuki, K., Sadée, W., Harding, B.W.: Endocrinology 99, 1097-1106 (1976)
8. Erbler, H.C.: Naunyn-Schmiedeberg's Arch. Pharmacol. 285: 395-401 (1974)
9. Conn, J.W., Hinerman, D.L.: Metabolism 26, 1293-1307 (1977)
10. Abshagen, U., Spore, S., Schöneshofer, M., L'Age, M., Rennekamp, M., Oelkers, W.: Clin. Sci. Mol. Med. 51, 307s-310s (1976)
11. Loriaux, D.L., Menard, R., Taylor, A., Pita, J.C., Santen, R.: Ann. Int. Med. 85, 630-636 (1976)
12. Schaue, H.P., Potts, G.O.: J. Clin. Endocrinol. Metab. 47, 691-694 (1978)
13. Menard, R.A., Stripp, B., Gillette, J.R.: Endocrinology 94, 1628-1636 (1974)
14. Taylor, A.A., Mitchell, J.R., Rollins, D.E.: Clin. Res. 24, 279 A (1976)
15. Rose, L.I., Underwood, R.H., Newmark, S.R., Kisch, E.S., Williams, G.H.: Ann. Int. Med. 87, 398-403 (1977)
16. Stalmann, S.L., Brooks, J.R., Morgan, E.R.: Steroids 14, 449-450 (1969)
17. Bonne, C., Raynaud, J.P.: Mol. Cell. Endocrinol. 2, 59-67 (1974)
18. Corvol,P., Michaud, A., Menard, J., Freifeld, M., Mahoudeau, J.: Endocrinology 97, 52-58 (1975)
19. Pita, J.C., Lippmann, M.E., Thompson, E.R., Loriaux, D.C.: Endocrinology 97, 1521-1527 (1975)
20. Rifka, S.M., Pita, J.C., Vigersky, A., Wilson, Y.A., Loriaux, D.C.: J. Clin. Endocrinol. Metab. 46, 338-344

(1977)

21. Cutler, G.B., Pita, J.C., Rifka, S.M., Menard, R.H., Sauer, M.A., Loriaux, D.C.:J. Clin. Endocrinol. Metab.47, 171-175 (1978)

22. Rousseau, G., Baxter, J.D., Funder, J.W., Edelman, I.S., Tomkins, G.M.: J. Steroid. Biochem. 3, 219-229 (1972)

23. Wambach, G., Higgins, J.R.: Endocrinology 102, 1686 (1978)

24. Wambach, G., Higgins, J.R.: In Antihormones (Ed. M.K. Agarwal) p. 167, Elsevier/North Holland Biomedical Press Amsterdam (1979)

25. Casals-Stenzel, J., Buse, M., Losert, W.: Eur. J. Pharmacol. 80, 37-41 (1982)

26. Funder, J.W., Feldman, D., Highland, E., Edelman, I.S.: Biochem. Pharmacol. 23, 1493-1507 (1974)

27. Sakauye, C., Feldman, D.: Am. J. Physiol. 231, 93-97 (1976)

28. Feldman, D.: In Antagonists in Clinical Medicine (Eds. Addison, G.M. et al.) p. 18, Excerpta Medica Amsterdam (1978)

29. Claire, M., Rafestin-Oblin, M.E., Michaud, A., Roth-Meyer, C., Corvol, P.: Endocrinology 104, 1194 (1979)

30. Peterfalvi, M., Torelli, V., Fournex, R., Rousseau, G., Claire, M., Michaud, A., Corvol, P.: Biochem. Pharmacol. 29, 353-357 (1980)

31. Karim, A., Zagarella, J., Hribar, J., Dooley, M.: Clin. Pharmacol. Ther. 19, 158-164 (1976)

32. Kagawa, C.M.: Anti-Aldosterones, in: Methods in Hormone Research, ed. R.J. Dorfman, Academic Press New York (1964)

33. Sadée, W., Abshagen, U., Finn, C., Rietbrock, N.: Naunyn-Schmiedeberg's Arch. Pharmacol. 283, 303 (1974)

34. Ramsay, L.E., Shelton, J.R., Wilkinson, D., Tidd, M.J.: Br. J. Clin. Pharmacol. 3, 607 (1976)

35. Casals-Stenzel, J., Brown, J.J., Losert, W.: Naunyn-Schmiedeberg's Arch. Pharmacol. Suppl. 316, R 49 (1981)

36. Sherman, M.R.: In Aldosterone Antagonists in Clinical Medicine (eds. G.M. Addison et al.) p. 1, Excerpta Medica Amsterdam (1978)

37. Raynaud, J. P., Bouton, M.M., Moguilewsky, M., Ojasoo, T., Philibert; D., Beck, G., Labrie, F., Mornun, J.P.: J. Steroid Biochem. 12, 143 (1980)

Acknowledgments

We would like to express our gratitude to Professor R. Wiechert, Dr. W. Eder and Dr. U. Kerb, Schering Aktiengesellschaft Berlin and Bergkamen, W-Germany, for the generous supply of the steroids used in this work. We thank Mr. M. Buse and Mrs. G. Suckau for excellent assistance. Part of the data are published in a contribution to Biochemical Pharmacology, 32, 1479-1485 (1983). These studies were supported by grants from Ministerium für Wissenschaft und Forschung des Landes Nordrhein-Westfalen.

CLINICAL PHARMACOLOGY OF THE SPIROLACTONES.

Lawrence E. Ramsay and Gordon T. McInnes
University Department of Therapeutics,
Royal Hallamshire Hospital,
Sheffield S10 2JF, UK

Introduction

Spirolactones are steroids with structures similar to that of aldosterone (Figure 1), and they generally have the property of specific competitive antagonism of aldosterone and other mineralocortocoids. The structures of the compounds which will be referred to in this review are shown in Figure 1. It should be noted that some are steroid lactones, for example spironolactone and SC 8109, whereas others are potassium salts of 21-carboxylic acids derived from spirolactones, for example SC 14266 (potassium canrenoate) and SC 23992 (potassium prorenoate). At the present time three spirolactones are in clinical use, namely spironolactone, canrenone and potassium canrenoate. These drugs have a valuable role in therapeutics but also have important drawbacks which have precluded their more widespread use. Briefly, they have to be given in large daily doses and are therefore relatively expensive drugs. In addition they cause endocrine side-effects such as gynaecomastia and impotence in men, and menstrual disturbance in women. These side-effects are generally acceptable to patients with disabling and life-threatening diseases such as heart failure or nephrotic syndrome, but they have proved a major disadvantage when treating conditions such as mild hypertension, in which therapy continues for many years and patients are otherwise completely healthy and symptom-free. There is therefore a need for new spirolactones which retain the therapeutic properties of spironolactone, but are more economic and free from unwanted endocrine side-effects. To this end numerous

Adrenal Steroid Antagonism

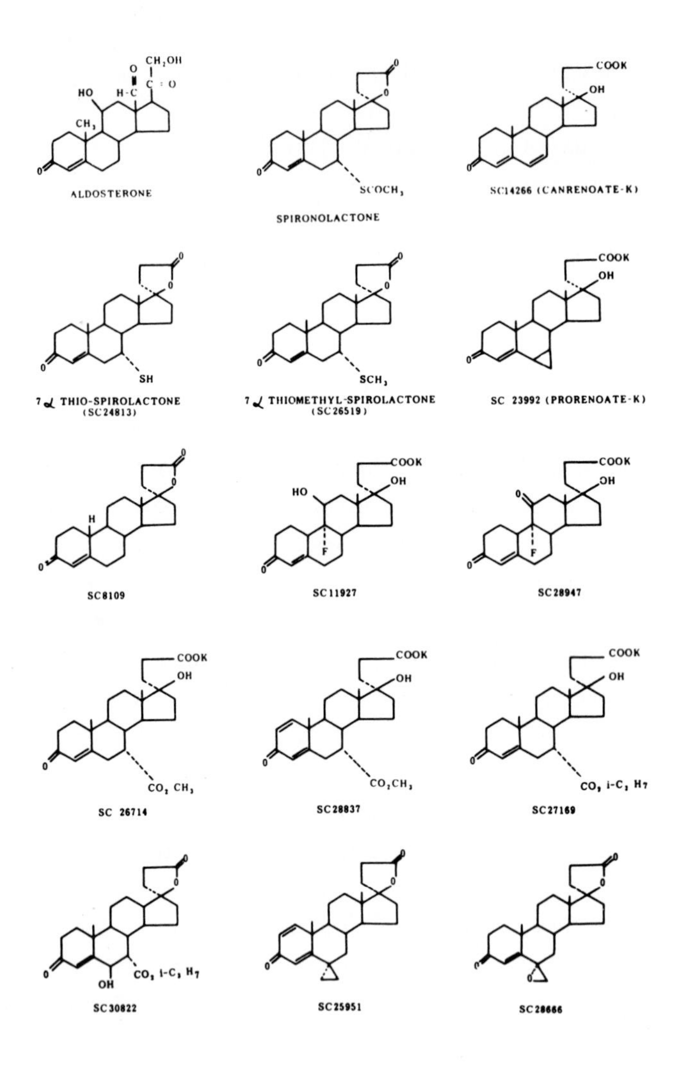

Figure 1. Structures of the spirolactones referred to in the text.

spirolactone structures have been synthesized and evaluated by several different methods. In this article we will describe and assess the various methods which have been employed in preclinical evaluation of new spirolactones, and advance what we believe to be a rational sequence of study. We will then review briefly the clinical pharmacology of the spirolactones which are currently in clinical use.

Evaluation of New Spirolactones.

Aldosterone receptor binding studies

In recent years the mode of action of aldosterone and other mineralocorticoids at the cellular level has been elucidated and it was a logical extension of this work to examine the binding of spirolactones to renal aldosterone receptors *in vitro*, and to explore their structure-activity relations at the cellular level.

The action of aldosterone is initiated by high-affinity stereo specific binding to receptor proteins in the cytoplasm of target cells (1-3) followed by time-dependent active transfer of the cytoplasmic complex to the nucleus to form a nuclear receptor complex. This complex interacts with chromatin to initiate synthesis of mRNA (4-6) and hence specific induced proteins which modulate the rate-limiting step in the physiological response. In the kidney there are probably independent mechanisms governing the kaliuretic and antinatriuretic actions of mineralocorticoids (1,7,8). The ability of a steroid to displace aldosterone from binding proteins correlates well with its potency as a mineralocorticoid (9). Specific high-affinity receptors for mineralocorticoids have been characterised in the cytoplasm and nuclei of rat kidney cells (1,3,9-12). The human kidney contains aldosterone receptors with very similar characteristics (13), suggesting that studies on rat kidney receptors will provide information on mineralocorticoids and mineralocorticoid antagonists which will be relevant to man(14), at least at the cellular level.

Spirolactones interact highly specifically (11,15) but relatively weakly (14,15) with these renal receptor proteins at concentrations which are associated with pharmacological activity _in vivo_ (1,9,10). However they do not activate the receptor (1,9-11) because the drug-receptor complex fails to translate to the nucleus and bind to chromatin (11,15,16). The chain of events which eventually results in the physiological response to aldosterone is thus interrupted.

The relative affinity of different spirolactones for rat kidney aldosterone-binding receptors has been reported in several studies (14-19). The relative affinities cited below are taken from the studies of Funder and his colleagues (14,18,19). The value and limitations of these _in vitro_ studies will be discussed later.

Bioassay in laboratory animals.

The pharmacological action of aldosterone antagonists has been studied widely in adrenalectomized rats maintained on a high sodium diet (20-24). Single doses of mineralocorticoids induce reproducible dose-dependent changes in urine electrolyte composition, namely decreased sodium and increased potassium excretion (23, 25-28). The responses to different mineralocorticoids are qualitatively similar (20-23,25,27) and deoxycorticosterone acetate (DOCA) has been used most widely. Simultaneous treatment with a competitive mineralocorticoid antagonist results in dose-dependent attentuation of the mineralocorticoid effects on urine composition by increasing sodium excretion and reducing potassium excretion (20, 22-28). Urinary $\log_{10}$ 10Na/K has proved the most useful single index of renal mineralocorticoid (21) and antimineralocorticoid (20, 22) activity because this ratio integrates the divergent electrolyte responses. The ratio is logarithmically transformed to normalise the distribution and stabilise the error variance, thus allowing analysis by parametric statistical methods, and it is multiplied by ten to avoid negative values. Using this method accurate and reproducible estimates of the relative potency of spirolactones in the rat have been obtained by standard parallel-line bioassays (29). More recently Hofmann and his colleagues have developed a similar bioassay to determine the renal antimineralocorticoid potency of spirolactones in rhesus monkeys (30-32). The monkeys are not adrenalectomized, and aldosterone is used as a mineralocorticoid agonist. The relative potency estimates for spirolactones in rat and monkey cited below are taken from several sources, both published and unpublished, and these are detailed in reference (33).

Bioassay in healthy man

To examine the pharmacological action of spirolactones satisfactorily in normal man it is necessary first to induce a state of mineralocorticoid excess. Several methods have been used, and they have been reviewed elsewhere (34). Briefly, endogenous mineralocorticoids may be stimulated by dietary sodium restriction or by administering a loop or thiazide diuretic, but the study conditions are not readily reproducible on different occasions. Exogenous aldosterone has been used, but it must be administered parenterally, preferably by intravenous infusion, and this is inconvenient. The simplest method involves treatment with a single oral dose of the synthetic mineralocorticoid fludrocortisone in the evening, followed by collection of the overnight urine. Fludrocortisone produces dose-dependent reduction of urine sodium excretion, increases potassium excretion, and reduces the urine Na/K ratio, and these changes are highly reproducible (35). The results obtained using this mineralocorticoid appear to be identical to those obtained using exogenous aldosterone, or with induced excess of endogenous mineralocorticoids (34). Single oral doses of spirolactones reverse the changes in urine electrolyte composition caused by fludrocortisone; they increase sodium excretion, reduce potassium excretion and elevate the urine Na/K ratio (35-37). As in the animal bioassays the urine $\log_{10}$ 10Na/K has proved the best single index of antimineralocorticoid activity.

Using this method it has been possible to define accurate and reproducible log dose-response curves for single doses of spironolactone and several other spirolactone analogues (38-41). Using spironolactone as the reference compound the relative potencies of new spirolactones have been defined with high precision by classic two-drug, three-dose parallel-line bioassays (38-41). The slope of the log dose-response curve for spironolactone has proved remarkably constant in several studies (41), and it is therefore possible to estimate the relative potency of spirolactones from data obtained at a single dose level, provided certain assumptions are accepted (41). Thus the relative potency of new spirolactones in healthy man can be estimated at a very early stage of development, using a simple study method which requires a minimal outlay on animal toxicology and formulation development. The methods used have been reviewed in detail elsewhere (34). To date we have determined the relative potency of fourteen spirolactones in healthy man (33), and the data presented below are drawn from this series of experiments.

Comparison of different methods

Table 1 shows the potencies of eight spirolactones and 21-carboxylic acid derivatives relative to spironolactone, comparing the findings in bioassays in healthy man, rats and monkeys, and from *in vitro* studies of binding to rat kidney aldosterone receptors. In man the potency of the new spirolactones varied from 0.08 to 2.7 compared to that of spirono-

Table 1 Relative potency (and 95% confidence limits) of single doses of nine spirolactones or 21-carboxylic acids in man, monkey and rat using urinary $\log_{10}$ Na/K as response, and relative binding affinity for aldosterone receptors *in vitro*. The structures are shown in Figure 1, and sources of the data are detailed in reference 33.

Compound	Healthy man	Monkey	Rat	In vitro
Spironolactone	1.0	1.0	1.0	1.0
SC 23992	2.7(2.1-3.5)	3.1	4.6	0.1
SC 25951	1.1(0.5-2.3)	-	4.3	0.4
SC 27169	0.6(0.5-0.8)	1.6	3.9	0.2
SC 26519	0.3(0.2-0.6)	0.1	1.1	2.0
SC 24813	0.3(0.1-0.5)	0.6	0.1	2.0
SC 28666	0.2(0.1-0.7)	1.1	0.9	0.7
SC 14266	0.3(0.2-0.4)	0.7	0.9	0.03
SC 8109	0.08(0.04-0.13)	-	0.2	1.3

lactone, with one compound significantly more potent than spironolactone (SC23992, prorenoate potassium), one approximately equipotent (SC 25951), and six significantly less potent than spironolactone. The relative affinity of these compounds for aldosterone receptors *in vitro* showed no relation to their renal potency in normal man. Indeed the estimates of relative affinity lay outwith the 95% confidence limit of potency in normal man for all compounds except SC 28666. The rank order of potency in healthy man was quite dissimilar to that at the receptor site. Results in the rat bioassay were also poorly predictive of potency in man, with the potency estimates in rats falling within the confidence limits in man for only one compound. The bioassay in monkeys was weakly predictive of the findings in man, with some similarity in the rank order of potencies. However even then the best estimate of potency in monkeys was inconsistent with that in man for four of the six compounds tested, as judged by the 95% confidence limits.

Superficially these data might be construed as indicating that *in vitro* studies of binding to aldosterone receptors are of little value in predicting the potency of spirolactones in man. In fact the reasons for the discrepancies are extremely instructive. Three of the compounds tested were 21-carboxylic acid derivatives rather than spirolactones, namely SC 14266 (potassium canrenoate), SC 23992 (potassium prorenoate) and SC 27169. All three had very low affinity for aldosterone receptors *in vitro* (0.03-0.2) while showing significant renal antimineralocorticoid activity in man (0.3-2.7). At physiological pH these acids exist largely in the ionized form which does not readily cross lipid membranes, and this is the cause of their weak *in vitro* binding properties. It has been shown in man that high plasma levels of canrenoate derived from SC 14266 (potassium canrenoate) do not contribute to renal antimineralocorticoid activity (39) indicating that the 21-

carboxylic acid form is also inactive in vivo. The 21-carboxylic acids are active when administered to intact animals or man because they are converted to the corresponding lactones in vivo. In fact lactones have much higher affinity for aldosterone receptors in vitro than the corresponding 21-carboxylic acid derivatives (14,42,43). For example prorenone, the lactone form of potassium prorenoate (SC 23992), has an affinity for aldosterone receptors 1.5-10 times that of spirondactone in vitro (42,43), findings which are more consistent with the potency estimates in man, monkey and rat (Table 1). The 21-carboxylic acid salts are in effect prodrugs for the lactones, selected for development because their water-solubility leads to more rapid and complete absorption from the gastrointestinal tract.

The findings for SC 8109 are also of interest. This compound had high affinity for aldosterone receptors in vitro but very little pharmacological activity in man or in the rat. SC 8109 has been shown to have partial agonist activity in laboratory animals (27), and partial agonism may well explain the discrepant findings in vitro and in vivo. This underlines an important consideration when interpreting receptor-binding studies, that such studies cannot differentiate between agonsts, antagonists and partial agonists (14,18).

The spirolactones SC-26519 and SC-24813 are believed to be intermediates in the metabolism of spironolactone, and it was thought that they might contribute significantly to the pharmacological activity of this drug, particularly after single doses (39,44). Their relative potency in healthy man (0.3) was much lower than would be predicted from their relative affinity for aldosterone receptors (2.0). The reason for this is not known, but we suspect that both spirolactones may be metabolized to canrenone when given orally to healthy man, either in the gut or during first pass through the liver. This highlights another major difficulty - perhaps the most important one - when extrapolating from receptor-binding studies to man, namely the metabolic transformation of drugs in vivo. Spironolactone itself undergoes extensive and complex biotransformation in man (45), and it is probable that its pharmacological and therapeutic effects are mediated entirely by canrenone and other active metabolites, and not by the parent drug (45). Canrenone is in fact a much less complex drug than spironolactone, and in retrospect it might have been better to use canrenone as the "standard" spirolactone when comparing results from human, animal and in vitro studies.

We conclude that aldosterone receptor binding studies cannot provide a simple direct prediction of the renal potency of spirolactones in man. However they have given valuable insight into the cellular mode of action of the spirolactones, and they have some general predictive value provided certain important points are noted. The lactones rather than 21-carboxylic acids should be examined in in vitro studies. Compounds showing high affinity in vitro must be examined in vivo to ensure that they are aldosterone antagonists, and not agonists or partial agonists. Finally, an antagonist exhibiting high affinity in vitro will be clinically useful only if it reaches its site of action in the human distal renal tubule unaltered. The ideal pharmacokinetic profile for a spirolactone in man might be complete absorption, absence of biotransformation, and a high renal clearance.

Receptor binding and endocrine side-effects

Gynaecomastia, impotence and menstrual disturbance are common and important side-effects of spironolactone (46). The precise mechanism is unclear, but spironolactone influences testosterone biosynthesis, plasma testosterone concentrations and the peripheral conversion of testosterone to oestradiol (45). It also binds to 5α-dihydrotestosterone receptors. In a detailed study Huffman et al (46) found no evidence for persistent alteration of androgen metabolism during prolonged treatment with spironolactone, and they concluded that competition by spironolactone or its metabolites at 5α-dihydrotestosterone receptors was the likely mechanism for gynaecomastia. Spironolactone and potassium canrenoate (SC 14266) do inhibit the binding of 5α-dihydrotestosterone to the cytosolic and nuclear receptors of the rat ventral prostate *in vivo* and *in vitro*, at doses or concentrations in the range required for antimineralocorticoid effect (47). Similar *in vitro* studies have suggested that potassium prorenoate (SC 23992) might have less antiandrogenic activity than spironolactone (15,19). These studies and others (48) raise the attractive prospect of selecting for development spirolactones with high affinity for aldosterone receptors but low affinity for 5α-dihydrotestosterone receptors hoping for a potent aldosterone antagonist which will prove free of unwanted endocrine effects. This elegant approach will prove very difficult to validate. Whereas the renal antimineralocorticoid activity of spirolactones can be measured very easily in man, as described above, evaluation of the endocrine side-effects of different spirolactones will need long-term studies in large numbers of subjects. The predictive value of 5α-dihydrotestosterone receptor binding studies is uncertain, and is likely to remain so.

Rational scheme of evaluation.

It is a sobering thought that several hundred spirolactone structures have been synthesized and studied over more than two decades, yet we still do not have a drug superior to spironolactone. Of those described above only potassium prorenoate (SC 23992) appeared to offer any potential advantage, but to our knowledge there is no intention to develop it for clinical use. Another drug, spirorenone, is currently under investigation (49). We believe that the chance of success in the future might be improved if the different methods of evaluation were used in the following manner;

1. Rat kidney aldosterone receptor binding studies are used to identify structures with high affinity, using the lactone form rather than the corresponding 21-carboxylic acid. Low affinity for 5α-dihydrotestosterone receptors may be an advantage, but should not be given undue weight.
2. Studies in rats and rhesus monkeys are used to confirm antimineralocorticoid activity and exclude agonist or partial agonist activity. The relative potency in monkeys may have weak predictive value for man.
3. Proceed to a single point bioassay in healthy man, to confirm activity and estimate the potency relative to spironolactone. Interpret these data in conjunction with the single-dose pharmacokinetics if possible. A salt of the 21-carboxylic acid, rather than the lactone, may be preferred for clinical development, as this form is likely to have superior bioavailability.

Spirolactones in Clinical Use.

Spironolactone has been in wide clinical use for twenty years. Potassium canrenoate (SC 14266, Figure 1) is used less widely, but can be administered intravenously because it is water soluble. The therapeutic action of both drugs is probably attributable to their common metabolite, canrenone, which has itself found limited clinical use.

Pharmacological action

The action of these drugs at the cellular level has been described above. In intact animals and man they satisfy the criteria for reversible specific competitive antagonism of aldosterone and other mineralocorticoids (29). Thus they have no important pharmacological action in the absence of mineralocorticoids, they reverse all the known effects of mineralocorticoids, and the responses to different doses of agonist and antagonist vary in accordance with the Law of Mass Action (29). Mineralocorticoids influence electrolyte transport in several organs other than the kidney, for example, the gut, sweat glands and salivary glands, and the spirolactones are active at all of these sites (50,51). However the kidney, and in particular the distal renal tubule (50), is the important target for aldosterone and its antagonists. As described above the spirolactones increase urine sodium excretion, decrease potassium excretion, and increase the urine Na/K ratio. In addition they diminish titratable acidity (51) and impair maximal acidification of the urine (52). Although the presence of a mineralocorticoid agonist is essential for the spirolactones to exhibit pharmacological activity, it is important to note that mineralocorticoids need not be present in excess. Indeed the spirolactones have weak but measurable natriuretic activity in healthy men even when endogenous mineralocorticoids have been suppressed by sodium loading (50).

Pharmacokinetics

Spironolactone is metabolized extensively when administered to healthy man. Abshagen et al (53) reported "considerable amounts" of unaltered spironolactone in plasma thirty minutes after dosing, but this is at variance with other studies (54, 55). The drug is probably metabolized completely on the first pass through the liver, with conversion of 80% of the dose to canrenone by removal of the 7α-acetylthio substituent (56). The remaining 20% undergoes partial metabolism at the 7α position, yielding a large number of sulphur-containing metabolites found mainly in the urine (54,57,58). The only sulphur containing compound which has been identified in plasma is the 6β-OH, 7α-thiomethyl derivative (54,59) with a concentration approximately 15% that of canrenone. Thus canrenone is the principal unconjugated metabolite of spironolactone in plasma, where it is in equilibrium with the 21-carboxylic acid form canrenoate (56). Further steps in the complex metabolic scheme include glucuronidation, hydroxylation and reduction (54, 59-61).

Activity of metabolites.

It has been assumed that canrenone was entirely responsible for the pharmacological action of spironolactone, but this is not the case, particularly after single doses (38,39,56,62, 63). When healthy men are given equal doses of spironolactone and potassium canrenoate the plasma canrenone concentrations are very similar, yet the renal antimineralocorticoid potency of potassium canrenoate is only one third that of spironolactone (39). Thus canrenone, as measured by fluorimetry, can account for only one third of the renal activity of single doses of spironolactone. However when healthy men are given repeated doses of spironolactone and potassium canrenoate until steady-state conditions are attained, canrenone is found to account for about 70% of the renal activity of spironolactone(64). It is probably true that canrenone is responsible for most of the therapeutic effect of spironolactone during chronic treatment.

The source of the two-thirds of the activity of single doses of spironolactone which cannot be attributed to canrenone is not known, but three suggestions have been advanced. It is possible that sulphur-containing metabolites may be highly active at the distal renal tubule because of their high renal clearance and high intrinsic activity, even though they are present in plasma only at very low concentrations (39). In support of this is the observation that oral canrenone and spironolactone have similar extrarenal antimineralocorticoid activity as measured by rectal potential difference (65), whereas spironolactone is markedly superior to canrenone as regards renal activity (66). A second possiblity is that spironolactone itself may contribute to the pharmacological activity, but the one report of unaltered spironolactone in plasma (53) has yet to be confirmed. Finally, studies measuring canrenone in plasma using HPLC suggest that the widely-used fluorimetric assay for canrenone is non-specific, and may measure (as canrenone) significant amounts of sulphur containing metabolites which are in fact present in plasma(56). If this is so, one or more of these metabolites may be responsible for the unexplained pharmacological action of single doses of spironolactone. It has been mentioned that canrenone exists in equilibrium with canrenoate in plasma, but it is clear that canrenoate does not contribute to the pharmacological effect (39).

Pharmacokinetics of canrenone

The considerations above apply only to single doses of spironolactone. On repeated dosing to steady-state canrenone is responsible for the major part of the pharmacological activity of spironolactone (64). In healthy men the half-life of canrenone in plasma is 17-22 hours (55,59,67-69) and with repeated dosing canrenone accumulates over 3-5 days (67). There is a linear relation between the dose of spironolactone (and of potassium canrenoate) and the plasma canrenone concentration at least up to doses of 800 mg (39,62,64,70). Canrenone is more than 90% protein-bound in plasma (59). Its renal clearance exceeds that of creatinine, suggesting active tubular secretion (71,72), but despite this renal excretion accounts for only a minor part of the total clearance of canrenone. Plasma canrenone concentrations do not vary much between healthy subjects (67,73). Hypertensive patients show only twofold variation in steady-state canrenone concentrations (70). This is related largely to body weight and presumably to the volume of distribution (70) and there is therefore surprisingly little variation between subjects in the metabolism of spironolactone despite its extensive biotransformation. On the other hand patients with heart failure and chronic liver disease show fifteenfold variation in canrenone concentrations for a given dose of spironolactone (68). This finding, and data from a study in rats (74), suggest that the disposition of spironolactone may be altered by liver dysfunction. In fact studies in patients with chronic liver disease are inconsistent, one showing prolongation of the plasma half-life of canrenone to 59 hours (69), whereas another showed no difference in half-life compared to healthy subjects (53). There is no evidence that the half-life of canrenone is shortened by long-term spironolactone treatment in man (68)as it is in the rat (74). There have been few attempts to correlate the therapeutic effects of spironolactone with the plasma canrenone concentration. There was a weak but significant correlation between steady-state canrenone concentrations and increments in plasma potassium in a group of thiazide-treated hypertensive patients (70).

Dose regimen

Spironolactone is usually prescribed three or four times daily, but this is unnecessary. The renal action of spironolactone reaches its peak only after seven hours and persists for 24 hours (75-77). Considering also the long half-life of canrenone (18 hours) and its accumulation over 3-5 days, single daily doses are sufficient and are more convenient for the patient. The antihypertensive action is maintained satisfactorily with once daily treatment (78,79). In theory a higher loading dose for one or two days might be useful when starting treatment (50,68) but this does not seem necessary in ordinary practice.

Bioavailability.

The absolute bioavailability of spironolactone has not been determined, but at least 75% of the dose is absorbed from an alcoholic solution (53). The extent of absorption from tablet formulations can be predicted from the in vitro dissolution rate, provided no major change is made to the excipients (73), and tablets with satisfactory dissolution rates are completely bioavailable when compared to solutions of spironolactone (80, 81). Micronization of spironolactone chemical increases the bioavailability (73,82,83) and pharmacological activity (83) of the drug, and the micronized formulations now available commercially probably have satisfactory and uniform bioavailability. The bioavailability of spironolactone is enhanced significantly when it is taken immediately after breakfast(84).

Therapeutic action.

The therapeutic effects of spironolactone are probably all related to mineralocorticoid antagonism, and in particular to its action on the kidney to promote sodium excretion and potassium conservation. The natriuresis is governed in part by the amount of sodium delivered to the distal renal tubule and the potency of spironolactone is therefore enhanced when it is used with thiazide or loop diuretics which act more proximally on the renal tubule. The need for the presence of mineralocorticoids for spironolactone to act has been mentioned before, but it is worth repeating that a state of mineralocorticoid excess is not necessary (85,86). In theory the therapeutic action of spironolactone might be blunted by very high levels of aldosterone. There is evidence that this is occasionally so (70) but in general this does not seem to be an important problem. From time to time there have been suggestions that some of the therapeutic effects of spironolactone cannot be attributed to its mineralocorticoid antagonist properties. In particular, it has been proposed that other mechanisms may be responsible for its antihypertensive effect and its influence on potassium homeostasis. These points have been discussed in detail elsewhere (45), with the general conclusion that there is in fact no need to invoke mechanisms of action other than mineralocorticoid antagonism.

Inhibition of aldosterone biosynthesis.

There is clear evidence that spironolactone (or its metabolites) causes a partial block of aldosterone biosynthesis in addition to antagonizing its peripheral effects. This has been shown *in vitro* (87-89), in healthy men (90-93), in hypertensive patients (85,94) and in primary hyperaldosteronism (95,96). Spironolactone bodies in the adrenal gland appear to be the morphological correlate of this phenomenon (96). It is important to note that the biosynthesis block is only partial and is countered by the increases in renin and angiotensin II stimulated by sodium loss, so that plasma aldosterone levels usually rise during spironolactone treatment. However the plasma aldosterone concentrations are approximately 20% lower than would be expected from the prevailing plasma angiotensin II and potassium concentrations (85). This inhibitory effect appears to persist for at least four weeks during spironolactone treatment (85). It is unlikely that the phenomenon has any clinical relevance, although it has been suggested as a possible mechanism for the selective hypoaldosteronism seen after surgical removal of aldosterone-producing adenomas (97).

Clinical uses.

The therapeutic use of spironolactone has been reviewed in

detail elsewhere (45,51). Briefly, it is a drug of first choice in treating primary aldosteronism, other uncommon forms of hypertension associated with mineralocorticoid excess, and fluid retention due to chronic liver disease and the nephrotic syndrome. In essential hypertension it is a useful alternative to the thiazides when the latter drugs are contra-indicated, and it also has a valuable role in the treatment of resistant hypertension. It is used as an adjunct to loop diuretics in patients with refractory heart failure. It is effective in the prevention or correction of hypokalaemia induced by the thiazide or loop diuretics, and this is in fact the most common indication for its use.

Drug interactions.

Spironolactone is a weak inducer of hepatic enzymes in man, but there is no evidence that this is clinically important. There is a suggestion that it may decrease the renal clearance of digoxin, but the evidence is conflicting. It is also unclear whether spironolactone cross-reacts in digoxin radioimmunoassays. Spironolactone reduces the anticoagulant action of warfarin significantly, but only slightly, and aspirin antagonises the renal action of spironolactone, but neither of these interactions is clinically important. Spironolactone blocks the ulcer-healing action of carbenoxolone, and this is a clinically significant drug interaction. These interactions are described in more detail and with references elsewhere (45).

Adverse effects.

Minor subjective side-effects are common when the dose of spironolactone exceeds 100 mg daily (85). Hyperkalaemia is the most dangerous adverse reaction, and is a serious hazard when spironolactone is prescribed with potassium supplements,other potassium sparing drugs, or captopril, or to patients with significant renal impairment. The risk of hyperkalaemia is negligible when ordinary doses of spironolactone are prescribed for patients with normal renal function, but high doses (eg 400 mg daily) have occasionally caused serious hyperkalaemia even when renal function was normal. The endocrine side-effects of spironolactone have been mentioned earlier. Gynaecomastia is common with prolonged treatment, even in healthy men. After ten months of treatment the incidence of gynaecomastia was 30% in men taking spironolactone 100 mg daily, and 62% in those taking 200 mg daily (46). It also occurs with lower doses, although relatively infrequently. More detail of these side-effects, and of the influence of spironolactone on several biochemical and hormonal measurements, can be found in reference (45).

References

1. Feldman, D., Funder, J.W., Edelman, I.S.: Amer.J.Med. 53, 545-560 (1972).
2. Sharp, G.W., Leaf, A.: Physiol.Rev. 46, 593-633 (1966).
3. Edelman, I.S., Fanestil, D.D.: Biochemical Actions of Hormones. Academic Press, London, 321-364 (1970).
4. Williamson, H.E.: Biochem. pharmacol. 12, 1449-1450 (1963).
5. Edelman, I.S., Bogoroch, R., Porter, G.A.: Proc. Nat. Acad. Sci. USA. 50, 1169-1177 (1963).
6. Rossier, B.C., Wilce, P.A., Edelman, I.S.: Proc. Nat. Acad. Sci. USA. 71, 3101-3105 (1974).
7. Forte, L.R.: Life Sci. 11, 461-473 (1972).
8. Fimognari, G.M., Fanestil, D.D., Edelman, I.S.; Amer. J. Physiol. 213, 954-962 (1967).
9. Herman, T.S., Fimognari, G.M., Edelman, I.S.: Biol.Chem. 243, 3849-3856 (1968).
10. Fanestil, D.D.: Biochem. Pharmacol. 17, 2240-2242 (1968).
11. Marver, D., Goodman, D., Edelman, I.S.: Kidney Int. 1, 210-223 (1972).

12. Funder, J.W., Feldman, D., Edelman, I.S.: Endocrinology 92, 994-1004 (1973).

13. Matulich, D.T., Spindler, B.J., Schambelan, M., Baxter, J.D.: J.Clin. Endocrinol. Metab. 43, 1170-1174 (1976).

14. Funder, J.W., Feldman, D., Highland, E., Edelman, I.S.: Biochem. Pharmacol. 23, 1493-1501 (1974).

15. Claire,M., Rafestin-Oblin, M.E., Michaud, A., Roth-Meyer, C., Corvol, P.: Endocrinology 104, 1194-1200 (1979).

16. Marver, D., Stewart, J., Funder, J.W., Feldman, D., Edelman, I.S.: Proc. Nat. Acad. Sci. USA. 71, 1431-1435 (1974).

17. Raynaud, J.P., Bonne, C., Bouton, M.M., Moguilewsky, M., Philibert, D., Azadian-Boulanger, G.: J. Steroid Biochem. 6, 615-622 (1975).

18. Sakauye, C., Feldman, D.: Amer. J. Physiol. 231, 93-97 (1976).

19. Funder, J.W., Mercer, J., Hood, J.: Clin. Sci. Mol. Med. 51, 333-345 (1976).

20. Kagawa, C.M. : In: Evaluation of Drug Activities: Pharmacometrics, Vol 2, eds, Lawrence, D.R., Bacharach, A.L.: Academic Press, London 1964: 745-762.

21. Johnson, B.B.: Endocrinology 54, 196-208 (1954).

22. Kagawa, C.M.: Endocrinology 67, 125-132 (1960).

23. Kagawa, C.M., Cella, J.A., Van Arman, C.G.: Science 126, 1015-1016 (1975).

24. Liddle, G.W.: Metabolism 10, 1021-1030 (1961).

25. Liddle, G.W.: Science 126, 1016-1018 (1957).

26. Kagawa, C.M., Sturtevant, F.M., Van Arman, C.G.:J.Pharmacol. 126, 123-130 (1959).

27. Kagawa, C.M.: In: The Clinical Use of Aldosterone Antagonists ed. Bartter, F.C.: Thomas, Springfield, 1960: 33-36.

28. Kagawa, C.M., Bouska, O.J., Anderson, M.L.: Acta Endocrinologica 45, 79-83 (1964).

29. Kagawa, C.M.: In: Methods in Hormone Research,vol III ed. Dorfman, R.I. Academic Press, New York and London, 1964, 351-414.

30. Hofmann,L.M.: In: Recent Advances in Renal Physiology and Pharmacology, eds. Wesson, L.G., Fanelli, G.M.: University Park Press, Baltimore, 1974: 205-216.

31. Hofmann, L.M., Chinn, L.J., Pedrera, H.A., Krupnick, M.I., Suleymanov, O.D.: J.Pharmacol.Exp.Ther. 194, 450-456 (1975)

32. Hofmann,L.M., Pedrera, H.A., Suleymanov, O.D.: J.Pharmacol Exp.Ther. 202, 216-220 (1977).

33. McInnes, G.T., Shelton, J.R., Ramsay, L.E., Harrison, I.R. Asbury, M.J., Clarke, J.M., Perkins, R.M., Venning, G.R.: Br. J. Clin. Pharmac. 13, 331-339 (1982).

34. McInnes, G.T., Shelton, J.R., Ramsay, L.E.: Meth. Find. Exp. Clin. Pharmacol. 4, 49-71 (1982).

35. Ross, E.J.: Clin. Sci. 23, 197-202 (1962).

36. Noel, P.R., Leahy, J.S.: Clin. Sci. 23, 477-483 (1962).

37. Gantt, C.L., Dyniewicz, J.M.: Metabolism 12, 1007-1011 (1963).

38. Ramsay, L.E., Shelton, J.R., Wilkinson, D., Tidd, M.J.: Br. J. Clin. Pharmac. 3, 607-612 (1976).

39. Ramsay, L.E., Shelton, J., Harrison, I., Tidd, M., Asbury M.: Clin. Pharmacol. Ther. 20, 167-177 (1976).

40. Ramsay, L.E., Harrison, I., Shelton, J., Tidd, M.: Clin. Pharmacol. Ther. 18, 391-400 (1975).

41. McInnes, G.T., Shelton, J.R., Asbury, M.J., Harrison, I.R. Clarke, J.M., Ramsay, L.E., Venning, G.R.: Clin. Pharmacol. Ther. 30, 218-225 (1981).

42. Claire, M., Rafestin-Oblin, M.E., Michaud, A., Corvol, P.: In: Aldosterone Antagonists in Clinical Medicine, ed. Addison, G.M. et al: Excerpta Medica, Amsterdam-Oxford, 1978, 65-69.

43. Wambach, G., Helber, A.: In: Hormone Antagonists, ed. Agarwal, M.K.: de Gruyter, Berlin and New York, 1982: 293-305.

44. McInnes, G.T., Asbury, M.J., Shelton, J.R., Harrison, I.R. Ramsay, L.E., Venning, G.R., Clarke, J.M.: Clin. Pharmacol. Ther. 27, 363-369 (1980).

45. Ramsay, L.E.: In: Hormone Antagonists, ed. Agarwal, M.K., de Gruyter, Berlin and New York, 1982: 335-363.

46. Huffman, D.H., Kampmann, J.P., Hignite, C.E., Azarnoff, D.L.: Clin. Pharmacol. Ther. 24, 465-473 (1978).

47. Corval, P., Michaud, A., Menard, J., Freifeld, M., Mahoudeau, J.: Endocrinology 97, 52-58 (1975).

48. Cutler, G.B.: J. Clin. Endoc. Metab. 47, 171-175 (1978).

49. Krause, W., Sack, C., Seifert, W.: Eur. J. Clin. Pharmacol 25, 231-236 (1983).

50. Streeten, D.H.: Clin. Pharmacol. Ther. 2, 359-373 (1961).

51. Manitius, A., Suchecki, T.: Mater.Med.Pol.4, 83-88 (1972)

52. Manuel, M.A., Beirne, G.J., Wagnild, J.P., Weiner, M.W.: Arch. Intern. Med. 134, 472-474 (1974).

53. Abshagen, U., Rennekamp, H., Luszpinski, G.: Naun. Schmied. Arch. Pharmacol. 296, 37-45 (1976).

54. Karim, A., Hribar, J., Aksamit, W., Doherty, M., Chinn, L.J.: Drug. Metab. Disp. 3, 467-478 (1975).

55. Sadee, W., Dagcioglu, M., Schroder, R.: J. Pharmacol. Exp. Ther. 185, 686-695 (1973).

56. Dahloff, C.G., Lundberg, P., Persson, B.A., Regardh, C.G.: Drug. Metab. Disp. 7, 103-107 (1979).

57. Karim, A., Brown, E.:Steroids 20, 41-62 (1972).

58. Abshagen, U., Rennekamp, H., Koch, K., Senn, M., Steingross, W.: Steroids 28, 467-480 (1976).

59. Karim, A., Zagarella, J., Hribar, J., Dooley, M: Clin. Pharmacol. Ther. 19, 158-169 (1976).

60. Karim, A.: Drug Metab. Rev. 8, 151-188 (1978).

61. Karim, A.: in Aldosterone Antagonists in Clinical Medicine ed. Addison, G.M. et al: Excerpta Medica, Amsterdam-Oxford 1978: 115-129.

62. Casals-Stenzel, J., Schmalback, J., Losert, W.: Eur. J. Clin. Pharmacol. 12, 247-255 (1977).

63. Ramsay, L.E.: in Aldosterone Antagonists in Clinical Medicine, ed. Addison, G.M. et al : Excerpta Medica, Amsterdam-Oxford, 1978: 199-206.

64. Ramsay, L.E., Asbury, M., Shelton, J., Harrison, I.: Clin. Pharmacol. Ther. 21, 602-609 (1977).

65. Huston, G. J., Al-Dujaili, E.A.S.: Br. J. Clin. Pharmacol. 7, 385-392 (1979).

66. Huston, G.J., Turner, P., Leighton, M: Br. J. Clin. Pharmacol. 3, 201-206 (1976).

67. Karim, A., Zagarella, J., Hutsell, T.C., Dooley, M.: Clin. Pharmacol. Ther. 19, 177-182 (1976).

68. Sadee, W., Schroder, R., Leitner, E., Dagcioglu, M.: Eur. J. Clin. Pharmacol. 7, 195-200 (1974).

69. Jackson, L., Branch R., Levine, D., Ramsay, L.E: Eur. J. Clin. Pharmacol. 11, 177-179 (1977).

70. Ramsay, L.E., Hettiarachchi, J.: Br. J. Clin. Pharmacol. 11, 153-158 (1981).

71. Hoffman, L.M., Polk, R.C., Maibach, H.I.: Clin. Pharmacol. Ther. 18, 748-756 (1975).

72. Ramsay, L.E., Harrison, I.R., Shelton, J.R., Vose, C.W.: Eur. J. Clin. Pharmacol. 10, 43-48 (1976).

73. Clarke, J.M., Ramsay, L.E., Shelton, J.R., Tidd, M.J. Murray, S., Palmer, R.F.: J. Pharm. Sci. 66, 1429-1432 (1977).

74. Solymoss, B., Toth, S., Varga, S., Krajny, M.: Steroids 16, 262-275 (1970).

75. Levine, D., Ramsay, L.E., Auty, R., Branch, R., Tidd, M.: Eur. J. Clin. Pharmacol. 9, 381-386 (1976).

76. Edmonds, C.J., Wilson, G.M.: Lancet 1, 505-509 (1960).

77. Ramsay, L.E., Shelton, J.R., Tidd, M.J.: Br. J. Clin. Pharmacol. 3, 475-482 (1976).

78. Henningsen,N.C.: In Aldosterone Antagonists in Clinical Medicine, ed. Addison, G.M. et al: Excerpta Medica, Amsterdam-Oxford, 1978: 227-233.

79. Schersten, B.: Hypertension 2, 672-679 (1980).

80. Karim, A., Zagarella, J., Hutsell, T.C., Chao, A., Baltes, B.J.: Clin. Pharmacol. Ther. 19, 170-176 (1976).

81. Tidd, M.J., Ramsay, L.E., Shelton, J.R., Palmer, R.F.: Int. J. Clin. Pharmacol. Biopharm. 15, 205-210 (1977).

82. Bauer, G., Rieckmann, P., Schaumann, W.: Arz. Forsch. 12, 487-489 (1962).

83. McInnes, G.T., Asbury, M.J., Ramsay, L.E., Shelton, J.R.: J. Clin. Pharmacol. 22, 410-417 (1982).

84. Melander, A., Danielson, K., Schersten, B., Thalin, T., Wahlin, E.: Clin. Pharmacol. Ther. 22, 100-103 (1977).

85. Ramsay, L.E., Hettiarachchi, J., Fraser, R., Morton, J.J.: Clin. Pharmacol. Ther. 27, 533-543 (1980).

86. Nicholls, M.G., Espiner, E.A., Hughes, H., Rogers, T.: Br. Heart J., 38, 1025-1030 (1976).

87. Erbler, H.C.: Naun. Schmied. Arch. Pharmacol. 277, 139-149 (1973).

88. Erbler, H.C.: Naun. Schmied. Arch. Pharmacol. 280, 331-337 (1973).

89. Aupetit, B., Bastien, C., Aubry-Marais, F., Legrand, J.C.: In Aldosterone Antagonists in Clinical Medicine, ed. Addison G.M. et al: Excerpta Medica, Amsterdam-Oxford, 1978: 36-40.

90. Erbler, H.C.: Naun. Schmied. Arch. Pharmacol. 285, 395-401 (1974).

91. Erbler, H.C.: Naun. Schmied. Arch. Pharmacol. 286, 145-156 (1974).

92. Erbler, H.C., Wernze, H., Hilfenhaus, M.: Eur. J. Clin. Pharmacol. 9, 253-257 (1976).

93. Abshagen, U., Sporl, S., Schoneshofer, M., L'Age, M., Rennekamp, H., Oelkers, W.: Clin. Sci. Mol. Med. 51, 307S-310S (1976).

94. Hoefnagels, W.H.L., Drayer, J.I.M., Smals, A.G.H., Kloppenborg, P.W.C.; Clin. Pharmacol. Ther. 27, 317-323 (1980).

95. Sundsfjord, J.A., Marton, P., Jorgensen, H., Aakvaag, A.: J. Clin. Endoc. Metab. 39, 734-739 (1974).

96. Conn, J.W., Hinerman, D.L., Cohen, E.L.: In Systemic Effects of Antihypertensive Agents, ed. Sambhi, M.P.: Stratton, New York, 1976: 359-382.

97. Bravo, E.L., Dustan, H.P., Tarazi, R.C.: J. Clin. Endoc. Metab. 41, 611-617 (1975).

THE USE OF STEROID INHIBITORS IN THE MANAGEMENT OF PATIENTS

Nicoletta Sonino, Gennaro Merola

Istituto di Semeiotica Medica - Università di Padova
Padova, Italy

Introduction

After the effect of the insecticide DDD on adrenal function had first been reported in 1949 (1), a large number of compounds were introduced, in the following twenty years, as inhibitors of corticosteroid production, affecting one or more steps of the biosynthetic pathways of steroid hormones from cholesterol (Fig. 1). These compounds can be generally grouped as follows: diphenylmethane derivatives such as o,p'DDD and amphenone B; pyridine derivatives such as metyrapone (SU-4885) and related compounds SU-8000, SU-9055, SU-10'603; disubstituted glutaric acid imides such as glutethimide and p-aminoglutethimide; other agents such as triazines, hydrazines, thiosemicarbazones (SKF 12185, cyanoketone). Some of these drugs turned out to be of great value to investigate the pathophysiology of the adrenal cortex and/or pituitary-adrenal axis both in laboratory and clinical studies, and some were employed in therapy of various diseases mainly of endocrine origin. Among compounds no longer in use: 1. cyanoketone, an inhibitor of the 3β-HSD enzyme system (2), administered to pregnant rats provided experi-

Abbreviations: 3β-HSD=3β-hydroxysteroid dehydrogenase; CAH=congenital adrenal hyperplasia; 17-OHCS=17-hydroxycorticosteroids; F=cortisol; B=corticosterone; DOC=deoxycorticosterone; 18-OH-DOC=18-hydroxydeoxycorticosterone; 18-OH-B=18-hydroxycorticosterone; DHEA-S=dehydroepiandrosterone-sulfate; PRA=plasma renin activity; SHBG=sex hormone binding globulin

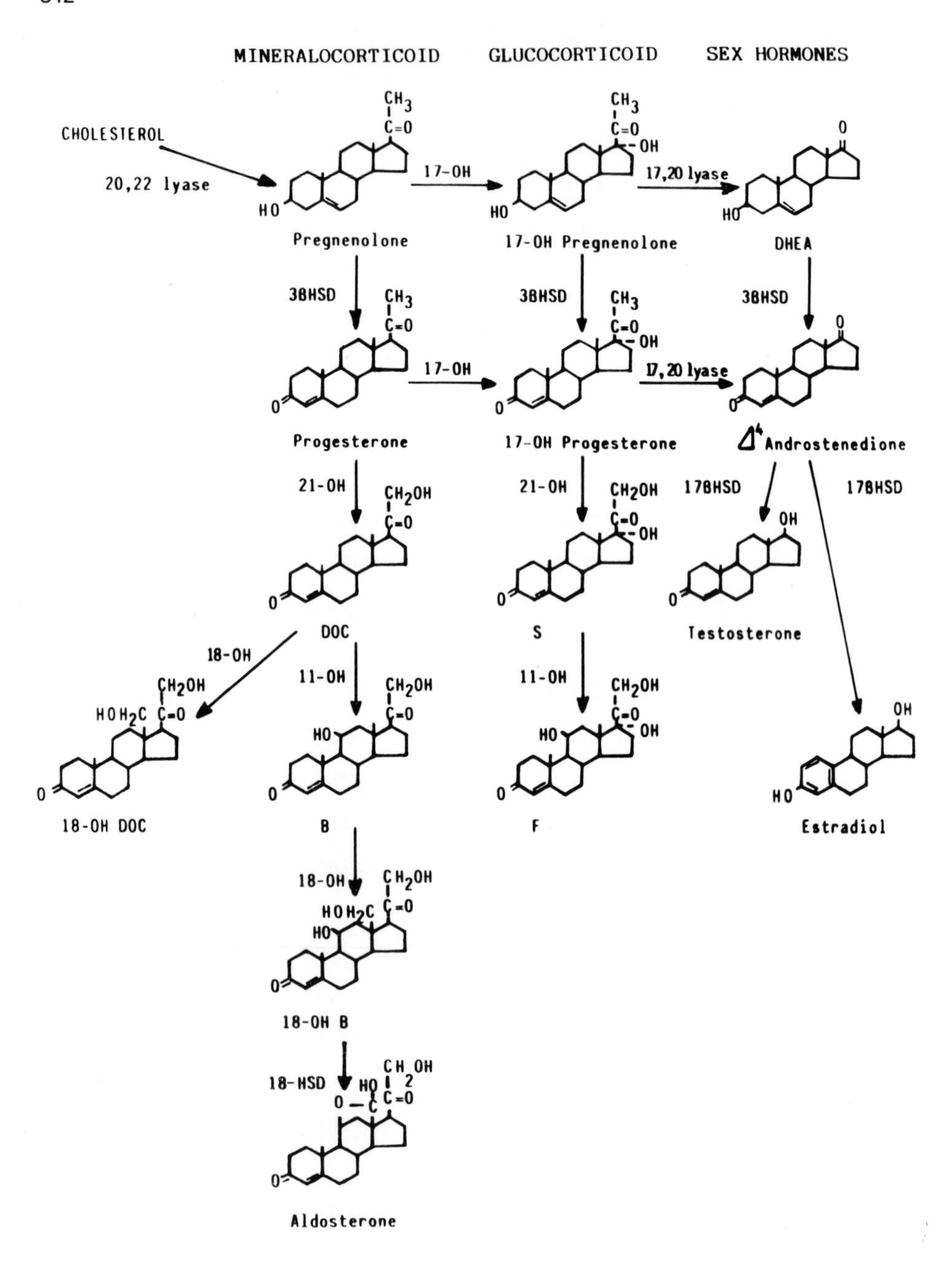

Fig. 1 Adrenal Steroidogenesis

mental models for the study of CAH due to 3β-HSD deficiency (3), while by the use of metyrapone in similar experiments the hypertensive form of CAH due to IIβ-hydroxylase deficiency was investigated (4). 2.amphenone B, shown to inhibit IIβ-, I7α- and 2I-hydroxylase (5) and pregnenolone formation from cholesterol (6), was administered to patients with adrenal hyperplasia (7), breast (7) and adrenal (8) carcinoma. 3. SKF I2I85, inhibitor of IIβ-hydroxylase, was used in the treatment of primary aldosteronism and Cushing's syndrome (9-II). On the other hand, aminoglutethimide (AG) and o,p'DDD were in clinical use, in spite of their toxicity, and metyrapone was routinely employed to assess pituitary function (I2). Thus, about fifteen years ago the research based on pharmacologic inhibition of adrenal steroid biosynthesis had already given most exciting results and still appeared as a promising field (I2). In I974 Gower thoroughly reviewed the chemistry, biochemistry and clinical applications of steroid-hormone metabolism modifiers adding new data on AG (I3). However, there have been no further significant developments over the years and the limited clinical use of adrenal inhibitors has mostly been based on the previously established. More recently, trilostane, a competitive inhibitor of the 3β-HSD system, has been introduced in clinical trials (I4,I5) mainly for the treatment of hypercortisolism. However, it seems to have no particular advantages compared with other available enzyme inhibitors. Finally, it has been lately reported on the effects of some clinically used fungicides, such as ketoconazole and chemically related compounds, on gonadal and adrenal steroid production (I6).
A serious clinical condition in the management of which these agents are still very valuable is Cushing's syndrome. Indeed, despite some advances in pharmacology (cyproeptadine, bromocriptine) and in surgical (microadenomectomy) and radiation (proton beam) procedures at pituitary level, no treatment has been proven fully satisfactory.
This review is concerned with those drugs that are in clinical use at the present time. Their structures are shown in Fig. 2.

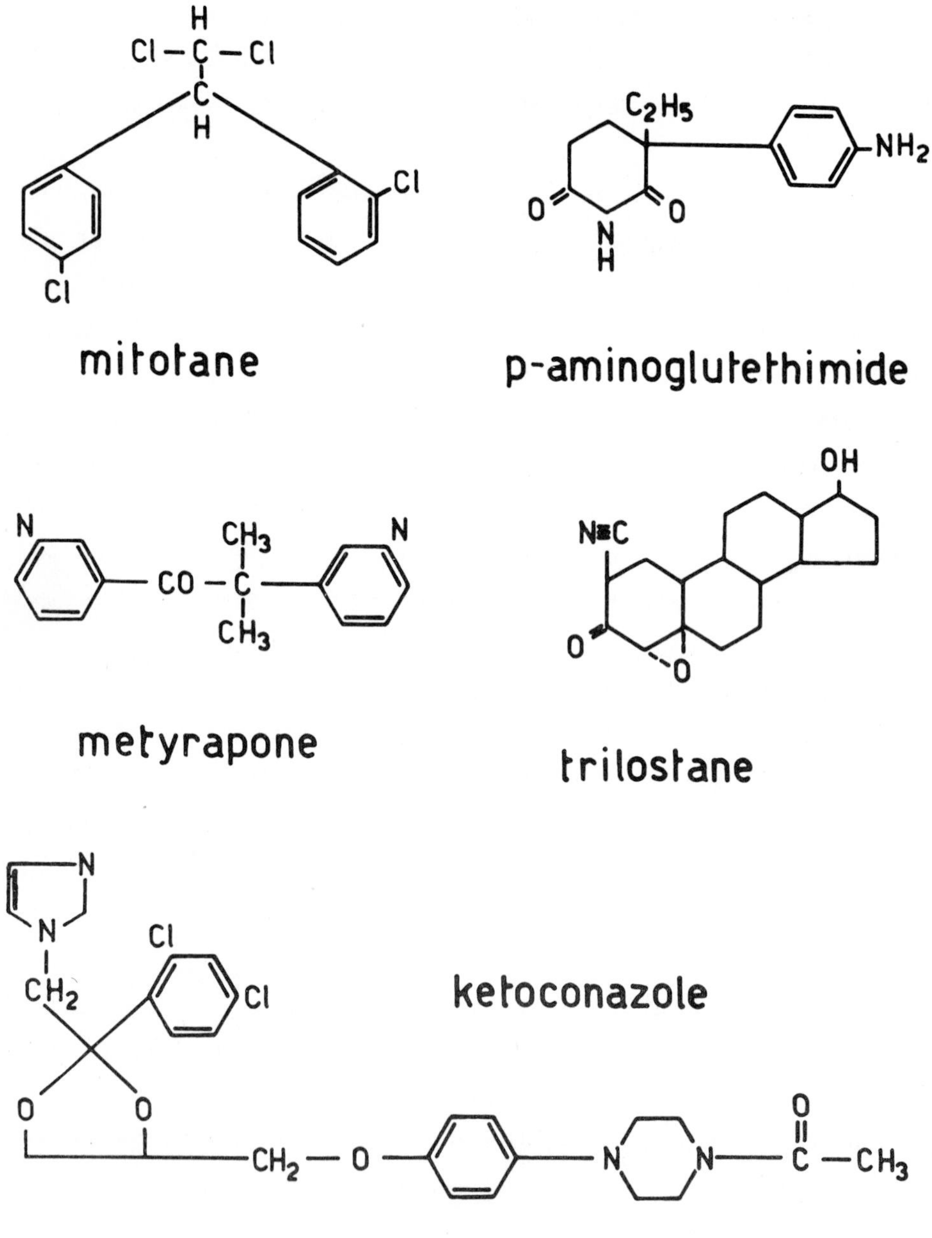

mitotane

p-aminoglutethimide

metyrapone

trilostane

ketoconazole

Fig. 2 Structures of adrenal inhibitors

Mitotane

o,p'DDD [2,2-bis(2-chlorophenyl-4-chlorophenyl)-I,I-dichloroethane]; Lysodren° (Bristol Lab.)

Mechanism of action. This drug acts by adrenocorticolytic effects, by modification of steroid peripheral metabolism and by direct inhibition of steroid synthesis. o,p'DDD was shown to cause cytotoxic atrophy of the adrenal cortex in dogs (I,I7), inducing degenerative changes of mitochondria as early as I2 hours after administration (I8). Changes were prominent in the zona fasciculata and zona reticularis, where most of the cells appeared destroyed within I2 days, while zona glomerulosa was relatively spared. In a study by Vilar and Tullner (I9), after prolonged treatment of dogs for 35 days (50 mg/Kg/d), the inner zones of the cortex had shrunk to a fibrous layer. Regeneration of the cortex was shown, after drug withdrawal, in case destruction had not been complete (20). Adrenal sensitivity to cytotoxic effects appears to be species dependent: while the dog is extremely sensitive, the rat, mouse, monkey and rabbit are resistant, showing only minor changes after prolonged treatment with high doses (I,20-22). Intermediate degree of susceptibility by the human adrenal has been demonstrated in several studies (23-26). After I month treatment with o,p'DDD 3 g/d, the adrenal glands of a patient with Cushing's disease were enlarged but grossly normal, with mitochondrial alterations at the electron microscopy similar to those observed in the dog (I8) in the zona fasciculata, and only minimal in the zona glomerulosa (27). Adrenal microadenomatosis was found in patients with Cushing's disease, who showed resistance to o,p'DDD (28). The biochemical mechanism of action of this compound is known to involve both the adrenal secretion and the peripheral metabolism of steroids. An inhibitory effect on steroid production was reported in early studies (I9,29), but it is still poorly understood. A decrease in urinary pregnanetriol by o,p'DDD was observed even at doses that would not affect other steroid metabolites (30). Later, inhibition of ACTH-mediated pregnenolone synthesis in adrenal cortex of dogs was reported by Hart and Straw (3I). More recently, Touitou et al. have shown inhibitory effect of o,p'DDD on II- and I8-hydroxylase activities by human adrenal incubations (32): a very significant decrease in the synthesis of F, B, I8-OH-B and aldosterone from radioactive precursors occurred in the adrenals removed from treated patients compared with untreated patients. However, addition of o,p'DDD or its metabolite o,p'DDE in the incubation medium of control human adrenals was ineffective in altering steroid synthesis, suggesting that peripheral metabolism is required for the drug to be active (32). o,p'DDD action is slow, requiring several months to reduce F secretion to the normal range (28,33). The striking fall in I7-OHCS excretion, that occur within a few days of treatment, is due to an altered extra-adrenal metabolism of cortisol and is not related to a decrease in plasma I7-OHCS and cortisol secretion rate nor to changes in serum transcortin (34). This is a major effect, which by itself accounts for a 50 to 80% decrease in urinary I7-OHCS. i .e., usual tetrahydrometabolites are formed less and there is a proportional increase in the excretion of 6β-OH-cortisol, a hydrophilic compound not extracted by the Porter-Silber method (34-36). In addition, o,p'DDD inhibits the peripheral conversion of 3β-hydroxy-Δ^5 steroids to their 3α-hydroxypregnane analogs (37).

Extra-adrenal effects. Peripheral effects on steroid metabolism by o,p'DDD may be explained by modification of liver enzyme activities, that apparently occur without gross alterations of hepatic function (34). Although found to be present, after treatment, in almost every tissue of the body (38,39), the drug does not seem to interfere with the function of endocrine glands other than adrenal cortex. i .e., no significant thyroid or gonadal alterations have been reported after prolonged administration in large groups of patients (28,33,40); rather, improvement of gonadal-pituitary function has been observed in both male and female patients (28). Transient gynecomastia is a relatively frequent side effect in males (28,40):inone case hormonal evaluation showed high estrogen excretion, high testosterone and Te-BG, and normal gonadotropins (4I). In two other cases (5 and I5 y.o.) there were no changes in testosterone, gonadotropins and prolactin compared to basal values (40). During prolonged treatment serum cholesterol significantly increased,and returned to basal values after drug withdrawal (28,4I). Increased urate excretion without changes in renal function has been observed in a patient (42).

Pharmacology. A study by Moy (38), in which o,p'DDD was administered at doses of 5, I0 or I5 g/d to patients with adrenal ca., showed that about 40% of the oral dose was absorbed by the gastrointestinal tract; I0% was recovered in the urine as the water-soluble metabolite o,p'DDA (dichlorodiphenylacetic acid); a small amount was excreted in the bile and the rest apparently stored in the tissues. Blood levels of the drug were measured by colorimetric method (38,39) or gas chromatography (43) and found to progressively increase during treatment. They ranged between 7 and 89 μg/ml when clinical effect became apparent (38). Severe neurologic disturbances were observed with serum concentrations higher than 20 μg/ml (43). Blood levels of the metabolite were 29-54 μg/ml (38). No correlation could be found between administered doses and blood levels, nor between blood levels and therapeutic or toxic effects (38,44). Following discontinuation of treatment, blood levels progressively decreased becoming undetectable after several weeks (38). Autopsy studies showed the highest concentrations in adipose tissues, followed by adrenal glands; detectable amounts were measured in the large number of tissues examined (38). Like the parent compound DDT, o,p'DDD is metabolized to o,p'DDA and unsaturated derivative DDE, both measurable in blood and tissues (38,43,45). Since o,p'DDD shows no direct effect on adrenal cortex *in vitro*, some sort of drug metabolism seems to be needed for its activity (32,44). Pharmacological antagonism between spironolactone and o,p'DDD has been reported both in man (46) and in dogs (47), suggesting a blocking effect of spironolactone on the action of o,p'DDD, the mechanism of which could be related to changes in cytochrome P-450 dependent enzymes (47).

Therapeutic use. o,p'DDD was first introduced for the treatment of metastatic adrenocortical ca. by Bergenstal et al., who in 1960 (23) reported the results of clinical trials in 18 patients. Since then, a large number of patients with adrenocortical ca. have been treated, in addition to surgical removal of the primary tumor, with o,p'DDD at various dosages (2-20 g/d) for several months or years (26,48-51). At an overall estimate, it appears that, as in the first report, more than two thirds of patients respond to o,p'DDD with decrease of high steroid levels. In one third,regression of tumor mass can also be observed. From various reports some conclusions can be drawn:

- There is no convincing evidence that treatment with o,p'DDD does influence the length of survival in adrenal ca. patients, except in some individual cases. However, the quality of life was improved in most patients by reversal of their endocrine manifestations.
- o,p'DDD treatment can suppress various active corticosteroids that a functioning adrenal ca. may produce, leading to complete reversal of Cushing's syndrome, virilization, feminization and hypermineralocorticism (49,50).
- The drug acts slowly, requiring a 4 week period or longer to evaluate a possible response. Since it can lead to adrenal insufficiency in an unpredictable way, simultaneous administration of substitutive glucocorticoids and mineralocorticoids may be appropriate. Interestingly, patients on such a regimen showed Addisonian features, probably due to alterations in exogenous steroid metabolism, as demonstrated for endogenous cortisol.
- Side effects are important and may be a limiting factor in the treatment. Most frequent adverse reactions are gastrointestinal (nausea,vomiting,diarrhea) and neurological (fatigue, lethargy, vertigo, ataxia, neuropathy). Nonetheless, abnormalities of routine laboratory testing, including hepatic and hematologic evaluation, were seldom reported.
- Patients in pediatric age do respond better to o,p'DDD treatment, showing sustained responses or even apparent cure (51-55).

After the first trial reported by Southren et al. (56) in 1966, o,p'DDD has been widely employed also in pituitary dependent Cushing's disease as a medical alternative to bilateral adrenalectomy. It seemed that this kind of treatment could result in selective destruction of zona fasciculata and zona reticularis, without impairment of zona glomerulosa (normal aldosterone levels), avoiding the need for mineralocorticoid substitution (33). However, in later reports, remission of Cushing's disease after o,p'DDD alone was not permanent in most cases, and eventually other forms of therapy were needed (40), even after doses of 6 to I2 g/d for several months (28). Nevertheless, the use of o,p'DDD allows long-term medical management of Cushing's disease. Apparent cure, with normal adrenal and pituitary function, 7 years after discontinuation of o,p'DDD was reported in a case (57). The drug had been administered over 3 years at high doses, as the only treatment.

Aminoglutethimide

[α-ethyl-α p-aminophenyl-glutarimide]; Orimeten° or Cytadren°
(Ciba - Geigy)

<u>Mechanism of action</u>. Introduced in I959 as an anticonvulsivant, aminoglutethimide (AG) was subsequently shown to be a potent inhibitor of adrenal steroidogenesis (58). After AG treatment the adrenal glands were found markedly enlarged with histologic changes, including cell vacuolation and excessive accumulation of lipid (59), similar to the lipoid adrenal hyperplasia described by Prader (60). Reversible mitochondrial alterations, that might be related to the drug mechanism of action, were also observed (6I). AG has been shown to inhibit the 20 α-hydroxylation of the cholesterol side chain, initial step for side chain cleavage, both in animal (62,63) and human studies (64). The desmolase complex required for the conversion is located in the mitochondria and is cytochrome P-450 dependent, like other hydroxylating enzymes affected by AG (I3). Inhibition of II β-hydroxylase has been shown (63,65). Touitou <u>et al</u>. (66) demonstrated a potent inhibitory effect of AG

on I8-hydroxylation of B by human and sheep adrenals *in vitro*, explaining the greater suppressibility of aldosterone compared to cortisol. This appears to be the most sensitive AG site of action, followed by the aromatizing system for the conversion of androgens to estrogens (67,68), the 20α-hydroxylation, and by the IIβ-hydroxylation. AG blocks such enzyme activities by binding to cytochrome P-450, thereby producing an altered ligand field and interferig with the substrate-hemoprotein interaction (I3,69,70). The effect of AG on androgen secretion is not completely clear, since androgen levels were found decreased, as might be expected from 20α-hydroxylase inhibition (7I,72) or either slightly increased (68,73). The latter finding could be attributed to aromatase inhibition (68) and/or to acceleration of the 3β-HSD, $\Delta_{5,4}$ enzyme system (74).

Extra-adrenal effects. In addition to peripheral aromatase inhibition, taking place in fat tissue, muscle, liver, breast (68), AG exerts several extra-adrenal actions. By inducing liver microsome enzymes, such as 6β-hydroxylase, the drug accelerates the metabolism of synthetic glucocorticoids (75), as well as of endogenous glucocorticoids and androgens (72,73). Thus,urinary free or plasma steroids should be measured to monitor the response of patients to AG treatment. In the thyroid AG interferes with the incorporation of iodine and may cause goiter and hypothyroidism, that actually accur in only 5% of chronically treated patients (76,77). AG has little clinical effects on ovarian function, causing no changes in the menstrual cycle (78). However, a decrease in estrogen production,minimal in the follicular phase and significant in the luteal phase, has been reported (79). Interference with progesterone synthesis in corpus luteum *in vitro* has been shown to occur by binding of AG to mitochondrial cytochrome P-450 (80). In addition to aromatization in the ovary, placental aromatization is also inhibited (8I). Enlargement of ovaries with cholesterol accumulation has been described in a patient who developed hirsutism during treatment (82), and in rat female foetuses with severe virilization after AG administration to their mothers (83).

Pseudohermaphrodism in a female child has also been described (84). Testis histology is not altered by AG (85). Reduction of testosterone production by acute AG administration to normal men has been demonstrated (86).

Pharmacology. After oral administration,AG is effectively absorbed and peak concentrations occur within 2 to 4 h (87). Blood levels have been measured by colorimetric method and found to range between 4.7 and 32.4 µg/ml in patients receiving AG I g/d, without significant changes in each patient throughout a long-term study (87). In the same study, AG half-life was initially I3.3 ± 2.6 h and fell significantly to 7.3 ± 2.I h during chronic treatment. The mechanism by which AG appears to accelerate its own metabolism might involve hepatic enzyme induction. These findings may provide an explanation for development of tolerance to side effects such as lethargy, dizziness, ataxia, rash and fever, that are prominent at the beginning of treatment and diminish with time (87). 36 h after discontinuation of long-term treatments serum levels of AG were undetectable (< I µg/ml)(88). About 50% of administered AG is excreted unchanged in urine (89), while 20-50% appears as a metabolite, aceto-AG,that is much less active, but might be responsible for side effects (90). Adrenal inhibition by AG is rapidly and fully reversible (72). Recovery of hypothalamic-pituitary-adrenal function, after administration of AG and hydrocortisone for 4-60 months, has been demonstrated by Worgul *et al.* to occur within 36 h (88), in contrast to previous observations in which children treated with AG without glucocorticoid replacement were unresponsive to ACTH up to 8 months after drug withdrawal (64). Pharmacologic interference by AG with anticoagulant therapy has been observed:in 3 patients coumarin doses had to be doubled to achieve the same anticoagulant effect 3-4 weeks after AG I g/d was started (9I).

Diagnostic use. As investigative and diagnostic tool, AG has been employed in patients with low renin essential hypertension (LREH) to assess the role of adrenal cortex in this condition (I2,92,93). In contrast to normal subjects and patients with essential hypertension and normal renin, patients with

LREH responded to AG with marked decrease in blood pressure, suggesting the role of some unidentified mineralocorticoid (93). On the other hand, in patients with low renin and normal aldosterone secretion rate, AG anti-hypertensive effect occurs without affecting DOC, I8-OH-DOC and I6β-OH-DEA secretion and drug action seems attributable to persistent suppression of aldosterone production, despite significant increase in PRA (94). Thus, the zona glomerulosa is more susceptible to I8-hydroxylase inhibition by AG, while mineralocorticoids from the zona fasciculata do respond to ACTH compensatory increase, unless prevented by glucocorticoid administration (95).

Therapeutic use. By administration of AG to subjects with normal function of pituitary-adrenal axis, only aldosterone is effectively suppressed. Cortisol secretion rate and plasma I7-OHCS are not significantly lowered, since ACTH increase can overcome enzyme blockade (72,73). Thus, to completely abolish measurable estrogen levels by "medical adrenalectomy" in postmenopausal breast ca. patients, the drug is currently administered in combination with hydrocortisone replacement therapy to avoid ACTH rise and prevent adrenal insufficiency as well (88,96-98). As a palliative treatment of Cushing's syndrome, this agent is more effective in cases of tumoral adrenal hyperfunction (72,78,99,I00) and ectopic ACTH production (33,I0I), in which it was also used in combination with other adrenal inhibitors (I0I), reversing symptoms such as diabetes, hypertension, poor wound healing, etc. Patients with pituitary dependent Cushing's disease appear to do less well when treated with AG alone (72,99). However, good control of Cushing's disease for long periods, before definitive treatment, has been reported by a combination of AG and metyrapone (MP), at lower doses of both drugs (I02). Similar therapy for I or 2 years in addition to megavoltage radiation to the pituitary was reported successful (I03). In our experience, a combination of AG and MP (AG 500-750 mg + MP 750 mg/d) was of great use in the management of patients with severe hypercortisolism. Six of such patients, after I month treatment, showed significant decrease in urinary free F (from 509.5 $\pm$ II4.5 to I2I.5 $\pm$

26.I µg/24 h; $p<0.05$) and plasma DHEA-S (from 2.2 ± 0.5 to I.3 ± 0.4 µg/ml; $p<0.05$). Aldosterone levels could not be evaluated for interference of anti-hypertensive medication. However, at I month, after anti-hypertensive drugs had been tapered or stopped, systolic (from I57.5 ± 7.5 to I40.8 ± 6.6 mmHg; $p<0.02$) and diastolic (from II0 ± 4.6 to 90 ± 3.4 mmHg; $p<0.02$) blood pressures were both significantly decreased. Serum K increased from 4.0 ± 0.I to 4.6 ± 0.I meq/l ($p<0.05$). Another patient showed no hormonal response to AG (750 mg/d) alone. Results are given as mean ± SE. We observed no side effects, nor hepatic or hematologic toxicity.
Finally, in both primary (72) and secondary hyperaldosteronism (I04), AG has been shown to cause a marked fall in aldosterone levels, increase in PRA and sodium diuresis, and lowering of blood pressure in the former.

Metyrapone

SU-4885 [2-methyl-I,2-bis (3 pyridil)-I-propanone]; Metopirone°
(Ciba - Geigy)

Mechanism of action. (See also ref. I05). Metyrapone (MP) was first introduced as a relatively specific inhibitor of IIβ-hydroxylase, affecting both the I7-hydroxy and I7-deoxy pathways (I06-I09). However, the following steps of steroidogenesis are also inhibited by MP: I8-hydroxylation of DOC (II0-II3); I8-hydroxylation of B (II3-II5); I9-hydroxylation (II4,II6-II8); cholesterol side chain cleavage (II9-I22). MP appeared to have greater affinity for the IIβ-hydroxylase than for cholesterol side chain cleavage enzyme system (I2I), and I8-hydroxylase was found to be more sensitive than IIβ-hydroxylase activity to MP (III). The I8-hydroxylation of B is a very sensitive site of MP action and is selectively inhibited *in vitro* at low concentrations of the drug (II4,II5). As a result, aldosterone and I8-OH-B are equally affected (II5). Since aldosterone is an II- and I8-hydroxylated compound and MP interferes with its biosynthesis in both sites with a primary effect (II3), its suppressibility is greater than that of other II- (F and B) or I8- (I8-OH-DOC) hydroxylated steroids. Although I9-hydroxylase inhibition by MP has been demonstrated in vitro in various experimental conditions (II4,II6-II8), clinical studies concerned with the effect of MP on androgen and estrogen production yelded contradictory results, probably due to the presence of too many variables (I23-I27). As AG, MP binds to cytochrome P-450, interfering with the substrate-hemoprotein interaction (I3,I28). In the adrenal cortex it seems to act selectively on mitochondrial and has no effect on microsomal cytochrome P-450, which differs in the type of electron transport protein (II7, I29). However, Greiner et al. (I30) found species differences in adrenal sites of action of MP and pointed out that its subcellular selectivity appears to be species dependent. i .e., in the guinea pig MP interacted with both mitochondrial and microsomal cytochrome P-450, inhibiting the IIβ- as well as the 2I- and the I7α-hydroxylase; whereas, it had no effect on 2I-hydroxylase activity of rat adrenal microsomes (I30).

Extra-adrenal effects. MP interferes with the oxidative metabolism of various substrates catalyzed by cytochrome P-450 dependent mixed function oxidation reactions in liver microsomes (I3I). It reduces cortisol half-life in plasma (I32,I33). A decrease of F binding capacity (I34) and alteration of steroid metabolizing enzyme activities (I35) have both been proposed as possible explanations for this effect. Investigating MP effect on gonads, no clear-cut results were obtained in vivo. In rat testis preparations MP decreased steroid production by inhibition of the mitochondrial cholesterol side chain cleavage (I36). A direct suppressive effect of MP upon plasma ACTH in the very early phase of drug i.v. infusion has been claimed in a recent report (I37). Normal plasma ACTH response to suppressed F levels, following either oral or i.v. MP administration, varied from 3 to I0 fold increase over baseline values in several studies (I38-I4I). Plasma levels of β-LPH and β-endorphin have been shown to change in parallel with ACTH (I42-I44). A significant rise of GH has been reported by some authors (I32,

I45), but not confirmed by others (I46,I47). In a recent report (I48), a rise in GH and prolactin right after a high MP dose was attributed to non-specific stimuli; while no changes in LH, FSH and TSH could be observed.

Pharmacology. After a 750 mg oral dose, peak levels of MP at I h (370 μg/I00 ml) and mean levels of 50 μg/I00 ml at 4 h were reported by *in vitro* bioassay (I49), while lower values were obtained by fluorimetric method (I50). Short half-life, between 20 and 26 min (I49), accounts for its relatively rapid disappearance from plasma. Cope *et al*. (I5I) found that the 750 mg oral dose every 4 h for 48 h produced 92-95% inhibition of II β-hydroxylase activity with only a moderate reduction of F secretion. MP metabolism takes place in liver microsomes (I52). The reduced, more polar derivative 2-methyl-I,2-bis (3 pyridil)-I-propanol (metyrapol) has been shown in rats to inhibit aldosterone and B synthesis as effectively as MP (I53). Simultaneous determination of MP and its reduced derivative has been performed by gas chromatography (I54) and RIA (I55). The formation of two N-oxide metabolites has also been demonstrated (I56).

Diagnostic use. Since its introduction in I958 (I06,I07), this agent has been widely employed as a diagnostic test of pituitary-adrenal reserve by virtue of its inhibitory action on II β-hydroxylase activity (I08). MP testing was then performed with numerous modifications. The huge amount of literature on this topic has been reviewed several times. See ref. I57, and also ref. I38-I4I.

Therapeutic use. As an inhibitor of F biosynthesis MP has been employed in the management of Cushing's syndrome of various etiologies (I0I-I03,I58-I60), for its ability to correct quickly the severe complications of the disease. Drug control of hypercortisolism is suitable for a better outcome of patients undergoing surgery or being treated with external pituitary radiation. Since MP is a relatively non-toxic drug, it has been used in combination therapy to lower the dose of more toxic agents such as AG capable of pharmacological adrenalectomy (I02,I03). Treatment with MP alone has been maintained

for 2 months to 5 years with no major side effects, except worsening of hirsutism in some cases (I58). Management with MP for I year has been successful in a case of severe Cushing's disease, who maintained spontaneous remission following discontinuation of the drug (I6I). Dramatic improvement of metabolic effects has been obtained in patients with ectopic ACTH syndrome (I0I,I59); whereas, MP was of little use in inoperable adrenal ca. (I02,I62).
Marked inhibition of aldosterone biosynthesis by MP occurs in normal subjects (I63,I64), and in both primary and secondary hyperaldosteronism (I63,I65). Electrolyte excretion, however, varies individually with the degree of aldosterone inhibition and DOC stimulation (I2,I63,I65). Although marked sodium and water diuresis could be achieved by addition of prednisone to MP treatment, such a therapy, in secondary hyperaldosteronism, has been considered of no practical importance, except for selected cases (I63,I65). Administration of MP for 5 to I0 days to normal children and children with dexamethasone suppressible and primary hyperaldosteronism, produced suppression of aldosterone in all cases without marked metabolic effects (I66). In conclusion, MP treatment might be beneficial in selected cases of hyperaldosteronism as it is in Cushing's syndrome.

Trilostane

WIN 24,540 [4α,5-epoxy-I7β-hydroxy-3-oxo-5α-androstane-2α-carbonitrile];
Modrenal° (Sterling Research Lab.)

Mechanism of action. This cyanoketone derivative (I4) is a competitive reversible inhibitor of the 3β-HSD, $\Delta_{5,4}$-isomerase enzyme system that converts pregnenolone to progesterone both in experimental animals (I5,I67,I68) and in humans (I69). Interference with the biosynthetic pathways of mineralocorticoids, glucocorticoids and androgens, results in decreased production of aldosterone, cortisol and androstenedione, respectively. By steroid measurements, dissimilar effects on 3β-HSD for pregnenolone than on that for

DHEA-S were observed, suggesting the existence of separate 3β-HSD enzymes (I69). The degree of steroid inhibition by trilostane (TL) appeared to be dose related (I67,I69), but very variable among individuals (I69). In addition to its inhibitory effect on 3β-HSD, TL has been reported to induce *in vitro* an increase of IIβ-HSD activity in sheep adrenals (I70). This effect was not observed by human adrenal incubations, thus suggesting species-related differences (I70). Whether TL does affect IIβ-HSD in the liver is not known. TL showed no effect on 2I-, II- and I7-hydroxylase activities (I70), nor on cholesterol conversion to pregnenolone in adrenal mitochondria (I67). Increase in adrenal weight with hypertrophy was seen in rats fed with high TL doses (40 mg/Kg) for 2 weeks (I67).

Extra-adrenal effects. TL was shown to affect gonadal steroidogenesis in both male and female rats at higher doses than those required to inhibit adrenal enzyme activity (I67). After TL administration to men, testosterone levels have been reported either decreased (I7I) or increased (I72). Since TL interferes with steroid assays (I72), to study the effect of the drug on the normal hypothalamic-pituitary-testicular axis, testosterone was measured after chromatography and found to fall significantly after treatment (I73). There was a concomitant rise in LH and no change in FSH nor in SHBG, while FSH and LH responses to LHRH were unaffected by TL administration for 4 weeks (I73). While causing adrenal enlargement, TL did not alter other organ weights at doses as high as 200 mg/Kg (I67). No effect on cortisol metabolism was observed (I69).

Pharmacology. Laboratory animals tolerated TL doses of up to 250 mg/Kg for I8 months (I67). In humans TL has been administered at doses of 240 to 960 mg/d up to I440 mg/d. Studies in monkeys showed that blood levels peak about 2 h after ingestion (I74). TL concentrations were measured by RIA, but no quantitative results were given since TL metabolites in plasma cross-reacted with TL antibody (I75). The drug is rapidly converted to several metabolites whose activity in man is not known (I74). Changes in TL metabolism or in adrenal 3β-HSD activity among individuals might account for conflicting results in human studies (I76,I77). TL interferes with testosterone RIA (I73) and with fluorimetric steroid assays (I78); no interference accurred in ACTH RIA (I79). Side effects include nausea, vomiting, diarrhea, flushing (I69, I75,I80), and appear to be mild. No alteration in blood chemistry has been reported. Male patients should be monitored for testicular function impairment (I73).

Diagnostic use. Among patients with LREH, those who responded to TL trial with decreases in blood pressure, were interpreted as producing a 4-ene steroid of uncertain identity responsible for the condition (93).

Therapeutic use. TL has been reported to afford reduction of cortisol levels in a number of patients with Cushing's syndrome (I69,I79,I80). However, in some studies it appeared of no use in the management of hypercortisolism, even at high doses (I76,I8I). Taking into account the great variability of response to TL among individuals (I69,I82), it seems not possible to rely on predictive criteria for treatment outcome. Elevated aldosterone levels (93,I80) as well as blood pressure (93) are lowered in patients with hyperaldosteronism. While minor or no effects on aldosterone or cortisol were seen in normal subjects (I75,I80). An advantage of this agent appears to be the low incidence of side effects.

Finally, ketoconazole (Nizoral°; Janssen Pharm.), an imidazole derivative, that interferes with the synthesis of ergosterol in fungi and of cholesterol in mammalian cells by blocking I4-demethylation of lanosterol (I83), is under investigation as a potent inhibitor of adrenal steroid synthesis (I6). In normal subjects, cortisol response to ACTH was significantly blunted 4 h after a 400-600 mg dose and the degree of inhibition was inversely related to drug serum concentrations (I6). Initial half-life of ketoconazole (KC) was calculated 2-3.3 h (I84). Its effects were rapidly and completely reversible (I84,I85). Recovery of steroidogenic blockade occurred 4-I6 h after the oral dose (I6). Cortisol levels were suppressed, after administration of KC 800 mg/d for 2 days, in a patient with a functioning adrenal adenoma (I85). After surgery, inhibition of F production by the drug was confirmed in tissue slices of the excised tumor _in vitro_ (I85).

Preliminary results in 3 of our patients with Cushing's disease, who received KC 800 mg in 24 h, showed a marked decrease in urinary free F in two (from 705 to I7I and from 353 to 28 μg/24 h) and no response (from 264 to 238 μg/24 h) in one.

In addition, KC has been shown to inhibit testosterone synthesis in men (I86, I87) and in Leydig cell coltures from rats (I86) and mice (I87). This finding could explain the occurrence of gynecomastia in a number of patients treated with KC (I6). Other side effects include gastrointestinal disturbances, rashes, and hepatotoxicity and seem to be infrequent (I88). The mechanism by which KC exerts its action on steroid biosynthesis is not fully elucidated, although inhibition of cholesterol formation by steroid-producing cells may be involved (I6). Inhibition of IIβ-hydroxylase has been shown in cell culture of mouse adrenal cortex tumors by Kowal (I89). In rat adrenal mitochondria, KC inhibited IIβ-hydroxylation by a competitive interaction with the mixed-function oxygenases involved in these activities (I89).

Antifungal imidazole derivatives chemically related to KC were also tested for interference with testosterone synthesis by Schürmeyer and Nieschlag (I87). These authors indicated a structure/activity relationship, since only compounds with a phenylated side chain (miconazole, econazole, isoconazole and clotrimazole) produced steroid inhibition (I87).

References

I. Nelson, A.A., Woodard, G.: Arch. Pathol. 48, 387-394 (I949).

2. Neville, A.M., Engel, L.L.: J. Clin. Endocrinol. Metab. 28, 49-60 (I968).

3. Goldman,A.S.: J. Clin. Endocrinol. Metab. 27, I04I-I049 (I967).

4. Goldman, A.S., Winter, J.S.D.: J. Clin. Endocrinol. Metab. 27, I7I7-I722 (I967).

5. Rosenfeld, G., Bascom, W.D.: J. Biol. Chem. 222, 565-580 (I956).

6. Kibelstis, J.A., Ferguson, J.J.Jr.: Endocrinology 74, 567-572 (I964).

7. Hertz, R., Pittman, J.A., Graff, M.M.: J. Clin. Endocrinol. Metab. I6, 705-723 (I956).

8. Gallagher, T.F.: J. Clin. Endocrinol. Metab. I8, 937-949 (I958).

9. Gabrilove, J.L., Nicolis, G.L., Gallagher, T.F.: J. Clin. Endocrinol. Metab. 27, I337-I340 (I967).

I0. Gabrilove, J.L., Nicolis, G.L., Gallagher, T.F.: J. Clin. Endocrinol. Metab. 27, I550-I557 (I967).

II. Gabrilove, J.L., Nicolis, G.L., Gallagher, T.F.: Metabolism I7, 936-942 (I968).

I2. Temple, T.E., Liddle, G.W.: Ann. Rev. Pharmacol. I0, I99-2I9 (I970).

I3. Gower, D.B.: J. Steroid Biochem. 5, 50I-523 (I974).

I4. Neumann, H.C., Potts, G.O., Ryan, W.T., Stonner, F.W.: J. Med. Chem. I3, 948-95I (I970).

I5. Potts, G.O., Ryan, W.T., Harding, H.R.: Endocrinology (Suppl.) 96, 58 (I975).

I6. Pont, A., Williams, P.L., Loose, D.S., Feldman, D., Reitz, R.E., Bochra, C., Stevens, D.A.: Ann. Int. Med. 97, 370-372 (I982).

I7. Cueto, C., Brown, J.H.U.: Endocrinology 62, 326-333 (I958).

I8. Kaminsky, N., Luse, S., Hartoft, P.: J. Nat. Cancer Inst. 29, I27-I59 (I962).

I9. Vilar, O., Tullner, W.W.: Endocrinology 65, 80-86 (I959).

20. Tullner, W.W.: Proceedings of the Chemotherapy Conference on ortho, para'DDD (eds L.E. Broder and S.K. Carter). National Cancer Institute, Bethesda. Nov. 6, I970, p. 7-I6.

2I. Nichols, J.: The Adrenal Cortex (ed. H.D. Moon) Paul B. Hoeber, Inc., New York, I96I, p. 84-87.

22. Fregly, M.J., Waters, I.W., Straw, J.A.: Can. J. Physiol. Pharmacol. 46, 59-66 (I968).

23. Bergenstal, D.M., Hertz, R., Lipsett, M.B., Moy, R.H.: Ann. Int. Med. 53, 672-682 (I960).

24. Danowsky, T.S., Sarver, M.E., Moses, C., Bonessi, J.V.: Am. J. Med. 37, 235-250 (I964).

25. Southren, A.L., Wisenfeld, S., Laufer, A., Goldner, M.G.: J. Clin. Endocrinol. Metab. 2I, 20I-208 (I96I).

26. Hutter, A.M., Kayhoe, D.E.: Am. J. Med. 4I, 58I-592 (I966).

27. Temple, T.E. Jr., Jones, D.J.Jr., Liddle, G.W., Dexter, R.N.: New Eng. J. Med. 28I, 80I-805 (I969).

28. Luton, J.P., Mahoudeau,J.A., Bouchard, Ph., Thieblot, Ph., Hautecouverture, M., Simon, D., Laudat, M.H., Touitou, Y., Bricaire, H.: New Eng. J. Med. 300, 459-464 (I979).

29. Tullner, W.W., Hertz, R.: Endocrinology 66, 494-496 (I960).

30. Verdon, T.A., Bruton, J., Hermon, R.H., Beisel, W.R.: Metabolism II, 226-233 (I962).

3I. Hart, M.M., Straw, J.A.: Steroids I7, 559-574 (I97I).

32. Touitou, Y., Bogdan, A., Luton, J.P.: J. Ster. Biochem. 9, I2I7-I224 (I978).

33. Orth, D.N., Liddle, G.W.: New Eng. J. Med. 285, 243-247 (I97I).

34. Bledsoe, T., Island, D.P., Ney, R.L., Liddle, G.W.: J.Clin. Endocrinol. Metab. 24, I303-I3II (I964).

35. Southren, A.L., Tochimoto, S., Isurugi, K., Gordon, G.G., Krikun, E., Stypulkowki, W.: Steroids 7, II-29 (I966).

36. Fukushima, D.K., Bradlow, H.L., Hellman, L.: J. Clin. Endocrinol. Metab. 82, I92-200 (I97I).

37. Bradlow, H.L., Fukushima, D.K., Zumoff, B., Hellman, L., Gallagher, T. F.: J. Clin. Endocrinol. Metab. 23, 9I8-922 (I963).

38. Moy, R.H.: J. Lab. Clin. Med. 58, 296-304 (I96I).

39. Cueto, C., Brown, J.H.U.: Endocrinology 62, 334-339 (I958).

40. Ridolfi, P., Boscaro, M., Masarotto, P., Scaroni, C., Mantero, F.: J. Endocrinol. Invest. 4 (Suppl. I), 27-30 (I98I).

4I. Luton, J.P., Rémy, J.M., Valcke, J.C.: Ann. Endocrinol. (Paris) 34, 35I-376 (I973).

42. Reach, G., Elkik, F., Parry, C., Corvol, P., Milliez, P.: Lancet i, I269 (I978).

43. Moolenar, A.J., Van Seters, A.P.: Acta Endocrinol. Suppl. I99, 266 (I975).

44. Touitou, Y., Bogdan, A., Legrand, J.C., Desgrez, P.: Ann. Endocrinol. (Paris) 38, I3-25 (I977).

45. Reif, V.D., Sinsheimer, J.E., Ward, J.C., Schteingart, D.E.: J. Pharm. Sci. 63, I730-I736 (I974).

46. Wortsman, J., Soler, N.G.: JAMA 238, 2527 (I977).

47. Menard, R.H., Cutler, G.B., Rifka, S.M., Gillet, J.R., Bartter, F.C.: The 59th Annual Meeting of the Endocrine Society, Chicago, June 8-I0, I977, I99:286.

48. Lubitz, J.A., Freeman, L., Okun, R.: JAMA 223, II09-III2 (I973).

49. Hoffman, D.L., Mattox, V.R.: Medical Clinics of North America 56, 999-I0I2 (I972).

50. Kelly, W.F., Barnes, A.J., Cassar, J., White, M., Mashiter, K., Loizou, S., Welbourn, R.B., Joplin, G.F.: Acta Endocrinol. 9I, 303-3I8 (I979).

5I. Greig, F., Oberfield, S.E., Levine, L.S., Ghavimi, F., Pang, S., New, M.I.: Clin. Endocrinol. 20, 389-399 (I984).

52. Fisher, D.A., Panos, T.C., Melby, J.C.: J. Clin. Endocrinol. Metab. 23, 2I8-22I (I963).

53. Helson, L., Wollner, N., Murphy, L., Schwartz, M.K.: Clin. Chem. I7, II9I-II93 (I97I).

54. Ostuni, J.A., Roginsky, M.S.: Arch. Int. Med. I35, I257-I258 (I975).

55. Rappaport, R., Schweisguth, O., Cachin, O., Pellerin, D.: Arch. Franc. Ped. 35, 55I-554 (I978).

56. Southren, A.L., Tochimoto, S., Strom, L., Ratuschni, A., Ross, H., Gordon, G.: J. Clin. Endocrinol. Metab. 26, 268-278 (I966).

57. Couzinet, B., Thomopoulos, P., Schaison, G.: Acta Endocrinol. I00, 63-67 (I982).

58. Huges, S.W.M., Burley, D.M.: Post-grad. Med. J. 46, 409-4I6 (I970).

59. Camacho, A.M., Cash, R., Brough, A.J., Wilroy, R.S.: J. Am. Med. Ass. 202, 20-26 (I967).

60. Prader, A., Gurtner, H.P.: Helv. Ped. Acta I0, 397-4I2 (I955).

6I. Racela, A. Jr., Azarnoff, D., Svoboda, D.: Lab. Invest. 2I, 52-60 (I969).

62. Kahnt, F., Neher, R.: Helv. Chim. Acta 49, 725-732 (I966).

63. Dexter, R., Fishman, R., Ney, R., Liddle, G.W.: J. Clin. Endocrinol. Metab. 27, 473-480 (I967).

64. Cash, R. Brough, A.J., Cohen, M.N.P., Satoh, P.S.: J. Clin. Endocrinol. Metab. 27, I239-I248 (I967).

65. Kowal, J.: Endocrinology 85, 270-279 (I969).

66. Touitou, Y., Bogdan, A., Legrand, J.C., Desgrez, P.: Acta Endocrinol. 80, 5I7-526 (I975).

67. Thompson, E.A. Jr., Siiteri, P.K.: J. Biol. Chem. 249, 5373-5378 (I974).

68. Santen, R.J., Santen, S., Davis, B., Veldhuis, J., Samojlik, E., Ruby, E.: J. Clin. Endocrinol. Metab. 47, I257-I265 (I978).

69. Uzgiris, V.I., Whipple, C.A., Salhanick, H.A.: Endocrinology I0I, 89-92 (I977).

70. Graves, P.E., Salhanick, H.A.: Endocrinology I05, 52-57 (I979).

7I. Cohen, M.P.: Proc. Soc. Exp. Biol. Med. I27, I086-I090 (I968).

72. Fishman, L.M., Liddle, G.W., Island, D.P., Fleischer, N., Küchel, O.: J. Clin. Endocrinol. Metab. 27, 48I-490 (I967).

73. Horky, K., Küchel, O., Gregorova, I., Starka, L.: J. Clin. Endocrinol. Metab. 29, 297-299 (I969).

74. Samojlik, E., Santen, R.J.: J. Clin. Endocrinol. Metab. 47, 7I7-724 (I978).

75. Santen, R.J., Lipton, A., Kendall, J.: J. Am. Med. Ass. 230, I66I-I665 (I974).

76. Rallison, M.L., Kumagai, L.F., Tyler, F.H.: J. Clin. Endocrinol. Metab. 27, 265-272 (I967).

77. Santen, R.J., Wells, S.A., Cohn, N., Demers, L.M., Misbin, R.I., Foltz, E.L.: J. Clin. Endocrinol. Metab. 45, 739-746 (I977).

78. Philibert, M., Laudat, M.H., Laudat, P.H., Bricaire, A.: Ann. Endocrinol. (Paris) 29, I89-2I0 (I966).

79. Santen, R.J., Samojlik, E., Wells, S.A.: J. Clin. Endocrinol. Metab. 5I, 473-477 (I980).

80. Inaba, T., Wiest, W.G.: Endocrinology I07, 578-583 (I980).

8I. Janne, O., Pastore, U., Timonen, H., Vihko, R.: Contraception 9, 239-247 (I974).

82. Cash, R., Petrini, M.A., Brough, A.J.: JAMA 208, II49-II52 (I969).

83. Goldman, A.S.: Endocrinology 87, 889-893 (I970).

84. Le Maire, W.J., Cleveland, W.W., Bejar, R.L., Marsh, J.M., Fishman, L.: Am. J. Dis. Child I24, 42I-423 (I972).

85. Morales, A., Connolly, J.G., Mobbs, B.G., Kraus, A.: Can. J. Surgery I4, I54-I60 (I97I).

86. Santen, R.J., Cohn, N., Misbin, R., Samojlik, E., Foltz, E.: J. Clin. Endocrinol. Metab. 49, 63I-634 (I979).

87. Murray, F.T., Santen, S., Samojlik, E.A., Santen, R.J.: J. Clin. Pharm. I9, 704-7II (I979).

88. Worgul, T.J., Kendall, J., Santen, R.J.: J. Clin. Endocrinol. Metab. 53, 879-882 (I98I).

89. Douglas, J.S., Nicholls, P.J.: J. Pharm. Pharmacol. I7 (Suppl.), II5 (I965).

90. Douglas, J.S., Nicholls, P.J.: J. Pharm. Pharmacol. 24 (Suppl.), I50 (I972).

9I. Bruning, P.F., Bonfrer, J.G.: Lancet i, 582 (I983).

92. Woods, J.W., Liddle, G.W., Stant, E.G. Jr., Michelakis, A.M., Brill, A.B.: Arch. Int. Med. I23, 366-370 (I969).

93. Liddle, G.W., Hollifield, J.W., Slaton, P.E., Wilson, H.M.: J. Ster. Biochem. 7, 937-940 (I976).

94. Taylor, A.A., Mitchell, J.R., Bartter, F.C., Snodgrass, W.R., McMurtry, R.J., Gill, J.R.Jr., Franklin, R.B.: J. Clin. Invest. 62, I62-I68 (I978).

95. Mancheno-Rico, E., Küchel, O., Nowaczynski, W., Seth, K.K., Sasaki, C., Dawson, K., Genest, J.: Metabolism 22, I23-I32 (I973).

96. Santen, R.J., Worgul, T.J., Samojlik, E., Interrante, A., Boucher, A.E., Lipton, A., Harvey, H.A., White, D.S., Smart, E., Cox, C., Wells, S.A.: New Eng. J. Med. 305, 545-55I (I98I).

97. Vermeulen, A., Paridaens, R., Heuson, J.C.: Clin. Endocrinol. I9, 673-682 (I983).

98. Samojlik, E., Santen, R.J., Worgul, T.J.: Clin. Endocrinol. 20, 43-5I (I984).

99. Smilo, R.P., Earll, J.M., Forsham, P.H.: Metabolism I6, 374-377 (I967).

I00. Misbin, R.I., Canary, J., Willard, D.: J. Clin. Pharmacol. I6, 645-65I (I976).

I0I. Carey, R.M., Orth, D.N., Hartmann, W.H.: J. Clin. Endocrinol. Metab.

36, 482-487 (1973).

102. Child, D.F., Burke, C.W., Burley, D.M., Rees, L.H., Russel Fraser, T.: Acta Endocrinol. 82, 330-341 (1976).

103. Ross, W.M., Evered, D.C., Hunter, P., Benaim, M., Cook, D., Hall, R.: Clin. Radiol. 30, 149-153 (1979).

104. Küchel, O., Horky, K., Gregorova, I.: Pharm. Clin. 2, 138-142 (1970).

105. Sonino, N.: in Hormone Antagonists (Ed. M.K. Agarwal) Walter de Gruyter & Co., Berlin.New York, 1982, p. 419-429.

106. Chart, J.J., Sheppard, H., Allen, M.J., Bencze, W.L., Gaunt, R.: Experientia 14, 151-152 (1958).

107. Liddle, G.W., Island, D., Lance, E.M., Harris, A.P.: J. Clin. Endocrinol.Metab. 18, 906-912 (1958).

108 Liddle, G.W., Estep, H.L., Kendall, J.W., Williams, W.C., Townes, A.W.: J. Clin. Endocrinol. Metab. 19, 875-894 (1959).

109. Diminguez, O.V., Samuels, L.T.: Endocrinology 73, 304-309 (1963).

110. Kraulis, I., Birmingham, M.K.: Can.J. Biochem. 43, 1471-1476 (1965).

111. Sanzari, N.P., Peron, F.G.: Steroids 8, 929-945 (1966).

112. Schöneshöfer, M., Schefzig, B., Oelkers, W.: Horm. Metab. Res. 11, 306-308 (1979).

113. Sonino, N., Levine, L.S., Vecsei, P., New, M.I.: J. Clin. Endocrinol. Metab. 51, 557-560 (1980).

114 Kahnt, F.W., Neher, R.: Experientia 18, 499-501 (1962).

115. Erickson, R.E., Ertel, R.J., Ungar, F.: Endocrinology 78, 343-349 (1966).

116. Griffiths, K.: J. Endocrinol. 26, 445-446 (1963).

117. Levy, H., Cha, C.H., Carlo, J.J.: Steroids 5, 469-478 (1965).

118. Sato, H., Ashida, N., Suhara, K., Itagaki, E., Takemori, S., Katagiri, M.: Arch. Bioch. Bioph. 190, 307-314 (1978).

119. Cheng, S.C., Harding, B.W., Carballeira, A.: Endocrinology 94, 1451-1458 (1974).

120. Carballeira, A., Cheng, S.C., Fishman, L.M.: Acta Endocrinol. 76, 689-702 (1974).

121. Carballeira, A. Cheng, S.C., Fishman, L.M.: Acta Endocrinol. 76, 703-711 (1974).

122. Carballeira, A., Fishman, L.M., Jacobi, J.D.: J. Clin. Endocrinol. Metab. 42, 687-695 (1976).

I23. Földes, J., Koref, O., Fehér, T., Steczek, K.: J. Endocrinol. 29, 207-208 (I964).

I24. Sfikakis, A.P., Ikkos, D.G., Diamandopoulos, K.N.: J. Endocrinol. 39, 6I-69 (I967).

I25. Tamm, J., Voigt, K.D.: Acta Endocrinol. 67, I5I-I58 (I97I).

I26. Nilsson, K.O., Hökfelt, B.: Acta Endocrinol. 68, 576-584 (I97I).

I27. DeLange, W.E., Sluiter, W.J., Pratt, J.J., Doorenbos, H.: Acta Endocrinol. 93, 488-494 (I980).

I28. Colby, H.D., Brownie, A.C.: Arch. Biochem. Biophys. I38, 632-639 (I970)

I29. Sweat, M.L., Dutcher, J.S., Youngh, R.B., Bryson, M.J.: Biochemistry 8, 4956-4963 (I969).

I30. Greiner, J.W., Kramer, R.E., Colby, H.D.: Biochem. Pharmacol. 27, 2I47-2I5I (I978).

I3I. Hildebrandt, A.G.: Biochem. Soc. Symp. 34, 79-I02 (I972).

I32. Bruno, O.D., Leclercq, R., Virasoro, E., Copinschi, G.: J. Clin. Endocrinol. Metab. 32, 260-265 (I97I).

I33. Blichert-Toft, M., Kehlet, H.: Clin. Endocrinol. 5, 295-298 (I976).

I34. Kehlet, H., Binder, C.: Acta Endocrinol. 8I, 787-792 (I976).

I35. Levin, J., Zumoff, B., Fukushima, D.K.: J. Clin. Endocrinol. Metab. 47, 845-849 (I978).

I36. Carballeira, A., Fishman, L.M., Durnhofer, F.: Metabolism 23, II75-II84 (I974).

I37. Schöneshöfer, M., Fenner, A., Claus, M.: Clin. Endocrinol. I8, 363-370 (I983).

I38. Jubiz, W., Matsukura, S., Meikle, A.W., Harada, G., West, C.D., Tyler, F.H.: Arch. Int. Med. I25, 468-47I (I970).

I39. Donald, R.A., Espiner, E.A., Beaven, D.W.: J. Endocrinol. 52, 5I7-524 (I972).

I40. Winkelmann, W., Allolio, B., Heesen, D., Hipp, F.X., Kaulen, D.: Acta Endocrinol., Suppl. 234, I40-I4I (I980).

I4I. Dolman, L.I., Nolan, G., Jubiz, W.: JAMA 24I, I25I-I253 (I979).

I42. Jeffcoate, W.J., Rees, L.H., Lowry, P.J., Besser, G.M.: J. Clin. Endocrinol. Metab. 47, I60-I67 (I978).

I43. Nakao, K., Nakai, Y., Oki, S., Horii, K., Imura, H.: J. Clin. Invest. 62, I395-I398 (I978).

I44. Wardlaw,S.L., Frantz, A.G.: J. Clin. Endocrinol. Metab. 48, I76-I80 (I979).

I45. Kunita, H., Takebe, K., Nakagawa, K., Sawano, S., Horiuchi, Y.: J. Clin. Endocrinol. Metab. 3I, 30I-306 (I970).

I46. Byyny, R.L., Orth, D.N., Nicholson, W.E., Liddle, G.W.: J. Clin. Endocrinol. Metab. 34, I093-I096 (I972).

I47. Lee, P.A., Keenan, B.S., Migeon, C.J., Blizzard, R.M.: J. Clin. Endocrinol. Metab. 37, 389-390 (I973).

I48. Schöneshöfer, M., Fenner, A.: Horm. Metab. Res. I3, 473-474 (I98I).

I49. Sprunt, J.G., Browning, M.C.K., Hannah, D.M.: Mem. Soc. Endocrinol. I7, I93-20I (I968).

I50. Meikle, A.W., Jubiz, W. West, C.D., Tyler, F.H.: J. Lab. Clin. Med. 74, 5I5-520 (I969).

I5I. Cope, C.L., Dennis, P.M., Pearson, J.: Clin. Sci. 30, 249-257 (I966).

I52. Kraulis, I., Traikov, H., Li, M.P., Lantos, C.P., Birmingham, M.K.: Can. J. Biochem. 46, 467-469 (I968).

I53. de Nicola, A.F., Dahl, V.: Endocrinology 89, I236-I24I (I97I).

I54. Hollands, T.R., Johnson, W.J.: Bioch.Med. 7, 288-29I (I973).

I55. Meikle, A.W., West, S.C., Weed, J.A., Tyler, F.H.: J. Clin. Endocrinol. Metab. 40, 290-295 (I975).

I56. De Graeve, J., Gielen, J.E., Kahl, G.F., Tüttenberg, K.H., Kahl, R., Maume, B.: Drug Metab. Dispos. 7, I66-I70 (I979).

I57. West, C.D., Meikle, A.W.: in Endocrinology (Ed. L.J. De Groot) Grune & Stratton, I979, vol. 2, p. II57-II77.

I58. Jeffcoate, W.J., Rees, L.H., Tomlin, S., Jones, A.E., Edwards, C.R.W., Besser, G.M.: Br. Med. J. 2, 2I5-2I7 (I977).

I59. Coll., R., Horner, I., Kraiem, Z., Gafni, J.: Arch. Int. Med. I2I, 549-553 (I968).

I60. Daniels, H., van Amstel, W.J., Shopman, W., van Dommenen, C.: Acta Endocrinol. 44, 346-354 (I963).

I6I. Kammer, H., Barter, M.: Am. J. Med. 67, 5I9-523 (I979).

I62. Kelley, W.F., Barnes, A.J., Cassar, J., White, M., Mashiter, K., Louizou, S., Welbourn, R.B., Joplin, G.F.: Acta Endocrinol. 9I, 303-3I8 (I979).

I63. Coppage, W.S. Jr., Island, D., Smith, M., Liddle, G.W.: J. Clin. Invest. 38, 2I0I-2II0 (I959).

I64. Sonino, N., Chow, D., Levine, L.S., New, M.I.: Clin. Endocrinol. I4, 3I-39 (I98I).

I65. Holub, D.A., Jailer, J.W.: Ann. Int. Med. 53, 425-444 (I960).

I66. Sonino, N., Levine, L.S., New, M.I.: Acta Endocrinol. 98, 87-94 (I98I).

I67. Potts, G.O., Creange, J.E., Harding, H., Schane, H.P.: Steroids 32, 257-267 (I978).

I68. Schane, H.P., Potts, G.O., Creange, J.E.: Fert, Steril. 32, 464-467 (I979).

I69. Komanicky, P., Spark, R.F., Melby, J.C.: J. Clin. Endocrinol. Metab. 47, I042-I05I (I978).

I70. Touitou, Y., Auzeby, A., Bogdan, A., Luton, J.P., Galan, P.: J. Ster. Biochem. 20, 763-768 (I984).

I7I. Belchetz, P.E., Davis, J.C., Diver, M.J., Hipkin, L.J.: Lancet i, 897 (I98I).

I72. Beastall, G.H., Ratcliffe, W.A., Thomson, M., Semple, C.G.: Lancet i, 727-728 (I98I).

I73. Semple, C.G., Weir, S.W., Thomson, J.A., Beastall, G.H.: Clin. Endocrinol. I7, 99-I02 (I982).

I74. Baker, J.F., Benziger, D., Chalecki, B.W., Clemans, S., Fritz, A., O'Melia, P.E., Shargel, L., Edelson, J.: Arch. Int. Pharmacodyn. Ther. 243, 4-I6 (I980).

I75. Semple, C.G., Thomson, J.A., Stark, A.N., McDonald, M., Beastall, G.H.: Clin. Endocrinol. I7, 569-575 (I982).

I76. Dewis, P., Anderson, D.C., Bu'lock, D., Earnshaw, R., Kelly, W.F.: Clin. Endocrinol. I8, 533-540 (I983).

I77. Dorrington Ward, P., Carter, G., Banks, R., MacGregor, G.: Lancet ii, II78 (I98I).

I78. Mattingly, D., Tyler, C.: Lancet i, 56I (I98I).

I79. Nomura, K., Demura, H., Imaki, T., Miyagawa, M., Ono, M., Yano, T., Shizume, K.: Acta Endocrinol. I05, 93-98 (I984).

I80. Scheinbaum, M.L., Blackmore, W.P., Potts, G.O., Boksay, J.E.: Clin. Pharmacol. Ther. 29, 28I: 3I (I98I).

I8I. Darling, J.A.B., Fraser, N.C.: Acta Endocrinol. Suppl. 243, 29I (I98I).

I82. Semple, C.G., Beastall, G.H., Gray, C.E., Thompson, J.A.: Acta Endocrinol. I02, I07-II0 (I983).

I83. Willemsens, G., Cools, W., Van den Bossche, H.: in The Host Invader

Interplay (Ed. H. Van den Bossche) Elsevier/North-Holland Biomedical Press, Amsterdam, I980, p. 69I-694.

I84. Brass, C., Galgiani, J.N., Blaschke, T.F., De Felice, R., O'Reilly, R.A., Stevens, D.A.: Antimicrob. Agents Chemother. 2I, I5I-I58 (I982).

I85. Engelhardt, D., Mann, K., Hörmann, R., Braun, S., Karl, H.J.: Klin. Wochenschr. 6I, 373-375 (I983).

I86. Pont, A., Williams, P.L., Azhar, S., Reitz, R.E., Bochra, C., Smith, E.R., Stevens, D.A.: Arch. Intern. Med. I42, 2I37-2I40 (I982).

I87. Schürmeyer, Th., Nieschlag, E.: Acta Endocrinol. I05, 275-280 (I984).

I88. Restrepo, A., Stevens, D.A., Utz, J.P.: Rev. Infect. Dis. 2, 5I9-699 (I980).

I89. Kowal, J.: Endocrinology II2, I54I-I543 (I983).

BLOOD PRESSURE PATTERNS AND CORTICOSTEROID CONCENTRATIONS IN RATS IMMUNIZED TO CORTICOSTERONE

Jürgen Wacker, Dietmar Uhler, Doris Haack, Klaus Lichtwald, Paul Vecsei
Pharmakologisches Institut der Universität Heidelberg
D-6900 Heidelberg

Introduction

Gless and Vecsei actively immunized rabbits against adrenal steroids and found signs indicating hypercorticism (1,5). The disadvantage of these experiments was the technical difficulty to measure the blood pressure of the rabbits. Therefore we decided to actively immunize rats instead of rabbits with adrenal steroids although it is known that this species is more difficult to immunize. We immunized to corticosterone, the main steroid hormone of rats. This was done as well with aldosterone, cortisol and with an antigen mixture of corticosterone, aldosterone, deoxycorticosterone and 18-hydroxydeoxycorticosterone.

Methods

White male Wistar rats were immunized to a corticosterone-21-hemisuccinate bovine serum albumin complex (BSA) (n=12), to cortisol-3-oxime-BSA (n=6) and aldosterone-3-oxime-BSA (n=21) at three to four week intervals. Six rats were immunized to a mixture of corticosterone-21-hemisuccinate-BSA complex, aldosterone-, deoxycorticosterone- and 18-hydroxydeoxycorticosterone-3-oxime-BSA complex (2), The control animals (n=12) were sham immunized. The antigen (0.5 mg) was

Adrenal Steroid Antagonism

suspended in 0.25 ml saline and 0.25 ml Freund's adjuvant and injected intradermally (30-45 different sides) under ether anaesthesia. In addition, the rats also received Pertussis vaccine suspension.
For practical reasons we conducted one series of animals immunized to cortisol and the mixture of four antigens, two series of animals immunized to corticosterone and three series of animals immunized to aldosterone.
The animals, weighing between 200-300 grams, were kept individually in metabolic cages. They received a laboratory diet with tap water ad libitum. The animal house had a constant temperature of 24±1°C with a twelve hour light/dark rhythm.
The volume of excreted urine was measured twice weekly, and the animals were weighed once a week.
The blood pressures were measured once a month by means of tail plethysmography carried out under slight ether aneasthesia. Blood samples were taken every 4-6 weeks to determine the hematocrit, the antibody titer and the plasma steroid levels. The concentration of aldosterone was determined by RIA after extraction and chromatography and of corticosterone by RIA after extraction (1,4).
Additionally, we determined the urinary excretion of unconjugated corticosterone, aldosterone and 18-hydroxydeoxycorticosterone by RIA after extraction and chromatography (3).
At the end of the experiments, the weight of adrenals, hearts and kidneys were determined. One half of the adrenals were used to study the steroid production in vitro. The adrenals were incubated at 37°C in Krebs-Ringer-bicarbonate-glucose solution. The aldosterone and corticosterone production was measured by RIA after extraction and chromatography.

Results

1) Body Weight, Urine Volume and Hematocrit
The initial weight of the animals was 200-300 g. At the end of the experiments the weight ranged from 360 to 440 g. There were no significant differences between immunized and sham immunized groups in body weight and hematocrit value. The urine volume did not change significantly.

2) Titer

a) Immunization against corticosterone
Figure 1 shows the titers determined during the first immunization series.

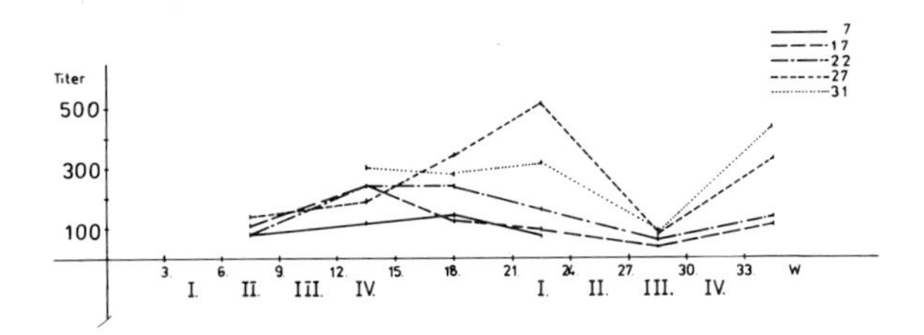

fig. 1: Titers of animals immunized to corticosterone (First series)

Titers above 1 : 80 were already observed in the eighth week after the second immunization. In the 22nd week the titers of two animals reached 1: 320 and 1: 520. In the second immunization series, all animals reached the highest titers after the fourth immunization (fig. 2).
Three animals had peak antibody titers over 1 : 300. Later, despite further immunization, the titers dropped and be-

came lower than 1 : 50. The specificity of the raised antisera was high showing negligible cross reactivity to 18-hydroxydeoxycorticosterone and deoxycorticosterone.

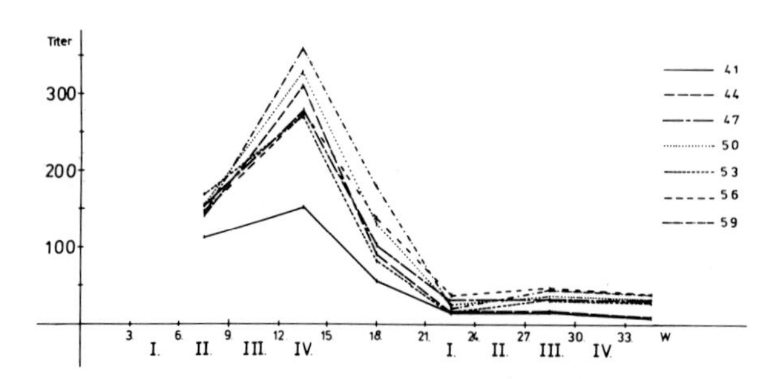

fig. 2: Titers of the animals immunized to corticosterone (Second series)

b) Immunization against the antigen mixture

The development of the corticosterone antibody titer of the animals which were immunized to the antigen mixture did not parallel the animals which were immunized to corticosterone only (fig. 3).

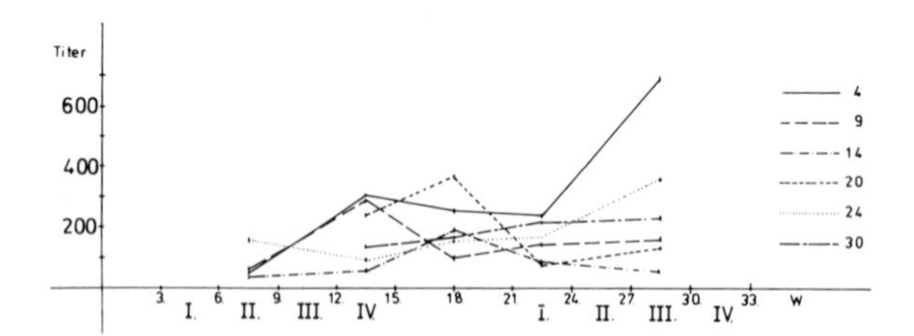

fig. 3: Corticosterone titers of the animals immunized to the antigen mixture

The aldosterone antibody titer became 1 : 50 after the immunization pause (fig. 4). The highest titer was greater than 1 : 800.

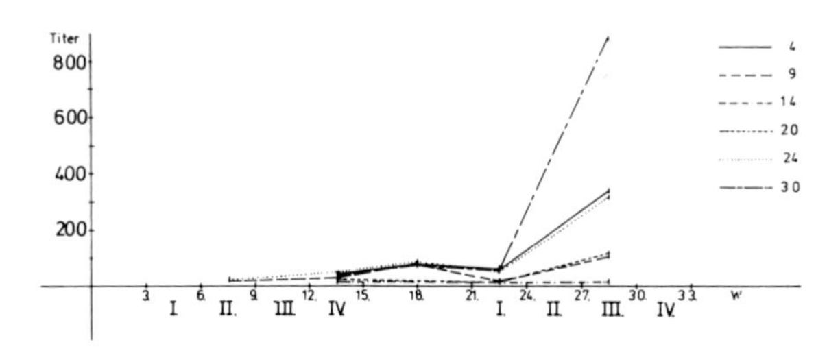

fig. 4: Aldosterone titers of the animals immunized to the antigen mixture

The antibody titer of deoxycorticosterone (fig. 5) and of 18-hydroxydeoxycorticosterone of the animals immunized to the antigen mixture reached values higher than 1 : 100 after the third booster immunization after the immunization pause (fig. 6).

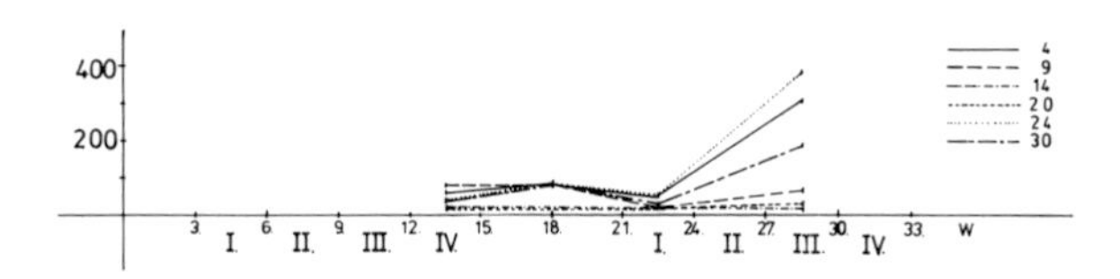

fig. 5: Deoxycorticosterone titers of the animals immunized to the antigen mixture

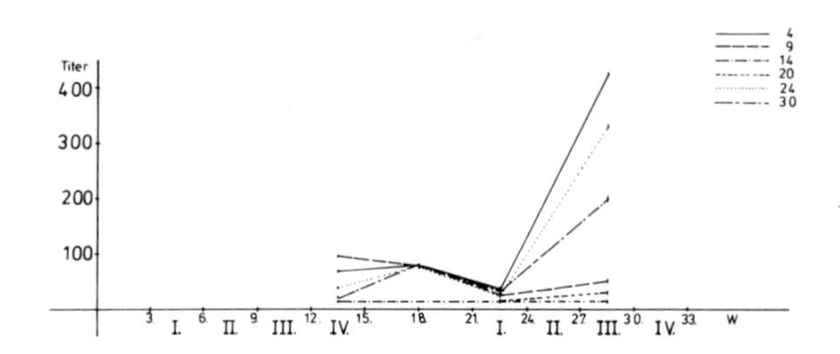

fig. 6: 18-Hydroxycorticosterone titers of the animals immunized to the antigen mixture

c) Immunization against aldosterone

The titers of the animals immunized with pure aldosterone antigen reached much higher values: up to 1 : 3000 (fig.12).

3) Blood Pressure, Heart and Kidney Weight

In the first immunization series, the blood pressure of the treated animals was significantly higher than the blood pressure of the control animals (151.3 ± 3.1 vs. 116.7 ± 3.8 mm Hg; $p<0.01$) (fig. 7).

In the second immunization series, the treated animals attained higher blood pressures (130.7 ± 2.3 mm Hg; $p< 0.01$) in the early phase of immunization. This elevated blood pressure remained high during the entire experiment and it was significantly higher than in the control animals after the fourth immunization (130 ± 10 vs. 115.8 ± 7.3 mm Hg; $p< 0.01$) (fig. 8).

The blood pressure curves of both immunization series demonstrate certain dynamics in the blood pressure elevation. During the first immunization series, blood pressures rose noticeably after the fifth booster immunization and were

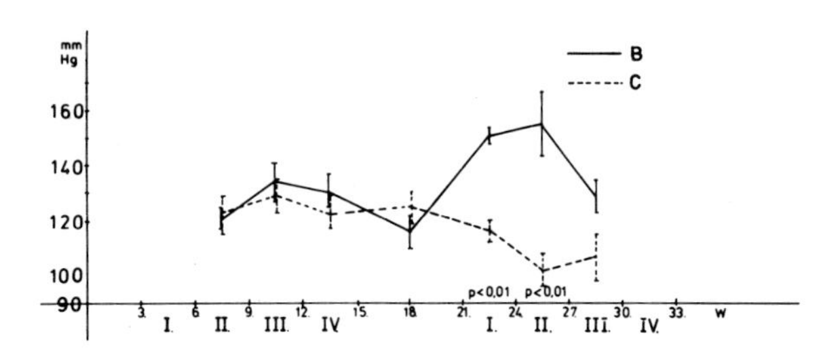

fig. 7: Blood pressure of the animals immunized to corticosterone (First series). Blood pressure values above 140 mm Hg were defined to be elevated

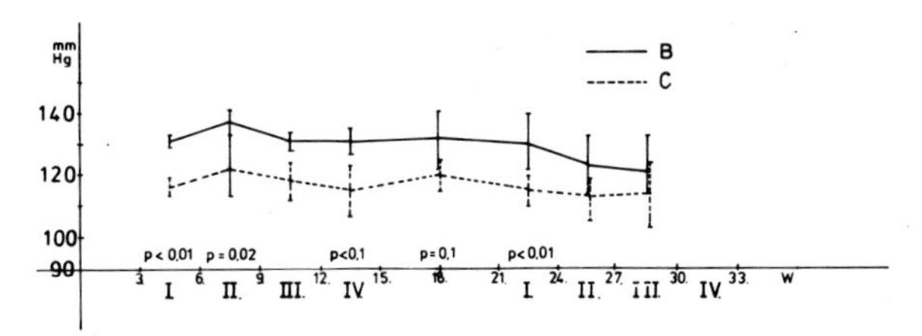

fig. 8: Blood pressure of the animals immunized to corticosterone (Second series)

significantly higher than those of the control animals. In the same experimental phase, the antibody titer also reached high levels (1 : 520, 1 : 320, 1 : 160), corresponding to the highest blood pressure values. In spite of a drop in all antibody titers after the sixth booster immunization, the blood pressures still rose slightly.

Fig. 9 shows the blood pressures of the animals immunized to the mixture of corticosterone, aldosterone, deoxycorticosterone and 18-hydroxycorticosterone. The blood pressures were significantly elevated in parallel to the antibody titers to corticosterone, which reached values above 1 : 5o after the fourth booster immunization.

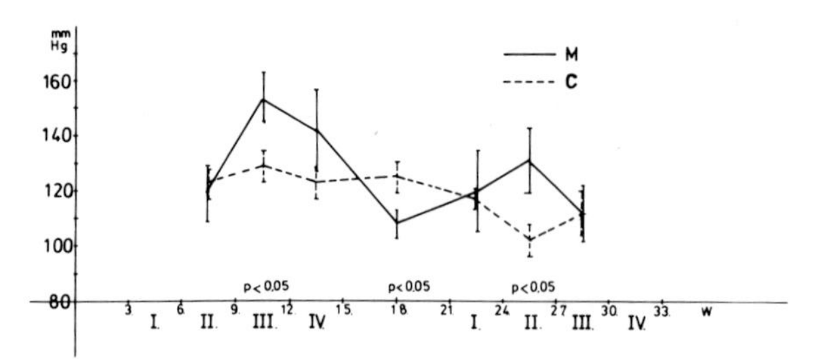

fig. 9: Blood pressure of the animals immunized to the antigen mixture

In fig. 10 the highest blood pressure values of different immunized groups are compared to the control values.
As the animals immunized to corticosterone had sustained elevated blood pressure the relative weights of heart and kidneys were determined.
The rat with the highest blood pressure (180 mm Hg) also had the heaviest heart (0.268 mg/100 g body weight as compared to 0.253 ± 0.012 mg/100 g body weight of the controls) and the highest kidney weight (0.694 mg/100 g body weight as compared to 0.669 ± 0.025 mg/100 g body weight). This animal also showed high antibody titer (1 : 320). The heart and kidney weights of the animals with the lowest blood pressure and the lowest antibody titer were noticeably lower.

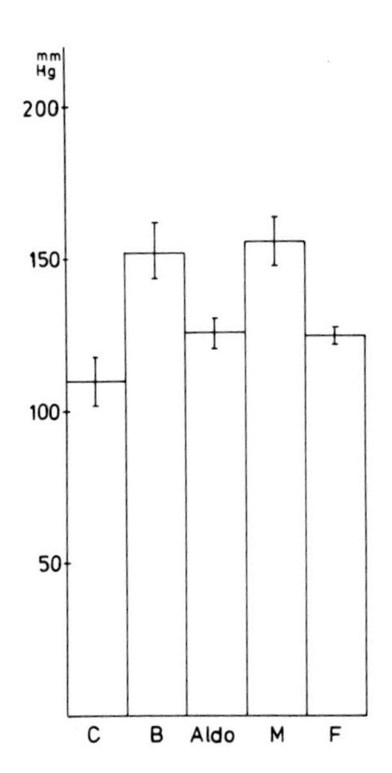

fig. 10: Blood pressure of the control animals (C), animals immunized to corticosterone (B), aldosterone (Aldo), antigen mixture (M) and cortisol (F).

4) Plasma Steroid and Renin Concentrations

In the first immunization series, the plasma corticosterone value of the immunized group (55.45 ± 0.08 µg/100 ml) was twice as high as that of the controls (25.08 ± 7.8 µg/100 ml; $p < 0.1$), whereas in the second immunization series no significant differences were measurable. At the end of the experiment, the plasma corticosterone concentration in all immunized animals was considerably higher as compared to the controls. This tended to be true as well for the first corticosterone immunization in relation to the animals immunized to cortisol and the antigen mixture (fig.11). However, the plasma aldosterone values of the animals immunized to corticosterone were significantly lower than the values of the controls (44.02 ± 3.50 ng/100 ml as compared to 80.22 ± 11.46 ng/100 ml; $p < 0.01$).

The third group of animals, immunized to aldosterone showed normal corticosterone values with three exceptions.

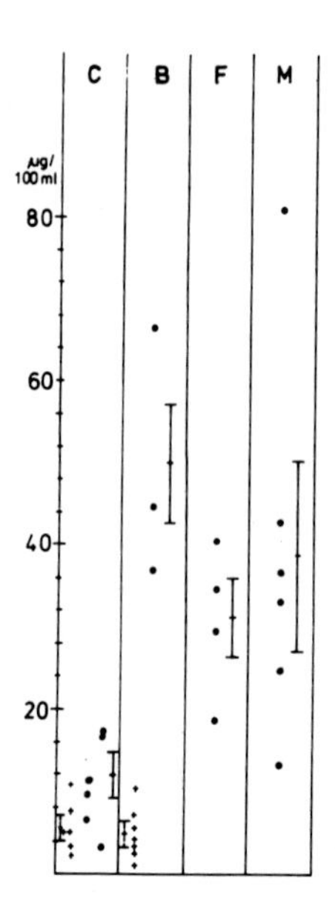

fig. 11: Plasma corticosterone at the end of the experiment: Controls (C), animals immunized to corticosterone (B), cortisol (F) and to antigen mixture (M) (First and second series)

Plasma aldosterone concentration was elevated in relation to the aldosterone titers, and concurrently plasma renin activity was lowered (fig. 12).

5) Steroids in Urine

The urinary corticosterone excretions of the animals immunized to corticosterone were higher than those of the controls in both immunization series (fig. 13).

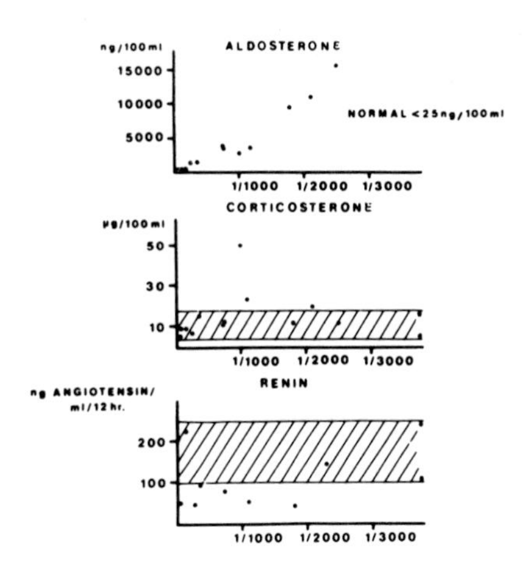

fig. 12: Plasma aldosterone and plasma corticosterone, renin activity; rats immunized to aldosterone (abscissa: antibody titer).

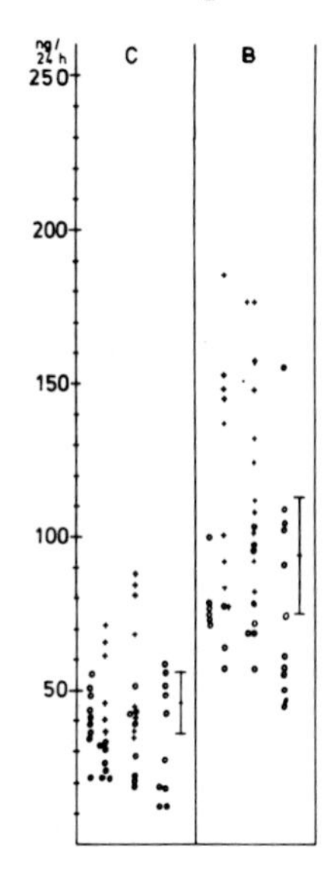

fig. 13: Urinary corticosterone excretion of animals immunized to corticosterone.

In the second immunization series, the urine corticosterone level lay well above the control group following the fourth booster immunization (69.9 ± 16.2 vs. 44.2 ± 10.75 ng/24h; $p < 0.05$).
In spite of the previously mentioned titer drop after the fourth immunization, urine corticosterone values continued to rise. Following the sixth booster immunization, corticosterone excretion in urine was 113.2 ± 21.05 ng/24h as compared to 67.2 ± 12.15 ng/24h ($p < 0.01$) of the controls. During this time, the corticosterone antibody titer of all animals was less than 1 : 50.
The high urine corticosterone level indicates increased activity of the adrenal cortex during this phase of the experiment.
In the first immunization series the urinary aldosterone excretion of the animals immunized to corticosterone were significantly higher ($p < 0.01$) than those of the control group (fig. 14). However, in the second immunization series no significant differences in aldosterone excretions could be found. There was no significant difference in the excretion of 18-hydroxydeoxycorticosterone in the urine between the animals immunized to corticosterone and the control group in the first and second series.
We did note, however, that during the experimental phase of the first series, when the excretion of unconjugated aldosterone rose, the values of 18-hydroxydeoxycorticosterone fell.

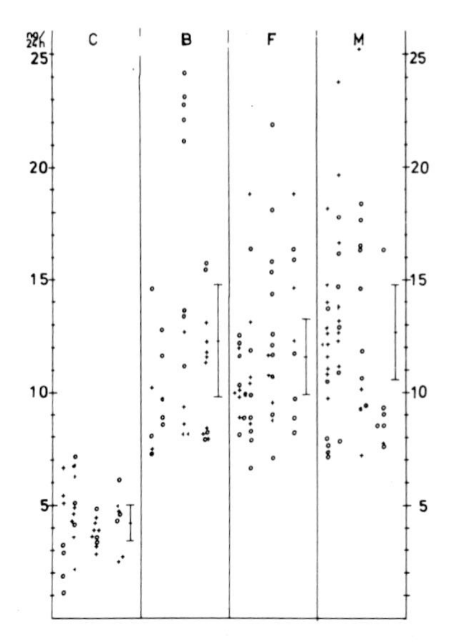

fig. 14: Urinary aldosterone excretion of controls (C), animals immunized to corticosterone (B), cortisol (F) and to antigen mixture (M) (First series)

6) Adrenal Weight and Steroid Production in Vitro

The adrenals of the animals immunized to corticosterone were heavier than those of the control animals in both immunization series (fig. 15).

In the first immunization series, the adrenals of the animals immunized to corticosterone weighed 10.4 ± 0.32 g body weight as compared to 8.8 ± 0.46 mg/100 g body weight of the controls. In the second immunization series, the adrenal weights of the animals immunized to corticosterone were only slightly elevated (9.4 ± 1.05 mg/100 g body weight).

In both immunization series, those animals which had the highest antibody titers during the immunizations also had the heaviest adrenals:

1 : 520 : 11.17 mg/100 g body weight
1 : 320 : 10.58 mg/100 g body weight
1 : 331 : 10.44 mg/100 g body weight

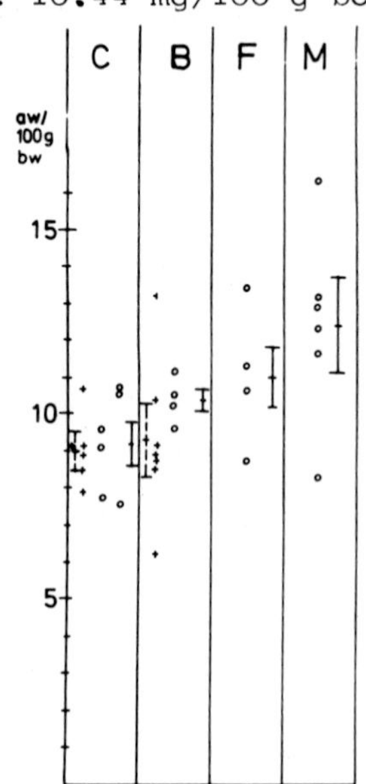

fig. 15: Adrenal weight of the controls (C), the animals immunized to corticosterone (B), cortisol (F) and the antigen mixture (M)

The weight of the adrenals of the animals immunized to the antigen mixture were 12.43 $\pm$ 1.3 mg/100 g body weight. The adrenals of this animal group showed the highest weight in average compared to all other groups.
At the end of the experiment the adrenals were removed and incubated in Krebs-Ringer-bicarbonate-glucose solution to determine the steroid production in vitro. The adrenals from the first corticosterone immunization group produced corticosterone (5.15 $\pm$ 1.4 μg/mg adrenal weight) and aldosterone (7.5 $\pm$ 1.1 ng/mg adrenal weight) in higher amounts than those of the control animals (corticosterone: 4.93 $\pm$ 0.6 μg/mg; aldosterone: 4.4 $\pm$ 0.1 ng/mg; $p < 0.1$). In the second corticosterone immunized group, elevated in vitro

corticosterone production was confirmed. The in vitro aldosterone production did not differ noticeably from that of the control animals. It is noteworthy that the animals with the heaviest adrenals, having also the highest antibody titer, showed an increased in vitro steroid production as well.

Discussion

There have been few reports of successful active immunization of rats with adrenal steroids. Therefore it was of interest to see if the individual animals had measurable antibody titers as a sign of an immune response at the end of the experiment.
Antibody titers were,indeed, ascertainable among all antigen groups we examined. The antibody titers of the animals immunized to corticosterone were considerably elevated after the second immunization (1 : 100). The antibody titers, however, dropped in the second series below 1:50 after an immunization pause and booster immunization. This variation between the first and the second series is not explainable by us. Perhaps the different rats or the different season are a reason for this variation.
The groups of rats immunized to the mixed antigen developed titers against corticosterone, aldosterone and deoxycorticosterone, but only the corticosterone titers were comparable to the corticosterone immunized rats with titers of 1 : 100 after the second immunization, while titers against aldosterone and deoxycorticosterone were only measurable after the 22nd week of immunization.
Determining the specificity of the corticosterone antisera (6) there are only minor cross reactions to 18-hydroxydeoxycorticosterone and to deoxycorticosterone.
Aside from the induction of antibodies to endogenous hormones, the biological effects of the antibodies in these

experiments were also of interest. The biological effects are a result of the level of the antibody titers and also of the affinity of the induced antibodies and the hormone receptors to the antibody hormone.
Elevated blood pressure values and hypertension were considered indicative of adrenal cortex hyperactivity among the animals immunized to corticosterone. Elevated heart weights were taken as signs of the organ manifestation of high blood pressure among the animals with hypertension. This is obviously a consequence of the active immunization to corticosterone i.e. the high degree of the antibody titer. The animal reaching the second highest antibody titer had the highest blood pressure and also the heaviest heart. Blood pressure in the group of rats immunized to the mixed antigen paralleled the titers against corticosterone, not to aldosterone or deoxycorticosterone titers.
Some authors (1,5) have recorded elevated plasma levels of the hormone against which the animals were immunized. This was confirmed in our study. These elevated levels could be the result of reduced plasma clearance, i.e. the association of the hormone to antibody prevented the hormone from being metabolized and excreted (7). This process has been referred by some authors as a "pooling effect" (8,9). On the other hand, stimulation of the endocrine organ by means of active immunization and the resulting increased production of the hormones may be responsible for the elevated plasma levels of the antibody hormones (1,5).
The unconjugated corticosterone urine values were also elevated. This can be interpreted as the result of adrenal cortex hyperactivity as well as increased adrenal weight and in vitro steroid production.
Endogenous aldosterone concentrations were lower in rats immunized to corticosterone as compared to control rats or to animals immunized to the mixed antigen.
The rise in blood pressure was related to the corticosterone

titers and also to elevated concentrations of endogenous corticosterone. The injection of exogenous corticosterone could induce blood pressure elevation besides a loss of sodium and water. The mechanism is still not clear, but vasoconstrictive factors as renin-angiotensin-system and catecholamines may play a role as well as redistribution between intra- and extracellular fluid volume (10).
In this study, biological effects appeared particularly noticeable when the antibody titer suddenly rose or fell. Of critical importance for the appearance of biological effects following active immunization are:

1. The antibody titer level,
2. The affinity of the antigen hormone to the induced antibody and to endogenous hormone receptors and their relationship to one another, and
3. The dynamics of the antibody titer progression.

This study discovered signs of increased adrenal cortex function following active immunization to corticosterone and with a mixture of corticosterone, aldosterone, deoxycorticosterone and 18-hydroxydeoxycorticosterone. This hyperactivity of the adrenal cortex has been described by Gless et al. (1,5) among rabbits and after active immunization of rats with aldosterone. In contrast, some authors reported diminished plasma levels of the hormone against which the animals were immunized (11). They found a hypofunction of the hormone producing organ. Rats immunized to testosterone showed sterility and animals immunized to vasopressin developed diabetes insipidus (12).
The presented data provide obvious indications for the presence of a stimulation of adrenal cortex activity following active immunization to corticosterone.

SUMMARY

White male Wistar rats (n=12) were immunized to corticosterone. Control animals (n=12) were sham immunized. The rats were individually kept in metabolic cages and got standard laboratory diet and tap water ad libidum.

The following values were measured: body weight, urine volume, antibody, titer blood pressure, hematokrit, urinary excretions of corticosterone, aldosterone and 18-hydroxy-DOC and plasma concentrations of corticosterone aldosterone.

Weight of adrenals, heart and kidneys were determined.

Blood pressure significantly increased in 10 of 12 animals (151.3 ± 3.1 mmHg vs. 116.7 ± 3.8 mmHg, $p < 0.01$).

The rat with the highest titer (1:520) had a blood pressure of 180 mmHg. The same rat had the greatest heart weight.

Plasma concentration of corticosterone was elevated (55.45 ± 5.08 µg/100ml vs. 25.08 ± 7.81 µg/100ml, $p < 0.01$), as well as urinary excretion of corticosterone was (105.3 ± 15.5 ng/24h vs. 43.7 ± 10.25 ng/24h, $p < 0.01$).

Plasma concentration of aldosterone was decreased (44.02 ± 3.5ng /100ml vs. 80.22 ± 11.45 ng/100ml, $p < 0.05$) while urinary excretion of aldosterone was elevated in most of the immunized animals.

The weight of the adrenals were higher in immunized animals (10.4 ± 0.32 mg/100g vs. 8.8 ± 0.46 mg/100g).

The corticosterone and aldosterone production in vitro was elevated.

References

1. Vecsei, P., Gless, K.H.: Aldosteron Radioimmunoassay, Enke Verlag, Stuttgart 1975.
2. Erlanger, B.F., Borek, F., Geiser,S., Liebermann, S.: J. Biol. Chem. 228, 713-727 (1957).
3. Connolly, T.C., Vecsei, P., Kohl, K.H., Abdelhamid S., Ammenti, A.: Klin. Wochenschr. 56 (Suppl. I), 173-181 (1978).
4. Vecsei, P.,: Methods of Hormone Radioimmunoassay, second edition, Academic Press, New York . London 1979.
5. Gless, K.H., Hanka, M., Vecsei, P., Gross, F.: Acta endocr. 75, 342-349 (1974).
6. Abraham G.E.: Acta endocr. 7, 183 (1974).
7. Vecsei, P., Haack, D., Gless, K.H.: Antihormones, Walter de Gruyter, Berlin . New York 1979.
8. Haack, D., Vecsei, P., Lichtwald, K., Vielhauer, W.: J. of Steroid Biochem. 11,971-980 (1978),
9. Sundaram, K., Tsong, Y., Hood, W., Brinsan, A.: Endocrinology 93, 843-847 (1973).
10. Haack, D., Möhring, J., Möhring, B., Petri, M., Hackenthal, E.: Amer. J. Physiol. 233 (50), F 403-F 411 (1977).
11. Nieschlag, E., Usadel, K.H., Schwedes, U., Kley, H., Schöffling, K., Krueskemper, H.L.: Endocrinology 92, 1142-1147 (1973).
12. Morton, J.H., Waite, M.A.: J. endocr. 54, 523 (1972).

AUTHOR INDEX

SUBJECT INDEX